W9-CGT-682

About Island Press

Island Press is the only nonprofit organization in the United States whose principal purpose is the publication of books on environmental issues and natural resource management. We provide solutions-oriented information to professionals, public officials, business and community leaders, and concerned citizens who are shaping responses to environmental problems.

In 1994, Island Press celebrated its tenth anniversary as the leading provider of timely and practical books that take a multidisciplinary approach to critical environmental concerns. Our growing list of titles reflects our commitment to bringing the best of an expanding body of literature to the environmental community throughout North America and the world.

Support for Island Press is provided by Apple Computer, Inc., The Bullitt Foundation, The Geraldine R. Dodge Foundation, The Energy Foundation, The Ford Foundation, The W. Alton Jones Foundation, The Lyndhurst Foundation, The John D. and Catherine T. MacArthur Foundation, The Andrew W. Mellon Foundation, The Joyce Mertz-Gilmore Foundation, The National Fish and Wildlife Foundation, The Pew Charitable Trusts, The Pew Global Stewardship Initiative, The Rockefeller Philanthropic Collaborative, Inc., and individual donors.

Special thanks to the foundations and individuals whose support made the development, publication, and distribution of this book possible:

Peter and Helen Bing
Heinz Foundation
Packard Foundation
Pew Charitable Trusts
W. Alton Jones Foundation
Winslow Foundation

Nature's Services

Nature's Services

Societal Dependence on Natural Ecosystems

Edited by Gretchen C. Daily

ISLAND PRESS
Washington, D.C. • Covelo, California

Library of Congress Cataloging-in-Publication Data

Nature's services: societal dependence on natural ecosystems/edited
by Gretchen C. Daily.
p. cm.
Includes bibliographical references and index.
ISBN 1-55963-475-8. — ISBN 1-55963-476-6 (pbk.)
1. Nature—Effect of human beings on. 2. Human ecology—Economic aspects. 2. Biotic communities—Economic aspects. 4. Ecosystem management. I. Daily, Gretchen C.
GF75.N37 1997
304.2—dc21 96-40401
CIP

Printed on recycled, acid-free paper

Manufactured in the United States of America

10 9 8 7 6 5 4 3 2 1

To all those blessed (or cursed) with
a passion to brighten humanity's future.

And to Chuck and Suzanne Daily,
with love and appreciation.

G.C.D.

CONTENTS

CONTRIBUTORS

SUSAN E. ALEXANDER
Earth Systems Science and Policy
100 Campus Center
California State University,
Monterey Bay
Seaside, CA 93955

KAMALJIT S. BAWA
Department of Biology
University of Massachusetts
Boston, MA 02125

STEPHEN L. BUCHMANN
USDA-ARS Carl Hayden Bee
Research Center
2000 East Allen Road
Tucson, AZ 85719

STEPHEN CARPENTER
Center for Limnology
680 North Park Street
University of Wisconsin
Madison, WI 53706

ROBERT COSTANZA
Maryland International Institute
for Ecological Economics
University of Maryland System
P.O. Box 38
Solomons, MD 20688

RICHARD M. COWLING
Institute for Plant Conservation
Private Bag
Rondebosch 7700
University of Cape Town
South Africa

GRETCHEN C. DAILY
Department of Biological Sciences
Stanford University
Stanford, CA 94305

PAUL DAYTON
Scripps Institution of
Oceanography
9500 Gilman Drive
University of California,
San Diego
La Jolla, CA 92093

PAUL R. EHRLICH
Department of Biological Sciences
Stanford University
Stanford, CA 94305

KATHERINE C. EWEL
USDA Forest Service
Pacific Southwest Research Station
1151 Punchbowl Street, Room 323
Honolulu, HI 96813

CARL FOLKE
Beijer International Institute of Ecological Economics
Royal Swedish Academy of Sciences
P.O. Box 50005
Stockholm S-104 05, Sweden

MADHAV GADGIL
Centre for Ecological Sciences
Indian Institute of Science
Bangalore 560 012, India

LAWRENCE H. GOULDER
Department of Economics and Institute for International Studies
Encina Hall, Room 200
Stanford University
Stanford, CA 94305

JOHN HARTE
Energy and Resources Group
310 Barrows Hall
University of California
Berkeley, CA 94720

STEVEN I. HIGGINS
Institute for Plant Conservation
Private Bag
Rondebosch 7700
University of Cape Town
South Africa

LES KAUFMAN
Department of Biology
5 Cummington Street
Boston University
Boston, MA 02215

DONALD KENNEDY
Institute for International Studies
Encina Hall, Room 200
Stanford University
Stanford, CA 94305

KALEN LAGERQUIST
Department of Biological Sciences
Stanford University
Stanford, CA 94305

JANE LUBCHENCO
Department of Zoology
Cordley 3029
Oregon State University
Corvallis, OR 97331

PAMELA A. MATSON
Environmental Science, Policy, and Management
University of California
Berkeley, CA 94720

NORMAN MYERS
Upper Meadow, Old Road
Headington, Oxford OX3 8SZ
England

HAROLD A. MOONEY
Department of Biological Sciences
Stanford University
Stanford, CA 94305

GARY PAUL NABHAN
Arizona Sonora Desert Museum
1201 North Kinney Road
Tucson, AZ 85743

ROSAMOND L. NAYLOR
Institute for International Studies
Encina Hall, Room 200
Stanford University
Stanford, CA 94305

JOSÉ M. PARUELO
Department of Ecology
Faculty of Agronomy
University of Buenos Aires
Avenida de San Martin 4453
Buenos Aires 1417, Argentina

CHARLES H. PETERSON
Institute of Marine Sciences
3431 Arendell Street
University of North Carolina, Chapel Hill
Morehead, NC 28557

SANDRA POSTEL
Global Water Policy Project
107 Larkspur Drive
Amherst, MA 01002

OSVALDO E. SALA
Department of Ecology
Faculty of Agronomy
University of Buenos Aires
Avenida de San Martin 4453
Buenos Aires 1417, Argentina

STEPHEN H. SCHNEIDER
Department of Biological Sciences and Institute for International Studies
Stanford University
Stanford, CA 94305

DAVID TILMAN
Department of Ecology, Evolution, and Behavior
100 Ecology Building
1987 Upper Buford Circle
University of Minnesota
St. Paul, MN 55108

PETER M. VITOUSEK
Department of Biological Sciences
Stanford University
Stanford, CA 94305

ANDREW WILCOX
Energy and Resources Group
310 Barrows Hall
University of California
Berkeley, CA 94720

PREFACE

The effort behind this book was initiated after dinner one night, under the Arizona desert sky, at an annual meeting of the Pew Fellows in Conservation and the Environment. A small group gathered informally to lament the near total lack of public appreciation of societal dependence upon natural ecosystems. This ignorance represents but one of a complex of interacting factors responsible for today's array of anthropogenic disruptions of the biosphere. Yet it clearly represents a major hindrance to the formulation and implementation of policy designed to safeguard earth's life-support systems. Moreover, lack of understanding of the character and value of natural ecosystems traces ultimately to a failure of the scientific community to generate, synthesize, and effectively convey the necessary information to the public.

A collective strategy to address this problem emerged from the group's discussion, the first phase of which consisted of producing a rigorous, detailed synthesis of our current understanding of a suite of ecosystem services and a preliminary assessment of their economic value. Thus, our first task was to assemble a broad, interdisciplinary group of natural and social scientists to undertake this work. The individuals we approached were extremely enthusiastic and remained so throughout the project development, reflecting a widely shared recognition of the need for such a book. After producing a first draft of the chapters, contributors met in Purity Springs, New Hampshire (as a special session of the next year's Pew Fellows meeting), to present and get feedback on their approach and analysis, and to discuss overarching issues pertaining to the whole book. This session was very productive, thanks in no small part to the participation of a large number of Pew Fellows not otherwise engaged in the undertaking. It led to the production of two additional chapters to make the book more comprehensive and coherent.

Coordinating this effort has been a great pleasure from the start, thanks to the support of the contributors, the Island Press staff, and the funders. I could not imagine a group of contributors more enthusiastic, timely, and responsive to queries, reviewers' suggestions, and general harassment. Nor could I conceive of more helpful and knowledgeable editors: Barbara Dean and Kristy Manning were fully engaged in every aspect of shaping the book. External reviewers of the chapters provided constructive criticism in the best sense; we were very fortunate to have the economics expertise of Michael Dalton and David Layton, who kindly reviewed the book end to end. The project was made possible by the generous support of the Packard Foundation, the Pew Charitable Trusts, and the W. Alton Jones Foundation; in addition, I was supported during the development of the book by the Winslow and Heinz foundations and by Peter and Helen Bing.

Scott Daily, Frédéric Lelièvre, and Kirsten Ziegenhagen were very helpful and encouraging with various aspects of the book. Jill Otto kindly tracked down obscure references, and Pat Browne and Steve Masley provided tremendous assistance with photocopying. I am grateful to Peter Bing, Sam Hurst, Donald Kennedy, Jonathan Lash, Peter Raven, Walter Reid, Kelsey Wirth, and Tim and Wren Wirth for freely offering advice and assistance with each phase of the group effort. Finally, I owe a special debt to Paul Ehrlich, Michael Kleeman, Jane Lubchenco, John Peterson Myers, Chuck Savitt, and Jeanne Sedgwick for providing extremely valuable insight and guidance on the overall course of this joint undertaking.

GRETCHEN C. DAILY
Stanford, California

Perspectives on Nature's Services

John Peterson Myers

Joshua S. Reichert

I fly window seat, incurably. It doesn't matter that I've been on countless flights over the past four decades and stared downward over numberless vistas of forest and wetland, prairie and river. When the flight attendants ask for people to lower the window shade so that you can see the movie better, that's my shade still up, my face pressed against the glass.

I revel in the details of geography and sweep of biology that can be seen from the air. I trace trails into box canyons and along mountain ridges; search tundra for oriented lakes; look at sediment roiling downcoast from a turbulent river mouth; scan bayous for cypress islands; enjoy the dendritic meanderings of channels through a tide flat; discover oxbow lakes along the fringes of a river basin.

Anyone who shares this affliction with me knows, however, that what one sees most of the time along too many common routes is the human footprint: dams and dikes, farmland, shopping malls and roads, hills bulldozed, forests flattened, massive works of civil (and uncivil) engineering.

The ubiquity of this footprint makes me think of Goldfinger. Bond's villain, you may remember, had a particularly twisted way of disposing of one of his enemies: he covered her entirely with gold paint. Up to a point, his victim could survive, but cover enough and her skin would no longer be able to breathe. However mythical might have been Goldfinger's technique, the image of that gilded body comes to mind when viewing the earth from the sky and seeing what a remarkable percentage of its surface has been churned or covered by human toil.

Metaphors leap from odd places (perhaps the strangest come from staring out of one too many airplane windows), but they can pose important questions. How much of the earth's ecological integrity can we disrupt before we pass a threshold in the loss of life-support services provided by natural ecosystems? Issues of ecosystem services—what they are, what they contribute to life on earth, their role in building and maintaining human prosperity, their vulnerability to disruption and the impact of their losses—are central questions for modern science. Or rather they should be, given the expanding scale of human impact upon the landscape.

That is why the work contained within this volume is so important. Its interdisciplinary, synthetic overview of the nature and value of ecosystem services reveals four things simultaneously. First, that we know enough now to understand in broad brush that ecosystem services are essential for human life as we live it today and as we would hope our children live it in the twenty-first century. These services are myriad. They include provision of clean water and flood control, creation of soil, pollination of crops, providing habitats for fisheries, and countless other benefits that underpin human well-being.

Second, the analyses presented herein show that for engineering to replace the services that ecosystems now provide is—if not completely beyond current technology—prohibitively expensive if implemented at anything but a trivial scale. Third, they make clear that our scientific and economic understanding of the true dimensions and details of these services is appallingly shallow.

Fourth—and with the greatest direct relevance for today's public debates about environmental protection—the combination of ignorance and import revealed dictate caution. We do not know where we are in Goldfinger's attack . . . how far we are today along the functions describing service output in relation to ecosystem disruption, nor the incremental effects accruing from each new degradation.

I am reminded of Wendell Berry in *Home Economics* (1987): "Acting on the basis of ignorance, paradoxically, requires one to know things, remember things—for instance, that failure is possible, that error is possible, that second chances are desirable (so don't risk everything on the first chance). . . ." The contents of this book teach that for ecosystem services, we need more than a second chance. We need a serious, unwavering commitment to science and policy that will ensure athe life-support processes flowing from natural ecosystems continue undiminished.

J.P.M

Human society has never had a more pressing need to understand its dependence on nature. From time immemorial we have too lightly valued some of the most basic resources on which we depend, including the air we breathe, the water we drink, and the ability of the earth to support a wide variety of life. The cumulative impact of human activity on the natural systems that produce these resources, particularly over the past one hundred years, and our rather recent understanding of the dramatic scope of that impact, make it impossible for us to take them for granted any longer.

We don't protect what we don't value. Over countless generations, people have tended to place relatively little emphasis on protecting the natural systems that surround them, in great part because they have neither understood nor appreciated their value. The goal of this book is to enhance that understanding and, in so doing, encourage greater efforts to protect the earth's basic life-support systems before it is too late.

The scientists who have contributed to *Nature's Services* represent an impressive array of disciplines and expertise. Equally important, the contributors share a commitment to applying their knowledge to solve our most serious environmental problems. That is also the central goal of the Pew Fellows Program in Conservation and the Environment, which provided the intellectual forum that led to the creation of this book, and which counts many of its contributors among its ranks.

The rapid deterioration of the global environment presents a set of new and urgent challenges to scientists worldwide. In addition to better understanding how natural ecosystems function, relate, and interdepend, there is a vital need to translate this information to the general public, as well as to policymakers, in ways that will prompt the action needed to preserve what remains of the planet's natural resources. This is not a traditional role for scientists and it is not an easy one for them to play. It calls on men and women who have typically defined themselves as observers and analysts to become more actively engaged in advocating the protection of those systems that are the objects of their study.

With such notable exceptions as Rachel Carson, scientists have generally avoided becoming involved in the rather messy process of shaping public policy. Indeed, they have often been criticized by their peers and penalized within their disciplines for such involvement as compromising their ability to remain sufficiently detached and, by implication, "objective."

Many of the men and women who have contributed to this volume and a small but increasing number of their peers in the scientific community have come to recognize that good science and sound environmental policy go hand in hand. Moreover, they realize that without the input of scientists in the public debate over the future of the global environment, we may fail to enact many of the policies needed to solve the world's most urgent environmental problems.

Nature's Services is of particular importance because it is one of the first efforts by scientists to provide insight to the many benefits and services that nature offers, and the extent to which the human race is vitally dependent on them. Without a firm understanding of the value of these systems to the quality of our lives and, in some cases, to our ability to maintain life at all, we are unlikely to make many of the hard choices and compromises needed to protect them.

Over the coming decade, policymakers in the United States and throughout the world will be forced to make very difficult decisions regarding the future of the earth's grassland, forest, and ocean systems, many of which are rapidly deteriorating. The continued destruction of these ecosystems holds enormous consequences not only for the men and women who depend directly on them for a livelihood, but also for much of the rest of humanity. Because it is not only the logger who has a stake in preserving the world's forests, the fisherman in maintaining a healthy population of fish in the sea, or the rancher in ensuring sufficient grasslands for his livestock. All of the world's population depends on these ecosystems—not simply for the goods they provide but also because of the critical role they play in maintaining the global atmosphere, regulating the earth's weather patterns, filtering much of the waste products produced by society, preserving watersheds, controlling soil erosion, and preventing floods and drought. Indeed, as *Nature's Services* points out, the value of these ecosystems far exceeds that of the commodities we commonly associate with them.

Yet because few people understand the broader importance of these systems, their worth is often expressed only in the most limited economic sense, i.e., the number of jobs or market goods they produce. Unless their true social and economic value is recognized in terms we all can understand, we run the grave risk of sacrificing the long-term survival of these natural systems to our short-term economic interests. This is the central message of *Nature's Services*. It is one we can ill afford to ignore.

J.S.R.

Chapter 1

Introduction: What Are Ecosystem Services?

Gretchen C. Daily

In the space of a single human lifetime, society finds itself suddenly confronted with a daunting complex of trade-offs between some of its most important activities and ideals. Recent trends raise disturbing questions about the extent to which today's people may be living at the expense of their descendents, casting doubt upon the cherished goal that each successive generation will have greater prosperity. Technological innovation may temporarily mask a reduction in earth's potential to sustain human activities; in the long run, however, it is unlikely to compensate for a massive depletion of such fundamental resources as productive land, fisheries, old-growth forests, and biodiversity.

On a global scale, different groups of people are now living at one another's expense, as is readily apparent in the disruption and overexploitation of earth's open-access resources and waste sinks. For example, whereas the levels of disruption caused by energy use were once small, local, and reversible, they have now reached global proportions and carry irreversible consequences. In fueling their industrialization historically and pursuing their activities today, the developed nations appear to have largely used up the atmosphere's capacity to absorb CO_2 and other greenhouse gases without risk of inducing climate change. In the process, they have foreclosed the option of safely using fossil fuels to sustain comparable levels of industrial activity by developing nations.

And, at the local scale, the tradeoffs between competing activities, and between individual and societal interests, are becoming ever more evident. In virtually any community, allocation of land or water to various activities often involves a zero sum game, as is apparent in the widespread loss of farmland and water to urban and industrial purposes. Thus, constraints on the scale of the human enterprise typically manifest themselves most tangibly not as absolute limits to a particular activity, but rather as tradeoffs, whose resolution is fraught with increasingly difficult practical and ethical considerations.

This book features contributions from a diverse group of natural and social scientists with expertise in different aspects of these issues, reflecting their own technical training, personal interests, and life experiences. Yet, as a whole, the contributors are oriented around a common set of fundamental premises. First, they share a conviction that, while civilization is presently careening along on a dangerous course, its fate is not sealed. The close of the twentieth century represents a period in history that demands not just a carefully tuned focus on crises of the moment, but also a long-term perspective on challenges to the human future. Second, by different paths they have reached the conclusion that society is poorly equipped to evaluate environmental tradeoffs, and that their continued resolution on the sole basis of the social, economic, and political forces prevailing today threatens environmental, economic, and political security. The chapter authors thus share a sense of urgency for developing analytical and institutional frameworks for the informed and wise resolution of these tradeoffs. Third, such decision-making frameworks must ensure the protection of humanity's most fundamental source of well-being: earth's life-support systems. A tremendous amount is known about the importance and value of the natural systems that underpin the human economy, but this information has neither been synthesized nor effectively conveyed to decision makers or to the general public.

The purpose of this book is to characterize the ways in which earth's natural ecosystems confer benefits on humanity, to make a preliminary assessment of their value, and to report this in a manner widely accessible to an educated audience. An ecosystem is the set of organisms living in an area, their physical environment, and the interactions between them. Although the distinction between "natural" and "human-dominated" ecosystems is becoming increasingly blurred, our focus is on the natural end of the spectrum, for three related reasons. First, the goods and services flowing from natural ecosystems are greatly undervalued by society. For the most part, the benefits those ecosystems provide are not traded in formal markets and do not send price signals of changes in their supply or condition. This is a major factor driving their conversion to human-dominated systems (e.g., agricultural lands), whose economic value is expressed, at least in part, in standard currency. Second, anthropogenic disruptions of natural ecosystems—such as alteration of the gaseous composition of the atmosphere, introduction and establishment of exotic species, and extinction of native species—are difficult or impossible to reverse on any time scale of relevance to society. Finally, if current trends continue, humanity will dramatically alter or destroy virtually all of earth's remaining natural ecosystems within a few decades.

What Are Ecosystem Services?

Ecosystem services are the conditions and processes through which natural ecosystems, and the species that make them up, sustain and fulfill human life. They maintain biodiversity and the production of *ecosystem goods*, such as seafood, forage, timber, biomass fuels, natural fiber, and many pharmaceuticals, industrial products, and their precursors. The harvest and trade of these goods represent an important and familiar part of the human economy. In addition to the production of goods, ecosystem services are the actual life-support functions, such as cleansing, recycling, and renewal, and they confer many intangible aesthetic and cultural benefits as well.

One way to appreciate the nature and value of ecosystem services (originally suggested by John Holdren) is to imagine trying to set up a happy, day-to-day life on the moon. Assume for the sake of argument that the moon miraculously already had some of the basic conditions for supporting human life, such as an atmosphere and climate similar to those on earth. After inviting your best friends and packing your prized possessions, a BBQ grill, and some do-it-yourself books, the big question would be, Which of earth's millions of species do you need to take with you?

Tackling the problem systematically, you could first choose from among all the species exploited directly for food, drink, spice, fiber and timber, pharmaceuticals, industrial products (such as waxes, lac, rubber, and oils), and so on. Even being selective, this list could amount to hundreds or even several thousand species. The space ship would be filling up before you'd even begun adding the species crucial to *supporting* those at the top of your list. Which are these unsung heroes? No one knows which—nor even approximately how many—species are required to sustain human life. This means that rather than listing species directly, you would have to list instead the life-support functions required by your lunar colony; then you could guess at the types and numbers of species required to perform each. At a bare minimum, the spaceship would have to carry species capable of supplying a whole suite of ecosystem services that earthlings take for granted. These services include:

- purification of air and water
- mitigation of floods and droughts
- detoxification and decomposition of wastes

- generation and renewal of soil and soil fertility
- pollination of crops and natural vegetation
- control of the vast majority of potential agricultural pests
- dispersal of seeds and translocation of nutrients
- maintenance of biodiversity, from which humanity has derived key elements of its agricultural, medicinal, and industrial enterprise
- protection from the sun's harmful ultraviolet rays
- partial stabilization of climate
- moderation of temperature extremes and the force of winds and waves
- support of diverse human cultures
- providing of aesthetic beauty and intellectual stimulation that lift the human spirit.

Armed with this preliminary list of services, you could begin to determine which types and numbers of species are required to perform each. This is no simple task! Let's take the soil fertility case as an example. Soil organisms play important and often unique roles in the circulation of matter in every ecosystem on earth; they are crucial to the chemical conversion and physical transfer of essential nutrients to higher plants, and all larger organisms, including humans, depend on them (Heywood 1995). The abundance of soil organisms is absolutely staggering: under a square yard of pasture in Denmark, for instance, the soil was found to be inhabited by roughly 50,000 small earthworms and their relatives, 50,000 insects and mites, and nearly 12 million roundworms. And that is not all. A single gram (a pinch) of soil has yielded an estimated 30,000 protozoa, 50,000 algae, 400,000 fungi, and billions of individual bacteria (Ehrlich et al. 1977; Overgaard-Nielsen 1955). Which to bring to the moon? Most of these species have never been subjected to even cursory inspection. Yet the sobering fact of the matter is, as Ed Wilson put it: they don't need us, but we need them.

Ecosystem services are generated by a complex of natural cycles, driven by solar energy, that constitute the workings of the biosphere—the thin layer near earth's surface that contains all known life. The cycles operate on very different scales. Biogeochemical cycles, such as the movement of the element carbon through the living and physical environment, are truly global and reach from the top of the atmosphere to deep into soils and ocean-bottom sediments. Life cycles of bacteria, in contrast, may be completed in an area much smaller than the period at the end of this sentence. The cycles also operate at very different rates. The biogeochemical cycling of carbon, for instance, occurs at orders of magnitude faster than that of phosphorus, just as the life cycles of microorganisms may be orders of magnitude faster than those of trees.

All of these cycles are ancient, the product of billions of years of evolution, and have existed in forms very similar to those seen today for at least hundreds of millions of years. They are absolutely pervasive, but unnoticed by most human beings going about their daily lives. Who, for example, gives a thought to the part of the carbon cycle that connects him or her to the plants in the garden outside, to plankton in the Indian Ocean, or to Julius Caesar? Noticed or not, human beings depend utterly on the continuation of natural cycles for their very existence. If the life cycles of predators that naturally control most potential pests of crops were interrupted, it is unlikely that pesticides could satisfactorily take their place. If the life cycles of pollinators of plants of economic importance ceased, society would face serious social and economic consequences. If the carbon cycle were badly disrupted, rapid climatic change could threaten the existence of civilization. In general, human beings lack both the knowledge and the ability to substitute for the functions performed by these and other cycles (Ehrlich and Mooney 1983).

For millennia, humanity has drawn benefits from these cycles without causing global disruption. Yet, today, human influence can be discerned in the most remote reaches of the biosphere: deep below earth's surface in ancient aquifers, far out to sea on tiny tropical islands, and up in the cold, thin air high above Antarctica. Virtually no place remains untouched—chemically, physically, or biologically—by the curious and determined hand of humanity. Although much more by accident than by design, humanity now controls conditions over the entire biosphere.

Interestingly, the nature and value of Earth's life-support systems have been illuminated primarily through their disruption and loss. Thus, for instance, deforestation has revealed the critical role of forests in the hydrological cycle—in particular, in mitigating flood, drought, and the forces of wind and rain that cause erosion. Release of toxic substances, whether accidental or deliberate, has revealed the nature and value of physical and chemical processes, governed in part by a diversity of microorganisms, that disperse and break down hazardous materials. Thinning of the stratospheric ozone layer sharpened awareness of the value of its service in screening out harmful ultraviolet radiation.

A cognizance of ecosystem services, expressed in terms of their loss, dates back at least to Plato and probably much earlier:

> What now remains of the formerly rich land is like the skeleton of a sick man with all the fat and soft earth having wasted away and only the bare framework remaining. Formerly, many of the mountains were arable. The plains that were full of rich soil are now marshes. Hills that were once covered with forests and produced abundant pasture now produce only food for bees. Once

> the land was enriched by yearly rains, which were not lost, as they are now, by flowing from the bare land into the sea. The soil was deep, it absorbed and kept the water . . . , and the water that soaked into the hills fed springs and running streams everywhere. Now the abandoned shrines at spots where formerly there were springs attest that our description of the land is true.
>
> —Plato (quoted in Hillel, p. 104)

Ecosystem services have also gained recognition and appreciation through efforts to substitute technology for them. The overuse of pesticides, for example, leading to the decimation of natural pest enemies and concomitant promotion of formerly benign species to pest status, has made apparent agriculture's dependence upon natural pest control services. The technical problems and cost of hydroponic systems—often prohibitive even for growing high-priced, specialty produce—underscore human dependence upon ecosystem services supplied by soil. Society is likely to value more highly the services listed above, and to discover (or rediscover) an array of services not listed, as human impacts on the environment intensify and the costs and limits of technological substitution become more apparent.

Organization of the Book

This introductory chapter is followed by a brief historical overview of modern concern for ecosystem services (chapter 2). Part I explores key philosophical and economic issues of valuation to provide a context for understanding the range of approaches employed in subsequent chapters to describe the importance of ecosystem services. The following two sections (parts II and III) examine a diversity of the major services operating in a variety of ecological systems over a spectrum of scales, from local to global. Part IV reports on a series of services whose nature and value are particularly well documented, typically by virtue of having been consciously exploited at the local level and, in some cases, marketed, at least informally. The book closes with a brief overview of our findings and a discussion of future challenges. Some parts of this structure—especially issues of valuation—merit further introduction, given below.

Valuation of Ecosystem Services

The disparity between actual and perceived value is probably nowhere greater than in the case of ecosystem services. If asked to identify *all* that goes into making a fine cake, a baker willing to share his or her secrets would most likely first identify its ingredients, and the knowledge and skill required to transform them into a culinary work of art. He or she might also describe the type of oven, pan, and various appliances and kitchen gadgets needed. If pressed further, an astute baker might also point out the need for capital infrastructure and human services to process, store, and transport the ingredients. With a helpful hint or two, he or she may even mention the cropland, water, chemical, and energy inputs to the whole process. However, the chances of the baker touching directly upon the natural renewal of soil fertility, the pollination of crops, natural pest control, the role of biodiversity in maintaining crop productivity, clean-up and recycling services outside the kitchen—or, indeed, upon *any* ecosystem service involved—are extremely remote. Ecosystem services are absolutely essential to civilization, but modern urban life obscures their existence.

Once explained, the importance of ecosystem services is typically quickly appreciated, but the actual assigning of value to ecosystem services may arouse great suspicion, and for good reason. Valuation involves resolving fundamental philosophical issues (such as the underlying bases for value), the establishment of context, and the defining of objectives and preferences, all of which are inherently subjective. Even after doing this, one is faced with formidable technical difficulties with interpreting information about the world and transforming it into a quantitative measure of value. Chapters 3 and 4 discuss these issues and describe alternative empirical valuation techniques, their applicability to different types of ecosystem services, and the advantages and limitations of their use. The final chapter (20) then reviews the major obstacles that contributors encountered.

This book does not attempt a comprehensive valuation of ecosystem services. Just as it would be absurd to calculate the full value of a human being on the basis of his or her wage-earning power, or the economic value of his or her constituent materials, there exists no absolute value of ecosystem services waiting to be discovered and revealed to the world by a member of the intellectual community. Contributors seek primarily to identify and characterize components of ecosystem service value and to make a preliminary assessment of their magnitude, as a prerequisite to their incorporation into frameworks for decision making.

Our concentration is on use values; aesthetic and spiritual values associated with ecosystem services are only lightly touched upon in this book, having been eloquently described elsewhere. The total value of ecosystem services may be best assessed in terms of physical magnitudes or proportions, such as the amount of human waste processed naturally, the amount of carbon sequestered in soils, the proportion of potential crop pests controlled naturally, and the proportion of pharmaceutical products derived from biodiversity. Where a technological substitute is available for an aspect of an ecosystem service, the market price of the substitute provides a lower-bound index of the value of the service (in terms of avoided costs). As a whole, ecosystem services have infinite use value because human life could not be sustained without them. The evaluation of the tradeoffs currently facing society, however, requires estimating the *marginal* value of ecosystem services (the value yielded by an additional unit of the service, all else held constant) to determine the costs of losing—or the benefits of preserving—a given amount or quality of services. The information needed to estimate marginal values is difficult to obtain and is presently unavailable for many aspects of the services. Nonetheless, even imperfect measures of their value, if understood as such, are better than simply ignoring ecosystem services altogether, as is generally done in decision making today.

Overarching Services and Services Supplied by Major Biomes

The next two sections provide an overview of some of the paramount services, from the perspective of those operating in most ecosystems, globally, and of those associated primarily with particular biomes. The section on overarching services (part II) opens with two fundamental issues, the interaction between climate and life (chapter 5) and the relation between biodiversity and aspects of ecosystem functioning important for the supply of ecosystem services (chapter 6). Subsequent chapters explore the services provided by soils, by pollinators, and by natural pest enemies (chapters 7–9). The section on services supplied by major biomes (part III) is structured around distinct services operating in four of the world's major biomes: marine, freshwater, forest, and grassland ecosystems (chapters 10–13). These two sections seek to identify the major components of ecosystem service value, to characterize them in terms of function, susceptibility to human disruption, amenability to repair, and societal importance.

Case Studies

This section of the book zeroes in on some of the more tangible and direct benefits derived from ecosystem services. The first two chapters (14 and 15) reveal the vast array of goods that societies extract from natural ecosystems and explain how ecosystem services sustain their production—on land and in the sea. Two subsequent chapters (16 and 17) explore the dependence of local economies—subsistence and modern—on ecosystem goods and services. The final chapters (18 and 19) take a detailed look at ways in which societies consciously exploit and manipulate particular ecosystem services to satisfy basic needs, such as for waste disposal and abundant clean water. This section thus illuminates the many connections between ecosystem services and aspects of daily human existence, from eating breakfast to shaping our traditions, values, and cultures.

The Policy Interface

Diverse human societies have now attained the status of ecological superpowers. That is, they have the capacity to seriously impair or destroy essential components of earth's life-support systems; moreover, they are currently using this capacity, almost without restraint. The persistence of all societies ultimately hinges upon those superpowers beginning to wisely coordinate and control the wielding of this power. This will especially be so if the magnitude of human influence continues to expand at unprecedented rates to unprecedented levels, through the momentum and inertia associated with population growth, expanding material desires, and the technical means by which fulfilling the latter is pursued. As the most accessible and suitable resources are sequentially exhausted, each additional person, all else equal, exerts greater per-capita impact in necessarily turning to lower-quality resources for the same end.

Historically, human societies have alleviated resource constraints primarily by pushing back intellectual and territorial frontiers. Yet, it would be difficult today for even the most optimistic rates of innovation and of adoption of improved technology (broadly defined) to offset the rates of increase in human disruption caused by rapid population growth and increases in per-capita impacts. Furthermore, opportunities for territorial expansion are now largely foreclosed—or never existed for inherently global impacts, such as those on the composition of the upper atmosphere.

The passage of time leaves in ever sharpening focus a daunting but critical need to tackle social and political frontiers with the same boldness and determination that took the first man to the moon. This will require not only strengthening existing institutions, but also creating entirely new regimes to manage globally human impacts on earth's life-support systems. It will also require an unprecedented level of international cooperation and coordination. It is at these policy frontiers that lie the brightest prospects for resolving the human predicament and converting the world's societies to new and sustainable resource management regimes.

Nature's Services represents an exploration of both the scientific and the policy frontiers. On the scientific front, the book provides a broad, preliminary characterization of the natural functioning of earth's systems; of the ways in which the human enterprise extracts material well-being from these systems; of the impacts exerted thereon by different human activities; and of the tradeoffs inherent in alternative courses of action. On the policy front, the book attempts an initial appraisal of the economic value of elements of earth's life-support systems. Of course, economic indices are likely to underestimate the total value of these systems. Nonetheless, economic markets play a dominant role in patterns of human behavior, and the expression of value—even if imperfect—in a common currency helps to inform the decision-making process. Making economic institutions sensitive and responsive to natural constraints and explicitly dealing with the limitations of such institutions in doing so are other requisites to effective earth management (Daily et al. 1996).

Present scientific understanding of ecosystem services is substantial, wide reaching, and extremely policy-relevant, and merits urgent attention by decision makers, since current patterns of human activity are unsustainable and threaten to impair critical life-support functions. Failure to foster the continued delivery of ecosystem services undermines economic prosperity, forecloses options, and diminishes other aspects of human well-being; it also threatens the very persistence of civilization. While the academic community remains a long way from a fully comprehensive understanding of ecosystem services, the accelerating rate of disruption of the biosphere makes imperative the incorporation of current knowledge into the policy-making process.

References

Daily, G., P. Ehrlich, and M. Alberti. 1996. "Managing earth's life support systems: The game, the players, and getting everyone to play." *Ecological Applications* 6:19–21.

Ehrlich, P., A. Ehrlich, and J. Holdren. 1977. *Ecoscience: Population, Resources, Environment.* San Francisco: Freeman & Co.

Ehrlich, P., and H. Mooney. 1983. "Extinction, substitution, and ecosystem services." *BioScience* 33:248–254.

Heywood, V., ed. 1995. *Global Biodiversity Assessment.* Cambridge, England: Cambridge University Press.

Hillel, D. 1991. *Out of the Earth: Civilization and the Life of the Soil.* New York: The Free Press.

Overgaard-Nielsen, C. 1955. "Studies on enchytraeidae 2: Field studies." *Natura Jutlandica* 4:5–58.

Chapter 2

Ecosystem Services: A Fragmentary History

Harold A. Mooney and Paul R. Ehrlich

While explicit recognition of ecosystem services is a relatively new phenomenon, the notion that natural ecosystems help to support society probably traces back to the time when our ancestors were first able to have notions. For example, Plato understood that the deforestation of Attica led to soil erosion and the drying of springs.

One might consider the origins of modern concern for ecosystem services to trace to George Perkins Marsh's publication of *Man and Nature* in 1864. The book was the first to attack the idea that America's resources (or the world's) were infinite, an error that persists among the scientifically ignorant.

Marsh, a lawyer, politician, and scholar, knew the Mediterranean well, having traveled there extensively and served as ambassador in Turkey and Italy. He noted that much of the once-fertile Roman Empire "is either deserted by civilized man and surrendered to hopeless desolation, or at least greatly reduced in both productiveness and population" (p. 9). He described the deterioration of the services of retaining soil and supplying fresh water: "Vast forests have disappeared from mountain spurs and ridges, the vegetable earth . . . [is] washed away; meadows, once fertilized by irrigation, are waste and unproductive, because . . . the springs that fed them dried up; rivers famous in history and song have shrunk to humble brooklets" (p. 9). He recognized the connections of deforestation to climate: "With the disappearance of the forest, all is changed. At one season, the earth parts with its warmth by radiation to an open sky—receives, at another, an immoderate heat from the unobstructed rays of the sun. Hence the climate becomes excessive, and the soil is alternately parched by the fervors of summer, and seared by the rigors of winter. Bleak winds sweep unresisted over its surface, drift away the snow that sheltered it from the frost, and dry up its scanty moisture" (p. 186).

Marsh was also quite aware of the waste-disposal service of natural ecosystems. For example, he wrote, "The carnivorous, and often the herbivorous insects render an important service to man by consuming dead and decaying animal and vegetable matter, the decomposition of which would otherwise fill the air with effluvia noxious to health" (p. 95). He noted the pest-control service: "man has promoted the increase of the insect and the worm, by destroying the bird and the fish which feed upon them" (p. 96). He was more aware in 1864 of the services performed by microorganisms than the average politician or economist in 1996:

> Earth, water, the ducts and fluids of vegetation and of animal life, the very air we breathe, are peopled by minute organisms which perform most important functions in both the living and the inanimate kingdoms of nature. It is evident that the chemical, and in many cases mechanical character of a great number of objects important in the material economy of human life, must be affected by the presence of so large an organic element in their substance, and it is equally obvious that all agricultural and all industrial operations tend to disturb the natural arrangements of this element. (p. 108)

Almost a century later, Aldo Leopold (1949) touched more poetically on ecosystem services. Writing of the loss of natural controls of herbivore herds, he said: "I now suspect that just as a deer herd lives in mortal fear of its wolves, so does a mountain live in mortal fear of its deer. . . . So also with cows. The cowman who cleans his range of wolves does not realize he is taking over the wolf's job of trimming the herd to fit the range. He has not learned to think like a mountain. Hence we have dustbowls, and rivers washing the future into the sea" (p. 132). Leopold recognized the basic impossibility of substituting satisfactorily for ecosystem services: "A land ethic changes the role of *Homo sapiens* from conquerer of the land community to plain member and citizen of it. . . . In human history we have learned (I hope) that the conquerer role is eventually self-defeating. Why? Because it is implicit in such a role that the conquerer knows, *ex cathedra*, just what makes the community clock tick, and just what and who is valuable, and what and who is worthless, in community life. It always turns out that he knows neither, and this is why his conquests eventually defeat themselves" (p. 204).

About the same time, two influential books appeared that helped reawaken interest in the sorts of ecosystem issues Marsh had discussed: Fairfield Osborn's *Our Plundered Planet* (1948) and William Vogt's *Road to Survival* (1948). Osborn summarized the situation simply and accurately: "As far as the habitable and cultivable portions of the earth's surface are concerned, there are four major elements that make possible not only our life but, to a large degree, the industrial economy upon which civilization rests: water; soil; plant life, from bacteria to forests; animal life, from protozoa to mammals" (pp. 48–49). Vogt pioneered the concept of natural capital. On the national debt he wrote: "By using up our *real* capital of natural resources, especially soil, we reduce the possibility of ever paying off the debt" (p. 44).

A little later, Paul Sears, distinguished botanist from Yale, explicitly recognized the recycling service: "Less obvious is the presence of a complex population of microorganisms and invertebrates which, among other functions, takes care of the breakdown of organic wastes and their return to chemical forms that can be reused to sustain life" (1956, p. 471).

Before the era of Leopold and Sears, the basic foundations for ecosystem ecology had been laid, providing a scientific basis for their views of the impact of human activities on earth's life-support systems. Those foundations can be traced as far back as Stephen Forbe's famous 1887 paper "The Lake as a Microcosm," which explicitly characterized one biological community within its physical context. But ecosystem ecology itself perhaps is best viewed as starting with the work of Henry Chandler Cowles (1899) on succession in the Indiana dunes. In that work the plant community and its physical environment were clearly tied together. The term *ecosystem* was first used by Tansley in his article (1935) "The Use and Abuse of Vegetational Concepts and Terms," a festschrift contribution for H. C. Cowles. The stature of Tansley as a scientist helped establish the ecosystem as a fundamental concept in ecology (Golley 1993).

The modern era of ecosystem ecology was ushered in by Raymond Lindeman's brilliant paper on a small lake ecosystem, published posthumously in 1942, shortly after his tragic death at the age of twenty-seven. He pointed out in his summary that, "Analysis of food-cycle relationships indicate that a biotic community cannot be clearly differentiated from its abiotic environment; the *ecosystem* is hence regarded as the more fundamental ecological unit" (p. 415).

The quantitative study of food chains was greatly stimulated during the early days of the nuclear age when intensive studies of the pathways of radionuclides in the environment were pursued. An energy-based approach to ecosystem studies was consolidated with the publication of Odum's classic textbook in 1953. Somewhat later, Bormann and Likens (1979) summarized their pioneering experiments that had begun in 1962 on whole watersheds. These studies demonstrated the crucial role of ecosystems in modulating the nutrient, sediments, and water budgets of landscapes. These studies, along with the efforts during the International Biological Program (IBP), in the late 1960s and early 1970s, to quantify the earth's productive capacity, firmly established the ecosystem as an important unit of study (Golley 1993). The investigation of the functioning of ecosystems centered primarily on the cycling of carbon, water, and nutrients between the biota and the soil and the atmosphere. During, and just subsequent to the IBP, enough momentum had been achieved in this area that research institutions were formed and government agencies began to organize in a manner that would allow long-term planning and funding of ecosystem research.

The environmental movement began with the publication of Rachel Carson's *Silent Spring* in 1962. Concern for preserving *ecosystem* functioning was expressed explicitly soon thereafter: "[Ecologists] realize how easily disrupted are ecological systems (called ecosystems), and they are afraid of both the short- and long-range consequences for these ecosystems of many of mankind's activities" (Ehrlich 1968, p. 47). In the first widely used environmental science text there is a chapter entitled "Ecosystems in Jeopardy," which defines ecosystems and then begins: "The most subtle and dangerous threat to man's existence . . . is the potential destruction, by man's own activities, of those ecological systems upon which the very existence of the human species depends (Ehrlich and Ehrlich 1970, p. 157).

As far as we can determine, the functioning of ecosystems in terms of delivering *services* to humanity was first described in the report of the *Study of Critical Environmental Problems* (SCEP 1970). It listed (pp. 122–125) the following "environmental services" that would decline if there were a "decline in ecosystem function":

- pest control
- insect pollination
- fisheries
- climate regulation
- soil retention
- flood control
- soil formation
- cycling of matter
- composition of the atmosphere.

This was expanded upon under the rubric "public-service functions of the global environment" (Holdren and Ehrlich 1974) to include:

- maintenance of soil fertility
- maintenance of a genetic library.

With this, the normally cited list of services was essentially complete. These were subsequently referred to as "'public services of the global ecosystem" (Ehrlich et al. 1977) and "nature's services" (Westman 1977) and elaborated upon simply as "ecosystem services" (Ehrlich and Ehrlich 1981).

Two questions about ecosystem services have been clear from the start (Ehrlich and Ehrlich 1981, pp. 95–96). One is how the loss of biodiversity will affect ecosystem services, and the other is whether it will be possible to find and deploy technological substitutes for the services. The first attempt to approach these questions systematically (Ehrlich and Mooney 1983) concluded: "The loss of services to humanity following extinctions ranges from trivial to catastrophic, depending on the number of elements (populations, species, guilds) deleted and the degree of control each exerted in the system. Most attempts to substitute other organisms for those lost have been unsuccessful, to one degree or another, and prospects for increasing the success rate in the foreseeable future are not great. Attempts to supply the lost services by other means tend to be expensive failures in the long run" (p. 248). Overall, however, the quantification of how ecosystems provide societal services has developed slowly, principally because ecosystem-level experiments are difficult, and costly, and need to be pursued for long periods of time (Carpenter et al. 1995).

Gradually, however, the extent to which species can compensate for one another in their roles in the delivery of ecosystem services has become an active area of ecological research (Ehrlich and Ehrlich 1981, B. Walker 1991, Schulze and Mooney 1993). A stimulus to this study area has been the recent concern about the consequences of the predicted massive losses of species in general but in particular on the functioning of ecosystems, and hence to the provision of ecosystem services. SCOPE (Scientific Committee on Problems of the Environment) launched a program in 1991 to assess our state of knowledge in this area in order to prepare the way for explicit experimentation. The initial activity of this assessment was a meeting in Bayreuth, Germany, in October 1991 (Schulze and Mooney 1993) where hypotheses were formulated and a plan to gather information on the following two issues was consolidated. The program focused on two basic but complex questions:

1. Does biodiversity "count" in system processes (e.g., nutrient retention, decomposition, production, etc.) including atmospheric feedbacks, over short- and long-term time spans and in face of global change (climate change, land-use change, invasions)?
2. How is system stability and resistance affected by species diversity and how will global change affect these relationships?

Vitousek and Hooper (1993) proposed a number of possible responses of ecosystem functioning to changes in species numbers in terms of model types. The data available were so poor that it was not possible to give support to any one of these models. However, they became the center of discussion and elaboration over the next several years.

The principal approach of the assessment was to look at the major biomes of the world and to examine surrogate data on the two questions posed by the program. Records are available, for example, on biological invasions, epidemics, and economic alteration of ecosystems to maximize harvesting, forces that tend to add or delete species. Such "experiments" could be used to assess the general impact of changing species diversity. The program broadened somewhat as it became part of the Global Biodiversity Assessment of UNEP (United Nations Environment Programme). New biomes were added to the assessment and the concept of diversity was broadened from looking at species only to also considering genetic, community, and landscape diversity and their roles in providing ecosystem services.

Detailed evidence was given of the ecosystem services provided by biodiversity in a number of biomes including arctic and alpine (Chapin and Körner 1995), Mediterranean (Hobbs 1992, Davis and Richardson 1995), Savannas (Solbrig et al. 1996), tropical forests (Orians et al. 1996), and islands (Vitousek et al. 1995). More comprehensive coverage of biomes, but in less detail, was given in Mooney et al. (in press). Even broader but less detailed information was given in two chapters of the Global Biodiversity Assessment (Mooney et al. 1995). This latter effort included the work of hundreds of scientists and was provided in a format that enabled cross-biome comparisons of processes. Clearly, the early ideas of ecosystem services had moved to the mainstream of ecological research.

The general conclusions of these assessments were that in many cases, we can make clear predictions of the ecosystem consequences of losses of certain types of species that possess specialized traits. For others such as keystone species, however, our knowledge base is limited and we have to rely on direct experimentation. Since keystones play such a vital role in ecosystem integrity, this calls for precaution in ecosystem management. Further, it was concluded that losses of populations reduce ecosystem flexibility to changing environments and to habitat rehabilitation. It was noted that species diversity is vital in the resilience of ecosystems to perturbation and presumably to changing environmental conditions. It was found that simple ecosystems, which have few representatives of major functional groups, such as the arctic and deserts, are particularly vulnerable to disruption from species losses. For virtually all ecosystem services it was found that species diversity was important although some services, such as primary productivity, were less sensitive to diversity than were other processes. It was concluded that our knowledge base is very poor at the moment yet suggestive of the fundamental requirement of diversity for providing ample free services to society. Moreover, since local diversity is very difficult to restore and global biodiversity loss is irreversible on a time scale of interest to humanity, we should exert great caution in our husbanding of our global biotic resources.

We are now entering a period of experimental refinement of our knowledge in the area of ecosystem functioning and biodiversity. The International Geosphere Biosphere Program is initiating a project on the role of ecological complexity in earth system functioning. Already there have been important contributions in this area utilizing model ecosystems (Naeem et al. 1994, Lawton 1995), natural climatic perturbations on a gradient of diversity imposed by nutrient variability (Tilman and Downing 1994), and most recently direct tests of ecosystem functioning in field gardens where species and functional type diversity has been manipulated (Tilman et al. 1996). All of these studies have supported the generalizations that arose from the assessment—diversity "counts" in ecosystem functioning (see also Perrings et al. 1995).

Conclusion

The scientific understanding of ecosystems and how they deliver essential services to humanity has advanced enormously since the day of George Perkins Marsh. But his view of humanity's role in the natural world was more accurate than that possessed by the average decision maker today. More than 130 years after the publication of *Man and Nature* most educated people remain sadly unaware of its basic message. Their inadvertent ignorance of the services that natural ecosystems supply to the human enterprise—of the reasons that the economy is a wholly owned subsidiary of those systems—amounts to a condemnation of schools, colleges, universities, and the print and electronic media. It also highlights the failure of professional ecologists to communicate their findings to the general public. We hope that this volume will help to end that sorry state of affairs.

References

Bormann, F.H., and G.E. Likens. 1979. *Pattern and Process in a Forested Ecosystem: Disturbance, Development, and the Steady State Based on the Hubbard Brook Ecosystem Study*. New York: Springer-Verlag.

Carpenter, S.R., S.W. Chisholm, C.J. Krebs, D.W. Schindler, and R.F. Wright. 1995. "Ecosystem experiments." *Science* 269:324–327.

Carson, R. 1962. *Silent Spring*. Boston: Houghton Mifflin.

Chapin, F.S., and C. Körner, eds. 1995. *Arctic and Alpine Biodiversity*, vol. 113. Berlin: Springer-Verlag.

Cowles, H.C. 1899. "The ecological relations of the vegetation on the sand dunes of Lake Michigan." *The Botanical Gazette* 27:95–117, 167–202, 281–308, 361–391.

Davis, G.W., and D.M. Richardson, eds. 1995. *Mediterrean-Type Ecosystems: The Function of Biodiversity.* Heidelberg: Springer-Verlag.

Ehrlich, P. 1968. *The Population Bomb.* New York: Ballantine.

Ehrlich, P., and A. Ehrlich. 1970. *Population, Resources, Environment: Issues in Human Ecology.* San Francisco: W.H. Freeman.

Ehrlich, P., and A. Ehrlich. 1981. *Extinction: The Causes and Consequences of the Disappearance of Species.* New York: Random House.

Ehrlich, P.R., and H.A. Mooney. 1983. "Extinction, substitution, and the ecosystem services." *BioScience* 33:248–254.

Ehrlich, P., A. Ehrlich, and J. Holdren. 1977. *Ecoscience: Population, Resources, Environment.* San Francisco: W.H. Freeman.

Forbes, S.A. 1887. "The lake as a microcosm." *Bulletin of the Peoria Scientific Association,* pp. 77–87. Reprinted in L. Real and J. Brown, 1991. *Foundations of Ecology.* Chicago: University of Chicago Press, pp. 14–27.

Golley, F.B. 1993. *A History of the Ecosystem Concept in Ecology.* New Haven: Yale University Press.

Hobbs, R.J. 1992. *Biodiversity in Mediterranean Ecosystems in Australia.* Chipping Norton, NSW, Australia: Surrey Beatty & Sons.

Holdren, J., and P. Ehrlich, 1974. "Human population and the global environment." *American Scientist* 62:282–292.

Hughes, J.D. 1975. *Ecology in Ancient Civilizations.* Albuquerque: University of New Mexico Press.

Lawton, J.H., 1995. "Ecological experiments with model systems." *Science* 269:328–331.

Leopold, A. 1949. *A Sand County Almanac and Sketches from Here and There.* New York: Oxford University Press

Lindeman, R.L. 1942. "The trophic-dynamic aspect of ecology." *Ecology* 23:399–418.

Marsh, G. P., 1864 (1965). *Man and Nature.* New York: Charles Scribner.

Mooney, H.A., J.H. Cushman, E. Medina, O.E. Sala, and E.-D. Schulze, eds. 1996. *Functional Roles of Biodiversity: A Global Perspective.* Chichester, England: John Wiley.

Mooney, H.A., J. Lubchenco, R. Dirzo, and O.E. Sala. 1995. "Biodiversity and Ecosystem Functioning." In *Global Biodiversity Assessment,* V. H. Heywood, ed. Cambridge: Cambridge University Press.

Naeem, S., L.J. Thompson, S.P. Lawler, J.H. Lawton, and R.M. Woodfin. 1994. "Declining biodiversity can alter the performance of ecosystems." *Nature* 368:734–737.

Odum, E.P. 1953. *Fundamentals of Ecology.* Philadelphia: Saunders.

Orians, G. H., R. Dirzo, and J.H. Cushman, eds. 1996. *Biodiversity and Ecosystem Processes in Tropical Forests.* Berlin: Springer-Verlag.

Osborn, F. 1948. *Our Plundered Planet.* Boston: Little, Brown and Company.

Perrings, C., K-G. Mäler, C. Folke, C.S. Holling, and B.O. Jansson. 1995. *Biodiversity Loss: Economic and Ecological Issues.* Cambridge: Cambridge University Press.

Schulze, E-D., and H. Mooney, eds. 1993. *Biodiversity and Ecosystem Function.* Berlin: Springer-Verlag.

Sears, P.B. 1955. "The Processes of Environmental Change by Man." In *Man's Role in Changing the Face of the Earth,* vol. 2, W. L. Thomas, ed. Chicago:University of Chicago Press.

Solbrig, O.T., E. Medina, and J.F. Silva, eds. 1996. *Biodiversity and Savanna Ecosystem Process: A Global Perspective.* Berlin: Springer-Verlag.

Study of Critical Environmental Problems (SCEP). 1970. *Man's Impact on the Global Environment.* Cambridge, Mass.: MIT Press.

Tansley, A.G. 1935. "The use and abuse of vegetational concepts and terms." *Ecology* 16:284–307.

Tilman, D., and J.A. Downing. 1994. "Biodiversity and stability in grasslands." *Nature* 367:363–365.

Tilman, D., D. Wedin, and J. Knops. 1996. "Productivity and sustainability influenced by biodiversity in grassland ecosystems." *Nature* 379:718–720.

Vitousek, P.M., and D.U. Hooper. 1993. "Biological Diversity and Terrestrial Ecosystem Biogeochemistry." In *Biodiversity and Ecosytem Function,* E.-D. Schulze and H.A. Mooney, eds. Berlin: Springer-Verlag.

Vitousek, P.M., L.L. Loope, and H. Adsersen, eds. 1995. *Islands. Biological Diversity and Ecosystem Function,* vol. 115. Berlin: Springer-Verlag.

Vogt, W. 1948. *Road to Survival.* New York: William Sloan.

Walker, B. 1991. "Biodiversity and ecological redundancy." *Conservation Biology* 6:18–23.

Westman, W.E. 1977. "How much are nature's services worth?" *Science* 197:960–964.

Part I

Economic Issues of Valuation

Chapter 3

VALUING ECOSYSTEM SERVICES: PHILOSOPHICAL BASES AND EMPIRICAL METHODS

Lawrence H. Goulder and Donald Kennedy

Societies often must choose between alternative uses of the natural environment. Should a given wetland be preserved, or should the land be drained and converted to agricultural use? Should a particular timberland be maintained in its current state, or should it be opened to forestry or other development? Should a certain park be maintained, or converted to a parking lot? These are difficult questions. The way they are answered has critical importance for the viability of species in the habitats involved as well as the performance of the complex ecosystems of which they are a part.

To make rational choices among alternative uses of a given natural environment, it is important to know both what ecosystem services are provided by that environment and what those services are worth. The first item lies in the realm of fact; the second, the realm of value. Societies cannot escape the value issue: whenever societies choose among alternative uses of nature, they indicate (at least implicitly) which alternative is deemed to be worth more. In many instances, environmentally concerned individuals sense that the wrong decision has been made—that society has imputed insufficient value to nature in its current state and has thereby permitted conversion to take place for the sake of an inferior alternative. Indeed, one may sense that nature routinely is undervalued. No matter how strong suspicions are along these lines, one cannot make a convincing case that nature is undervalued without having a philosophical and empirical framework for assessing nature's values. The philosophical element seeks to identify the ethical or philosophical basis of value, that is, articulate what constitutes the source of value. The empirical element aims to find techniques for the measurement of value, as defined according to a given philosophical notion.

This chapter considers both components in offering a framework for valuing ecosystem services. While most of the other chapters in this volume examine valuation issues as they apply to particular ecosystem services (soil conservation, pest control, pollination, etc.), this chapter is more philosophical and broader in its focus. Our attention to philosophical underpinnings helps clarify the ethical issues underlying different approaches to value. And our general approach to empirical valuation methods helps convey the range of empirical approaches available to researchers, as well as the strengths and limitations of these approaches.

The Philosophical Basis of Value

From what do nature's values derive? When we claim that a given living thing or species or habitat is worth such and such, what is the basis of that claim?

Competing Approaches

A broad class of approaches to value is represented by anthropocentric viewpoints: elements of nature are valuable insofar as they serve human beings in one way or another. Within the anthropocentric group is utilitarianism, which maintains that natural things (indeed, all things) have value to the extent that they confer satisfactions to humans. Economists endorse the utilitarian viewpoint; as we will discuss later, this approach is inherent in benefit-cost analysis.

At first blush, it might seem that a utilitarian basis for value cannot be consistent with safeguarding the planet or protecting "lower" forms of life. But utilitarianism does not necessarily imply a ruthless exploitation of nature. On the contrary, it can be consistent with fervently protecting nonhuman things, both individually and as collectivities. After all, we may feel that the protection of certain forms of life is important to our satisfaction or well-being, and thus be led to place a high value on these forms. Utilitarianism doesn't rule out making substantial sacrifices to protect and maintain other living things. But it asserts that we can assign value (and therefore help other forms of life) only insofar as we humans take satisfaction from doing so. The notion of satisfaction here should be interpreted broadly, to encompass not only mundane enjoyments (as with consuming plants or animals for food) but also more lofty pursuits (such as marveling at the beauty of an eagle).

The utilitarian approach allows value to arise in a number of ways. It embraces both direct use values (for example, the satisfaction from eating fish) and indirect use values (for example, the value that can be attached to plankton because it provides nutrients for other living things that in turn feed humans). This approach does not restrict value to forms of nature that are consumed: there are both consumptive and nonconsumptive use values. An example of the former are the values that might be attached to ducks insofar as they provide food. An example of the latter are the values we attribute to ducks that provide pleasure to bird watchers. This approach also includes non-use values: values that do not involve any actual direct or indirect physical involvement with the natural thing in question. The most important value of this type may be existence value (or passive use value)—the satisfaction one enjoys from the mere contemplation of the existence of some entity. For example, a New Jersey resident who has never seen the Grand Canyon and who never intends to visit it can derive satisfaction simply from knowing it exists.[1] The array of services provided by ecosystems spans all of these categories of values. The pest-control and flood-control services they offer have direct use value to nearby agricultural producers.[2] Their provision of habitats for migratory birds implies an indirect use value to people who enjoy watching or hunting these animals; depending on whether such birds are hunted or just observed, the indirect use value may be consumptive or nonconsumptive. Ecosystems also yield an existence value: wetlands, for example, provide such value to people who simply appreciate the fact that wetlands exist.

One can distinguish weak and strong forms of utilitarianism. The weak form asserts that the value of a given species or form of nature to an individual is entirely based on its ability to yield satisfaction to the person (directly or indirectly). The stronger form makes an assertion about the value of a species (or other natural thing) to society. It claims that the value to society of the natural thing is the sum of the values it confers to persons.

This stronger form of utilitarianism is inherent in benefit-cost analysis. An attraction of strong utilitarianism is that it provides a rather convenient way of ascertaining social values of alternative policies and thus offers a way to make difficult decisions. Benefit-cost analysis seeks to ascertain in monetary terms the gain or loss of satisfaction to different groups of human beings under each of various policy alternatives. Under each alternative, it adds up the gains and subtracts the losses, and then compares the net gains across policy options. Importantly, benefit-cost analysis doesn't cast judgment on the differences between one person's valuation of a given species and another's. Each person's valuation receives the same weight. It makes no attempt to correct for differences in awareness, education, or "enlightenment" among individuals. The preferences of people who have no concern for future generations, or who have no sense of the ecological implications of their actions, count the same as those of people who are more altruistic or who recognize more fully the fragility of ecosystems. Benefit-cost analysis is nondiscriminating, perhaps to a fault.[3]

Among ecologists' criticisms of benefit-cost analysis, this is one of the most important. When philosophers argue with economists about the use of benefit-cost analysis, a critical underlying issue is whether some preferences are better than others and ought to count more.[4] Note that it is perfectly consistent to uphold the weaker form of utilitarianism—to believe that human satisfaction is the source of all value—while rejecting the stronger form, that is, while maintaining that some persons' valuations ought to take precedence over others'. In this case, it is the strong utilitarian assumption that social value is just the sum of the individual valuations that is objectionable.

Some would argue that the fate of other species becomes too precarious when it must depend on a link to human satisfactions. The utilitarian view contends that if a species doesn't convey satisfaction—either directly or indirectly—to human beings, it should be given no value, and thus no sacrifice to protect this species is warranted.[5] Many philosophers are uncomfortable with these implications; some have embraced, as an alternative, an intrinsic rights approach to dealing with other species. According to the intrinsic rights view, species and other natural things have intrinsic rights to exist and prosper, independent of whether human beings derive satisfactions from them. Many animal rights advocates appeal to certain intrinsic rights. In *Animal Liberation* (1975), ethicist Peter Singer argues that nonhuman animals have the basic right to be spared of suffering that is deliberately caused by humans. This argument is grounded in the notion that, like humans, other animals are sentient creatures, capable of experiencing pleasure and pain, and that there is something fundamentally wrong about causing pain to any creature.[6]

The intrinsic rights approach falls within the category of biocentric (as opposed to anthropocentric) approaches. It puts other living things on a moral plane comparable to that of human beings. Defenders of anthropocentrism point out that since human beings are the dominant species on the planet, they are obliged to define ethical principles in terms of human wants and needs (see Watson 1983). But biocentrists can counter by pointing out the following implication of anthropocentric logic: Suppose that representatives of another species should arrive from outer space, a species that is clearly superior to human beings in intelligence, perceptiveness, and technological know-how. To the extent that defenders of anthropocentrism have invoked the "dominant species" argument, consistency would seem to require them to yield the central moral status to this new, superior species.

Consistency would require humans to abide by whatever decisions are made by this other species, since humans would no longer have any moral authority.[7] This may be troubling to many of us. This *reductio ad absurdum* argument can be invoked to support biocentric approaches that are more generous in the allocation of moral status.

A similar rights-based approach is given by the Kantian categorical imperative. The essence of Kantian justice is the notion that each human being should only act in ways that are able to be universalized in the sense that they would seem appropriate for any human being in the comparable situation. To determine the right action in a situation involving alternative choices, one should first remove from consideration one's own stake in the outcome, and imagine what action one would be willing to uphold for any individual facing the same circumstances.[8]

Kant's approach gives rise to certain rights and obligations, including certain duties to "lower" animals. This may seem to imply a biocentrism in Kantian justice. But in fact Kantian justice is anthropocentric in that it confers no moral status to nonhumans, as the following passage (Kant 1963, 239–241) suggests: "so far as animals are concerned we have no direct duties. Animals are not self-conscious and are there merely as a means to an end. That end is man. Our duties toward animals are merely indirect duties towards humanity."

Ethical Bases and Social Decisions

It's easy to get entranced by philosophical nuance, but our main concerns here are practical—recognizing values in order to make ethical collective decisions about the preservation of nature. What does this brief excursion into the bases of value imply about valuing ecosystems in practice?

First, it establishes the utilitarian—indeed, strong utilitarian—basis of benefit-cost analysis. As traditionally practiced, benefit-cost analysis not only regards human satisfaction as the source of the value of every natural thing, but also gives the same ethical status to every person's valuation. Some ecologists might concede the economists' claim that value is sourced in individual satisfactions yet insist that social value, or just decision making, should not be determined simply by adding up individual values. Some economists might counter by defending strong utilitarianism. But there is an intermediate position that we find attractive. In cases where various policy alternatives are related to the use of a given habitat, we would recognize that the results of a benefit-cost study are not sufficient to settle the question of which policy is best. There are some ethical dimensions that benefit-cost analysis cannot consider. Here is where the nondiscriminating element of benefit-cost analysis falls short. To the extent that ecologists can show that the general public was unaware of significant ecological issues in forming their own valuations, this seems relevant to decision making. Moreover, even if individual valuations were based on very good information, there is an ethical dimension to the decision—associated with how the benefits and costs are distributed across affected parties or generations—that is not addressed by the simple adding up of individual benefits and costs. At the same time, we would affirm that benefit-cost information—in particular, the aggregate net benefits from various alternatives—remains useful in weighing the various policy alternatives.

A second main insight is that the leading alternatives to utilitarianism (and benefit-cost analysis) usually do not deal with "values" at all! The exercise of imputing values to different elements of nature is part and parcel of utilitarianism, but is not an essential ingredient of intrinsic rights or Kantian approaches to decision making. If one adopts an intrinsic rights or Kantian approach, the choice as to how to choose among policy alternatives usually reflects issues of whether fundamental rights are violated or whether the action in question is able to be universalized. Values may not be a central part of the consideration. This is important, because it suggests that making arguments for social policy by referring to the "value of ecosystem services" is to conform, to a degree, to the utilitarian approach. Ecologically concerned individuals should recognize that this is the case, and realize what issues can and cannot be addressed by a focus on values.

Measuring Ecosystem Values

It is difficult enough to agree on a philosophical basis for value. Further difficulties arise in attempting to measure nature's values (after assuming a given basis for value). This section presents some important measurement methods. Considerable progress has been made over the years in developing such methods. But the science is far from perfect. Controversies persist.

Ecosystem services are especially difficult to measure for the same reason that ecosystems themselves are threatened. Many of the services provided by ecosystems are positive externalities. The flood-control benefits, water-filtration services, and species-sustaining services offered by ecosystems are usually external to the parties involved in the market decision as to whether and at what price a given habitat will be sold. As a result, the habitats that support complex ecosystems tend to be sold too cheaply in the absence of public intervention, since important social benefits are not captured in the price. Public attention to the values of these (largely external) benefits is important to provide support for reasonable public policies to protect important habitats. This makes it all the more important to determine the values of these services. At the same time, it explains why gauging these values is so difficult: in many cases the values of these services are not directly expressed in market prices.

The prevailing approach to ascertaining value is benefit-cost analysis. As indicated, benefit-cost analysis implicitly adopts the utilitarian basis for value. The value of a given living thing is the amount of human satisfaction that thing provides. How could such satisfaction be measured? Nearly every empirical approach assumes that the value of a given natural amenity is revealed by the amount that people would be willing to pay or sacrifice in order to enjoy it. Willingness to pay is thus regarded as the measure of satisfaction.

It is important to note that willingness to pay is not always an actual, expressed willingness; it is not restricted to what we observe from people's actual payments in market transactions. Rather, it expresses how much people would be willing to pay for a given good or service, whether or not they actually have opportunities to do so. Market behavior often gives evidence of willingness to pay, but in many instances researchers must rely on other, more indirect methods to fathom it.

Ecosystem Services and Valuation Methods

The myriad services offered by ecosystems can be divided into three main categories: (1) the provision of production inputs, (2) the sustenance of plant and animal life, and (3) the provision of non-use values, which include existence and option values. Different types of valuation techniques are called for, depending on the category of service involved. Table 3.1 shows the relationships between service types and valuation methods.

Valuing Production Inputs

Table 3.1 lists four examples of production inputs from ecosystems: pest control, flood control, soil fertilization, and water filtration. These services are inputs to the sustained production of agricultural products in the sense that it would be difficult to maintain agricultural production without pest control, flood control, fertile soil, or (at least in some cases) relatively pure water.

One can place a value on these production inputs by recognizing what costs or expenditures agricultural producers manage to avoid by virtue of the availability of these inputs. For example, where ecosystems provide effective pest control, farmers can avoid undertaking expenditure on alterna-

Table 3.1. Ecosystem services and valuation methods

Service	Valuation Method
Provision of Production Inputs	
Pest Control	Avoided Cost
Flood Control	Avoided Cost
Soil Fertilization	Avoided Cost
Water Filtration	Avoided Cost
Sustenance of Plant and Animal Life	
Plants/Animals with Direct Use Values	
• consumptive uses	Direct valuations based on market prices
• nonconsumptive uses	Indirect valuations (travel cost method, contingent valuation method)
Plants/Animals with Indirect Use Values	(No valuations necessary if plants/animals with direct use values are counted)
Provision of Existence Values	Indirect valuations (contingent valuation method)
Provision of Option Values	Empirical assessments of individual risk-aversion

tive pest-control methods such as the use of synthetic pesticides (Naylor and Ehrlich, this volume). To the extent that data are available on expenditures on synthetic pesticides, they provide an indication of the value of the pest-control services provided by ecosystems.[9]

Similarly, the flood-control services offered by ecosystems eliminate farmers' needs to undertake alternative flood control expenditures. The avoided costs of flood-control again indicate the value of the services provided by ecosystems; here the cost may be avoided by taxpayers (who otherwise pay for flood-control projects), rather than farmers, but the principle still applies. The same logic applies to soil fertilization and water filtration services.

Of course, farmers' circumstances vary, and the avoided costs associated with these ecosystem services will therefore vary for different farming enterprises. Given this heterogeneity, it becomes difficult to pinpoint the ecosystem values. Nevertheless, attention to avoided costs offers a very useful gauge of the values of the production inputs supplied by ecosystems.

Valuing Plant and Animal Life

As suggested by table 3.1, a second main type of service provided by ecosystems is the sustenance of plant and animal life. In choosing a method for valuing this type of service, it helps to distinguish living things with direct use values from those with indirect use values. Examples of the former are plants or animals that are consumed as food or that directly offer recreational values (sightseeing, nature watching, etc.). Examples of the latter are plants and animals (such as organisms that are lower on the food chain) that help sustain other plants and animals that we enjoy directly. To give specific examples: ecosystems generate direct use values by supporting the various types of birds that we enjoy either nonconsumptively as bird watchers or consumptively as bird hunters. They generate indirect use values by supporting the life of various plants or insects that in turn enable birds to thrive.

Direct, Consumptive Use Values. When direct use values are involved, two main valuation methods may apply. In the case of direct consumptive use values, it may be possible to employ direct valuation methods based on market prices. When natural ecosystems provide a habitat for animals that are harvested and sold commercially, the commercial market value provides a gauge of the value of the habitat services. For example, part of the value of marine ecosystems is conveyed by the value of the commercial fish that they help sustain. Of course, this only represents a portion of the value of the ecosystem—namely, the value of the ecosystem's potential to sustain fish with a market value.

There is an important difference between the marginal and the total value associated with market prices or the willingness to pay of consumers in markets. Economists regard the prices that people are willing to pay as indicators of the marginal value—the value they place on the last unit purchased. Consider what a homeowner would be willing to pay for residential water in a given month. He might be willing to pay a huge sum for the privilege of consuming the first ten cubic feet, because doing without them would deprive him of even the most fundamental (and valuable) uses of water for that month: drinking water, the occasional shower, etc. The next ten cubic feet would probably not be worth quite as much. They would allow him additional opportunities to fill a glass from the faucet, and an extra shower or two, but these would not be as critical to him (or to the people with whom he associates!) as the first ten cubic feet. Thus the marginal value of water—

the amount one is willing to pay for each successive increment—falls steadily.

Figure 3.1 displays a typical willingness-to-pay schedule. The first cubic foot is shown to be worth a great deal more than the fiftieth, which in turn is worth much more than the hundredth. In reality, of course, households don't have to purchase each unit of water at its marginal value. If they did, they would be charged larger amounts for the first increments than for later ones. Instead, households generally pay a given price per unit of water, regardless of how much they consume.[10]

In figure 3.1, the horizontal line at $0.02 represents the price charged for the water. (We use this number arbitrarily.) The standard economic assumption is that users will continue to purchase water until the marginal value of the water (or marginal willingness to pay) is equal to the marginal sacrifice (or price). Under these circumstances, the price is an expression of the marginal willingness to pay, or marginal value. (In the example of figure 3.1, the user would demand four hundred cubic feet of water per month at this price.)

The total value of the water consumed is much more than the price, however. The total value is the area under the marginal willingness-to-pay schedule (the sum of areas I and II in the diagram). Note that to ascertain total value (as opposed to marginal value), researchers need to have infor-

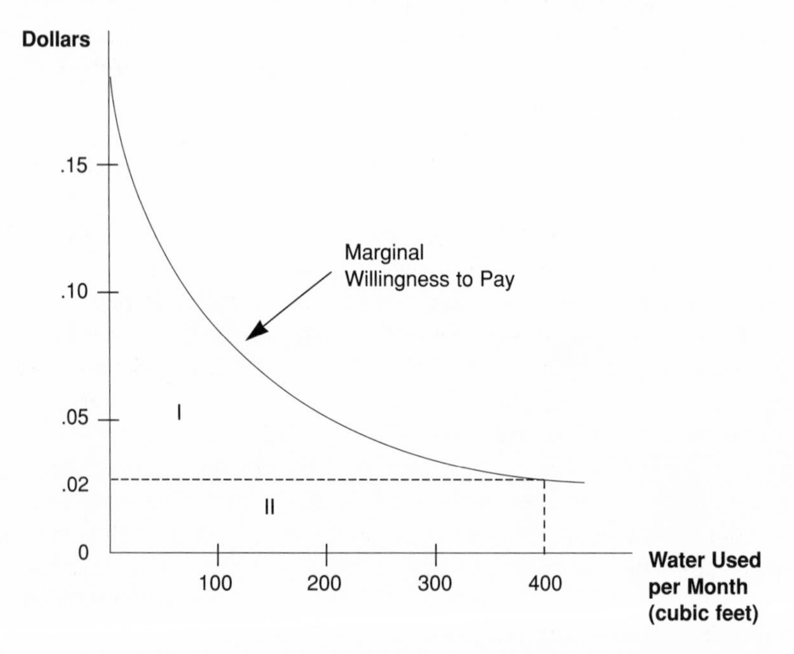

Figure 3.1. Relationship between water use and marginal willingness to pay.

mation on the entire marginal willingness-to-pay schedule (or demand curve), not just the price paid.[11] A main challenge of empirical valuation techniques is to trace out marginal willingness-to-pay schedules.

In the context of commercial products of ecosystems, this means that market prices represent only the marginal value of these products. The value of the total sales of these products corresponds to area II in figure 3.1. Note that this is less than the total value to consumers, which is the sum of areas I and II. Thus market sales understate the overall value of the commercially viable forms of life supported by ecosystems.

Direct, Nonconsumptive Use Values. Within the category of direct use values from living things maintained by ecosystems, we have another case to consider. This is the case where the life forms are used nonconsumptively. For such uses, the relevant markets do not usually arise, and thus it is not possible to gauge values directly by observing market prices.[12] For example, there usually are no markets for the bird-watching opportunities that ecosystems provide by offering suitable habitats. In these cases, it is necessary to apply more inferential methods to ascertain the relevant values.

The travel-cost method is a widely used inferential approach. The method has been applied to ascertain some of the values provided by parks, lakes, and rivers—or, equivalently, the costs that result from the loss of these elements of nature. The nonconsumptive uses are not directly bought or sold in markets; prices are not usually charged for their use. And in those instances when use prices are charged (through entry fees, etc.), the prices are unlikely to be good indicators of (marginal) value. That is because the users of these resources actually "pay" more than the entry fees to use them. For example, the cost of the family visit to Yosemite National Park is much greater than the $15-per-day use fee. The travel-cost method recognizes that by adding to the entry fee (if any) the transportation cost and time cost expended to visit a particular site, one can ascertain the overall travel cost. This method regards the overall travel cost as a measure of the marginal willingness to pay by a visitor to the park; this is considered to be the same as the marginal value of the park to the visitor. The underlying assumption is that people will continue to visit the park until the value of the last unit (that is, the marginal value) is just equal to the travel cost.[13]

It is also possible to use survey methods, such as the contingent valuation method, to determine how much value people place on the nonconsumptive uses.[14] Many economists distrust results from survey approaches, claiming that individuals' asserted preferences in the hypothetical circumstances posed by surveys bear no systematic relationship to their true preferences. Defenders of survey methods counter that, in many cases, surveys are the only method available. This "only game in town" argument may have force when existence values are involved, as discussed below.

Indirect Use Values. Ecosystems support many "lower" forms of life that provide only indirect use value. It is sometimes argued that the value of ecosystem services should include the values of the services provided by these life forms. But in fact there is no need to include the values of these services in an accounting of the overall value of an ecosystem. The values of these services are already captured in the values attached to the life forms that humans enjoy directly. Consider the value of certain plants whose fruits are eaten by birds and other "higher" life forms; assume humans obtain no direct use value from these plants. If we abide by the utilitarian approach to value, then there is no value to these plants over and above the value that we attach to the higher life forms to which they contribute.[15] To add their indirect use values to the direct use values would be double-counting.[16]

Non-Use Values

Some of the values from ecosystems do not involve direct or indirect uses of the good or service in question. These are non-use values. There are two main types of non-use value.

Existence Value. This is the value that derives from the sheer contemplation of the existence of ecosystems—apart from any direct or indirect uses of goods and services they provide.

Survey approaches such as contingent valuation assessments may be the only way of ascertaining existence value, since actual market and nonmarket behavior gives little hint of its magnitude. As mentioned, survey approaches are controversial. Yet they may be the only way of measuring existence values because people's actions do not leave a "behavioral trail" from which their valuations can be inferred. In this limited space we cannot offer an appraisal of survey approaches.[17] But we can point out what seems to be the key underlying question: whether the information obtained from surveys, however imperfect, is better than no information at all. In the next section we revisit issues of uncertainty and imperfect information.

The existence value could include a pure biodiversity component. This is the appreciation for the variation or richness we observe in the ecosystem; it is based on the contemplation of the ecosystem as an ensemble of life forms, as contrasted with an appreciation for each of its members individually. Although we mention this value in connection with existence value, the pure biodiversity value may also have a use-value component: we take pleasure in the ecosystem's heterogeneity when we visit the habitat in question and observe the diversity of life forms that reside there.

Option Value. As developed in the economics literature,[18] the term "option value" refers to a premium that people are willing to pay to preserve an environmental amenity, over and above the mean value (or expected value) of the use values anticipated from the amenity.[19] This premium reflects individual risk-aversion: in the absence of risk-aversion, people's willingness to pay would equal the mean use value (its expected value), and option value would be zero. It is much easier to define option value than to measure it. Its measurement requires a gauging of individuals' risk-aversion, and this may depend on the specific context: persons are not equally averse to different types of risk. For an empirical assessment of option value, see Fisher and Hanemann 1986.

Marginal vs. Total Value

In much of the preceding discussion, we have concentrated on measurement of the total value of ecosystems. But in many real-world circumstances, the policy debate concerns the change in value or marginal loss of value that results from alteration or conversion of a part of the region that occupies an ecosystem. In benefit-cost analyses, when a portion of the ecosystem is threatened with conversion, it may be more important to know the change or loss of ecosystem value associated with such conversion than to know the total value of the entire original ecosystem. Does a "minor" encroachment on the land area of an ecosystem generate small losses in ecosystem value, or do small encroachments precipitate large damages?

To examine this issue, we can begin with a very large area of a (relatively) undisturbed ecosystem.[20] The value we place on a given amount of area lost to other uses depends on the area of this system.[21] Let A represent the land area of our ecosystem, and suppose that the initial area is A_0. This ecosystem, valued for its natural beauty and its biological diversity, is being decreased marginally in area by being converted to farmland. Suppose first (counter to fact) that this decrease takes place without changing the ecosystem's character through species loss. Since a larger area is worth more than a small one, the marginal value of each withdrawn unit rises gradually as the area (A) decreases. But in the limit, an area of size zero is worthless, and tiny areas are less attractive because they have a rather zoo-like character. Thus at small values of A, the marginal value begins to fall again. This relationship is shown in the path marked "1" in figure 3.2. The relationship between area and value expresses the pure ecosystem-scale effect.

In fact we know that the biological diversity of the ecosystem—one of the features contributing to its value to nature lovers—is not area-independent. The relationship, established mainly in studies on islands and (to a more limited extent) on tropical forests, is a nonlinear one. The precise form varies, but in a variety of studies the number of species lost is slight until a quarter to a half of the area is lost, and rises precipitously after about three-

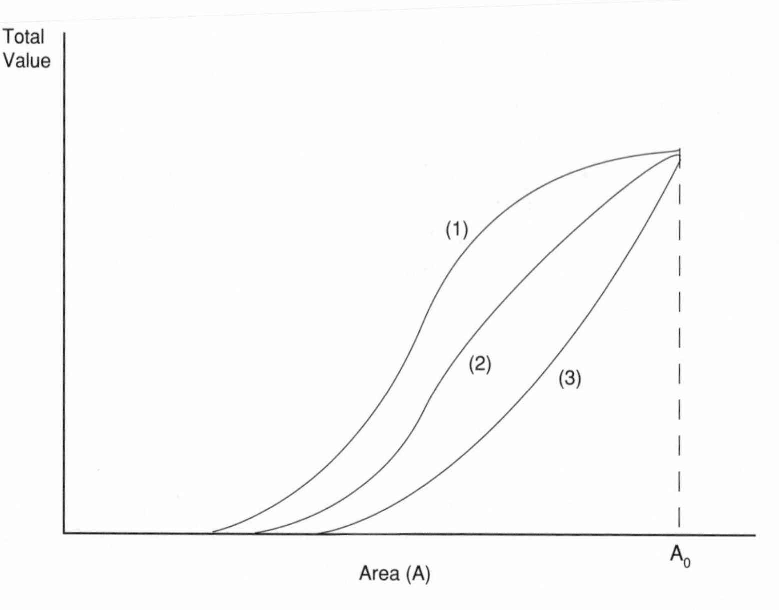

Figure 3.2. Habitat area and ecosystem value: ecosystem-scale, diversity, and species-composition effects.

quarters of the area is lost. As A is reduced the effect on marginal value is to exaggerate the loss of ecosystem value. The impact of the loss in numbers of species as A is reduced may be termed the diversity effect. This effect is taken into account in the path marked "2" in figure 3.2. As indicated by the differences between paths 1 and 2, this intensifies the marginal loss of value from a given reduction in A.

There is a third effect that needs to be considered. The species in ecosystem A are not considered to be of equal value to humans. People seem to care more about eagles and panthers than about mosses and bacteria. We also know that species are related to one another in a complex, co-evolved web of dependencies: prey and predator, plant and pollinator. Trophic relationships are also vitally important. Often, higher-order species on the food chain have the most exacting environmental requirements and are thus valuable indicators of the health of the entire ecosystem; they or others may also be critical "keystone" species because they are located at the center of a network of interdependencies. Thus, as a practical matter, species values become proxies for ecosystem values: the Endangered Species Act in the United States is an embodiment of this principle in policy. And of course we regularly justify large expenditures to save some species (e.g., the African rhinoceros) but not others (there is no Save the Furbish Lousewort Society).

On what basis do we assign value to species? The following are some axes along which different people make selections.

Taxonomic Proximity. We like animals that are like us. Primates attract human attention not only because there may be utility in the relationship ("animal models" for human disease) but because we respond to their quasi-human qualities.

Rarity. All other things being equal, we have more interest in rare things than in common ones. This is not simply a matter of vulnerability, although it is true that rare organisms are more vulnerable to extinction than abundant ones. Rarity itself can be the attraction; in some sense animals and plants in nature are "collectibles," if only in the sense of finding and listing them, and collections of the rare are more desired than collections of the commonplace. Indeed, "collecting" in the form of listing is a motive with powerful economic consequences. Many bird watchers will undertake extreme expenditures to visit ecosystems harboring rare species for the purpose of expanding their "life-lists."

Genetic Uniqueness. If a species represents a unique evolutionary line—is, for example, the only extant member of its genus or family—then it may be entitled to higher value than it would otherwise. Scientists especially would favor the use of this criterion.

Importance to Ecosystem Function. Certain species (such as "dominant" and "keystone" species) create conditions that permit the maintenance of the entire ecosystem. The dominant trees in a forest, or birds that dig nest holes in trees that are used by other species, or insects vital to the pollination of a dominant plant, would be examples.

How can these preferences be related to the marginal value calculation? Biological diversity is reduced as A shrinks, but species do not fall out randomly; certain kinds tend to drop out relatively early, others only when A becomes quite small. For conservation biologists and others, this means that wise policies cannot be made unless some value is attached to the different kinds.[22]

Obviously the number of possible criteria is large enough to prohibit development of a precise relationship among area, species loss, and value. But larger organisms with broad ranges that are especially area-sensitive would be likely to be rarer, on average closer taxonomically to humans, favored for "charm," and important to ecosystem function. Thus it is reasonable to assume a species-composition effect: that as A is reduced, the species lost early in the reduction are more valuable than those lost later. When this effect is taken into account, the marginal loss from a reduction in species area is even greater than indicated by path 2. Path 3 incorporates this effect (and the others).[23]

Obviously some of these relationships are uncertain, and the exercise could be applied to real natural areas only after substantial research. But it points up the importance of thinking about value in marginal rather than aggregate terms and suggests a discipline that could be applied in the framing of general conservation policy.

Uncertainty and Policy Making

It is evident that precise information on the values of ecosystem services will often be lacking. That is a fact of life, yet we still need to make choices. How can they be made as rationally as possible? What is the right framework for decision making in a realm of uncertainty?

An important first principle emerges from the still developing discipline of risk assessment, best known in the context of efforts to make quantitative estimates of the risks (to health, safety, and the environment) of pollutants. Plainly the same analytical format is applicable to other kinds of global change, including loss of ecosystem services. The frequent criticisms of the use of risk assessment in toxic substances regulation have usually singled out uncertainties in the extension of animal data to humans and in the extrapolation of results from high dosages to low. But a more serious problem has been a tendency to deliver assessments in the form of point estimates rather than probability distributions.

In the even more uncertain domain of ecological risk assessment, the density of probabilities around any estimate may be as important to the formation of policy as the point estimate itself. This is the case for several reasons.

First, many people, including those involved in the policy process, are risk averse and likely to concentrate attention on the unfavorable or high-cost side of the distribution of possible outcomes. If society as a whole approaches risk from a conservative or risk-averse position, the information contained in a distribution may be vitally important; that is especially true if the variance among estimates is high, that is, if the distribution is broad.

Second, the distribution of risk estimates is sometimes skewed, often toward the downside. The asymmetry is often not revealed until independent estimates are pooled. Furthermore, experience often modifies our view of a certain class of risk. For example, the adverse environmental consequences of introducing exotic organisms for control purposes have nearly always been unforeseen—so much so that it is probably now prudent to assume the worst.

Third, ecological risk factors are frequently multiple and interconnected. The leverage provided by such relationships is difficult to predict, but it is clear that apparently independent events may summate to produce levels of effect much greater than the sum of disaggregated risk estimates. Thus the distribution of outcomes for the system as a whole is much broader than one would expect based on risk assessments that concentrate on the individual risk factors separately. For example, a major environmental issue concerns the state of Everglades National Park in Florida, where a unique wetland ecosystem is threatened because historic flows of fresh water into the area have been slowed by human activity. It is interesting to note that in the complex history of development that has led to the present state of affairs in the Everglades, no single development—the Tamiami Canal, the engineering projects on the Okechobee drainage, or the intensification of agriculture—would by itself have been predicted to interrupt sheet-flow into the central Everglades and thereby disrupt the entire ecosystem.

Finally, the time dimension is often ignored in traditional risk assessment, yet the dynamic character of ecological risk often raises the time issue in a way that should amplify our policy concerns. In the first place, ecological change often shows a strong hysteresis: restoration processes work slowly, and intense perturbations may exact costs over a very long period. Second, and perhaps more important, human preferences—in this case, our interest in natural ecosystems—have changed with industrialization and affluence. Such changes pose a challenge for traditional benefit-cost analysis, since our assumptions about future preferences may err in understating value for future generations. Problems of intergenerational equity, difficult as they are to resolve, are at the heart of ecosystem valuation.

Examples of Real-World Valuation Challenges

The challenge in real-world decision making about land use is to evaluate the costs of altered use against the benefits. The latter are relatively easy to measure, but the costs of conversion—that is, the value of the loss associated with the "native" ecosystem—are much more difficult. The following examples illustrate some of the problems.

Wetlands

Consider the following situation applying to Jason Shifflet, a hypothetical farmer in the lower Mississippi valley. Shifflet has on his property a fifty-acre forested wetland. He wishes to drain this wetland, harvest the trees, and convert it to productive cropland. The parcel is connected (barely) to a larger swamp on state land; the entire wetland has been used heavily by local duck hunters and bird watchers.

In order to accomplish the conversion, Shifflet must follow provisions under two different federal laws. Under the terms of the "swampbuster" section of the 1990 farm bill, he would become ineligible for Department of Agriculture farm program benefits. Shifflet is not bothered by this, since his operation has been subsidy free, but he is concerned with meeting the requirements of Section 404 of the Clean Water Act. In order to ditch and drain the property, he is required under that law to obtain a permit from the U.S. Army Corps of Engineers. The law requires that steps be taken to minimize or avoid impacts on wetlands and to provide compensation for unavoidable impacts by other activities to restore or create wetlands.

Shifflet applies, emphasizing the care with which he proposes to accomplish the drainage. He will leave the portion of the property closest to the state land untouched and create a new wetland of nearly equal area on another piece of land he owns. His application will be examined not only by the Corps of Engineers but by the Environmental Protection Agency as well. They will look carefully at other values of the wetland parcel that Shifflet proposes to convert. They may apply value estimation methods that would encompass both consumptive (duck-hunting) and nonconsumptive (bird-watching) uses, by using travel-cost and other measures. Existence values may or may not be considered; contingent valuation techniques have been applied to some situations.[24]

In the end, Shifflet's application is denied; when added to the state parcel, his wetland generated substantial recreational values and represented—in the view of EPA reviewers consulted by the Corps of Engineers—an unacceptable loss. The Corps was inclined to agree, since Shifflet's drainage plan would have altered the remaining wetland area in unpredictable ways.

Shifflet has since become an active member of the Lower Mississippi Property Rights Forum, an organization dedicated to lobbying in favor of the application of "takings" provisions to lands devalued as a result of regulatory decisions.

The Galapagos Islands

A second example, international in character, is provided by the Galapagos Islands. This archipelago, located six hundred miles west of the Ecuador coast, consists of thirteen large islands and a number of smaller ones. All are of recent volcanic origin (100,000 to a million years old), and they contain a unique assemblage of plants and animals. They were visited by Charles

Darwin during the voyage of the *Beagle*, and now are an important site for contemporary studies of evolutionary biology—many of them carried out under the auspices of the Darwin Research Station located on the largest island.

Managed as a national park by Ecuador since the 1950s, the islands have also become a favorite destination for tourists, who explore the islands from boats and debark on the islands to follow carefully marked trails in the company of trained naturalist-guides. With the growth in popularity of "ecotourism," the Galapagos now attract about fifty thousand visitors each year.

There is a resident population on several of the larger islands, with a few service industries and a subsistence economy that depends on agriculture and fishing. These have been augmented by other direct uses that compete with the "natural" state of the islands. A significant fishery for sea cucumbers, a delicacy prized in Asian and French cuisine, has developed. Not only does it threaten the rich intertidal fauna; it has posed significant risks to the terrestrial ecosystem, through the introduction of "exotic" species and destructive camping on some islands. Other extractive industries are either established or in prospect.

Arrayed against these direct, consumptive use values are two other values. The first is the direct, nonconsumptive use value from ecotourism, which brings significant revenue. A sample calculation of this value would estimate that the average visitor spends the equivalent of $3,000 (a week on a boat is a typical excursion). If the visitor is from the United States, additional revenue will accrue to the Ecuadorian economy through accommodations on the mainland, the flight to the islands, and (if a national carrier is used) the flight to Quito or Guayaquil. A total per-visit value of $5,000 would be a reasonable figure for the "overseas" visitor: if half were Ecuadorian nationals and half from elsewhere, the value of the industry would be $200 million annually.

Local residents, however, would make quite a different calculation. The T-shirt shops and restaurants at Puerto Ayora collect some money, and the support of the Darwin Station by tourists flows into the local economy. Some boat operators are islanders, and some services for all vessels are locally provided. But the vast majority of the revenue flows to tour operators, many of them non-Ecuadorian, and to other off-island entities.

Thus it is not surprising that a sometimes violent controversy has arisen over the protection of the islands. When the government closed the sea-cucumber fishery in 1994 because the catch limit was being vastly exceeded, fishermen and some other local residents seized the Darwin Station and took scientists hostage. In a political controversy over a bill that would have given the fifteen thousand islanders more local autonomy (and relaxed many of the ecological protections) there was another takeover. The tense contest between extraction and conservation in the Galapagos is, at least with respect to this particular indirect use value, the result of distributional effects. The economic potential of ecotourism is almost certainly greater than that of the resource-extraction uses. Yet the residents retain most of the rents from the second and little from the first.

A second use value stems from the (uncertain) future benefits that would emanate from the scientific research under way on the Galapagos. The large number of endemic species found there, and the recency of their evolutionary divergence from mainland relatives, make the islands a living laboratory for studies of species formation. Important recent work (see Grant 1986) depends on the integrity of the ecosystems of certain islands. Calculating its value, of course, would be extremely difficult.

Finally, there are two important non-use values. First, as in the case of the wetland example, people who have never been to the Galapagos and never expect to, may experience a loss of existence value that they would willingly pay to avoid. The unique quality of the islands and the considerable publicity they have received as a mecca for naturalists gives this consideration a weight it might lack in less special areas. In addition, in the presence of uncertainty, people might be willing to pay a premium (over and above the expected future use value) to ensure the preservation of the unique flora and fauna of the islands. This is the option value.

Conclusions

To assess the value of ecosystem services we must choose among alternative philosophical bases of value as well as alternative measurement techniques. Philosophers will continue to debate the relative merits of alternative philosophical approaches, and we cannot hope to settle this debate here. We have given special emphasis to the utilitarian basis for value, in part because it underlies nearly all empirical assessments of value, including all benefit-cost analyses. Selecting the utilitarian approach does not eliminate from consideration nonconsumptive enjoyments of nature, nor does it disregard satisfactions that do not entail direct or indirect use, such as existence value.

The problems of measurement are at least as daunting as the problems of selecting or justifying a philosophical basis. Empirical assessments tend to disfavor the "natural" state in comparison with economically desired alternative uses for the same land, simply because the benefits from the alternative uses are usually more easily measured than are the benefits of ecosystem services. Many of the most important beneficial services of ecosystems are public (that is, jointly enjoyed) goods whose values are not expressed in market prices. The values of these services are therefore especially difficult to measure. Evaluators often concentrate on the most easily measured impacts and ignore the difficult ones. As a consequence, the unacknowledged pathways may fade from view, yielding an overall value estimate of ecosystem value that is far too low. This indicates the critical importance of developing and improving measurement techniques oriented toward those ecosystems services whose values are not expressed directly in markets.

We have noted the importance of distinguishing carefully between aggregate value and marginal value. In many instances public projects or private developments encroach on portions of ecosystems, rather than the entire system, and in these cases the relevant question is the change in ecosystem value (or marginal loss of value), not the overall ecosystem value. Whether a particular ecosystem is or is not "worth saving" may depend critically on how much of the total area devoted to that ecosystem is still extant. Earlier, we illustrated the dependence of marginal value on total area, taking account of area-diversity relations and the differential value of species. That same kind of analysis could be modified to extend beyond the continuous-patch model we used to apply to the global distribution of certain plant and animal assemblages. Such considerations could help develop a more comprehensive strategy for allocating scarce conservation resources among competing needs.

Even granting our fondest hopes for success in this venture, however, for some time the values we can associate with natural ecosystems will be full of uncertainty. That uncertainty has to be incorporated into the estimates that serve decision makers; point estimates without probability distributions often lead to wrong conclusions, especially when—as often happens—the unstated distributions are skewed toward the more costly outcomes.

Although our discussion acknowledges a key role for benefit-cost analysis in the valuation of ecosystem services, we would emphasize that such analysis does not yield a sufficient criterion for deciding policy. Fundamental issues of fairness or distribution are ignored in benefit-cost assessments. At best, benefit-cost analyses yield useful information on aggregate net benefits under alternative policy scenarios. This information needs to be accompanied by a recognition of the distribution of the gains and losses, both across the current generation and between current and future generations. When the distributions of benefits and cost differ, the ethical issue of "who decides" becomes central to policy making. How much weight should we give to the well-being of future generations, as compared to that of current inhabitants of the planet? And how can we gauge the preferences of future generations in attempts to ascertain the gains or losses they might experience under different policies? Among the members of the current generation, do we give preference to particular members of society? Are sophisticated ecologists worth more votes than city dwellers who evidence neither knowledge of nor interest in "nature"? And what do we do about the enormous variation in risk aversion among our citizens? Do we owe extra deference to those who truly believe that we are threatening our very futures?

These questions reach to the very center of our views about the design of society and the appropriate relationship between state and citizen. The fact that they have no easy answers need not make us pessimistic about the prospects for sensible public policy. We can go a long way toward improving policy making simply by calling attention to the underlying philosophical questions, by developing empirical methods that generate better information about the gains and losses at stake under alternative public policies, and by developing channels for communicating this information to the general public.

Acknowledgments

We are grateful to Gretchen Daily, Michael Dalton, David Layton, Jason Robinson, and anonymous referees for helpful comments on an earlier draft.

Notes

1. As another example, many people experienced a loss of satisfaction or well-being upon learning of the ecological damage resulting from the 1989 *Exxon Valdez* oil spill. The spill caused a lot of existence value.

2. These are direct, nonconsumptive use values in that the enjoyment of the wetland's flood-control or pest-control services does not use up the potential of the wetland to continue to provide these services.

3. A further, and related, issue is that preferences change. They may change for a given person over his or her lifetime, or from generation to generation. To impute values for future generations (such as the value that future generations might place on ecosystem functions), benefit-cost analysis must impute preferences to these generations. Clearly, this can only involve guesswork. Usually, benefit-cost analyses assume that future generations' preferences are similar to those of the current generation. Costanza, Norton, and Bishop (1995) indicate that preferences seem to evolve toward an increasing concern for sustainability. They consider the notion that this natural evolution of preferences ought to be accounted for in social decisions—that more evolved, developed preferences deserve greater weight in analyses of policy options.

4. Costanza, Norton, and Bishop (1995) and the chapter by Costanza and Folke in this volume consider this issue in some detail.

5. This organism must produce no use value, either directly or indirectly. Thus it must be something we don't enjoy eating (there is no consumptive use value) and something we don't enjoy observing (there is no nonconsumptive use value). In addition, the organism must not serve any positive ecosystem function (there must be no indirect use value). And it must be the case that we're certain that human's tastes and ecosystem function won't change to give rise to a future use value. To complete the picture, the organism must also have a zero existence value—humans must not enjoy contemplating this thing. Is there any real-world organism that fits this picture? Perhaps some lowly species of cockroach comes close. Whether it exactly fits the picture isn't important. The key point is that such a creature would be given virtually no value in a benefit-cost analysis. This means that if we are considering a development project that threatens its existence, this threat does not cause us (as utilitarians) to refrain from undertaking the project. As long as there are some benefits from the project and no other, "significant" form of life is put at risk, we would not prevent the loss of the particular species.

6. The animal rights position is sometimes extended to embrace other "rights" such as freedom; hence the occasionally observed bumper sticker, "Pet Breeders are Pimps."

7. We thank Partha Dasgupta for pointing this idea out to us.

8. The Kantian emphasis on removing one's own identity from the consideration is inherent in John Rawls's notion of the original position. This notion gives rise to a Rawlsian conception of justice that is close in many respects to the Kantian conception. See Rawls 1973.

9. If the pest-control services provided by the ecosystem in question are perfect substitutes for the pest-control services offered by the alternative (e.g., synthetic substitutes), then the avoided expenditure is a fairly good measure of the pest-control benefit provided by the ecosystem. However, if the services are imperfect substitutes for one another, the avoided costs can significantly understate the value of pest-control services generated by ecosystems. For details on this issue see Freeman 1993.

10. There are exceptions. In some cases, there is one unit rate or price for up to a certain quantity of water, then another unit for consumption in excess of that quantity. This is a case in which two prices are charged, but it does not constitute a charge based on willingness to pay for each unit. That would require a multitude of prices.

11. It may be noted that the total value or benefit from the water consumed (areas I and II) exceeds the sacrifice associated with paying for the water (area II). Thus there is a consumer surplus given by area I.

12. Markets tend to arise for goods or services that are excludable: the failure to pay for the good or service implies an inability to enjoy or consume the good. For nonconsumptive use values (like bird watching) it is difficult to establish a market because people cannot easily be excluded from enjoying the good or service.

13. For an illustration of the use of the travel-cost method, see Goulder and Kennedy 1995. For a detailed exposition, see Freeman 1993.

14. In contingent valuation assessments of value, interviewees are asked what they would be willing to pay in order to provide some real or hypothetical amenity.

15. The accounting here is perfectly analogous to the economic valuation of net economic output, which disregards the value of intermediate inputs, that is, inputs that are used up in the process of producing final goods such as consumer goods and capital goods.

16. Our attention to the possibilities for double-counting should not be misinterpreted. We do not mean to suggest that there is a general tendency to overvalue ecosystems services. To the contrary, these services are often undervalued because important direct use values and production services are ignored. But we do wish to indicate that if these types of services are valued correctly, there is no need to add further values attributed to indirect contributions by various life forms.

17. A collection of thoughtful examinations of the contingent valuation method is provided in the fall 1994 issue of the *Journal of Economic Perspectives*.

18. For a detailed discussion, see Bishop 1982. A closely related concept is that of the quasi-option value, which relates to the value of flexibility in situations involving the irreversibilities; on this see, for example, Dasgupta (1982, ch. 10). We follow general practice in subsuming option value is so closely connected with (potential) use that it should be placed in the use-value category.

19. For example, suppose a habitat is threatened with destruction. Suppose that, if the habitat is preserved, there is a 50 percent chance you would visit it, you would derive a use value of 10; if you didn't, you would enjoy no use value. In this case the expected value of the use value is 5. But you might be willing to pay, say, 7 to ensure the preservation of the habitat. If so, your option value is 2 (7–5).

20. The degree of historical disturbance, of course, is difficult to estimate. It is usually underestimated by human observers, whose decisions often are based on what they believe the ecosystem was like in their grandfather's time.

21. The same principle applies to other resources: as indicated earlier, the marginal value of water to households dwindles as the amount of water consumed increases. Working in the other direction, the marginal value rises the lower the amount of water available for consumption.

22. If, for example, the ones we view as most valuable did well in relatively small areas, we might argue for a patchwork of little parks; whereas, if the opposite were true, we would insist on large refuges.

23. Indeed, our analysis applies specifically to the simple case in which A is reduced by shrinkage from the outside edges. In many situations, the reduction occurs by fragmentation—a patch here, a patch there, leading to a checkerboard of "natural" and "modified" areas. The new habitats provided by "edge effects" can raise local biodiversity (at least transiently). In the longer run the area/diversity rule will apply over the entire region, but the value of species lost may differ. In recent studies of plant diversity in grassland patches, the first species lost are the most effective, narrow-niche competitors: fragmentation gives an advantage to those species adept at dispersal and at rapid colonization. (See, for example, the results discussed in the chapter by David Tilman in the volume.)

24. A technical discussion of these models is in Bergstrom and Stoll 1993.

References

Bergstrom, J. C., and J. R. Stoll. 1993. "Value estimation models for wetlands-based recreational use values." *Land Economics* 69(2), 1993.

Bishop, Richard C., 1982. "Option value: An exposition and extension." *Land Economics* (February): 1–15.

Costanza, Robert, Bryan Norton, and Richard C. Bishop. 1995. "The Evolution of Preferences: Why 'Sovereign' Preferences May Not Lead to Sustainable Policies and What to Do About It." Paper presented at the Swedish Collegium for Advanced Study in the Social Sciences (SCASSS) workshop on Economics, Ethics and the Environment, Uppsala, Sweden, August 25–27.

Dasgupta, Partha. 1982. *The Control of Resources*. Cambridge, Mass.: Harvard University Press.

Fisher, Anthony C., and W. Michael Hanemann. 1986. "Option value and the extinction of species." In V. Kerry Smith, ed., *Advances in Applied Microeconomics* 4:169–190.

Freeman, A. Myrick. 1993. *The Measurement of Environmental and Resources Values: Theory and Methods*. Washington, D.C.: Resources for the Future.

Groombridge, Brian, ed. 1992. *Global Biodiversity: Status of the Earth's Living Resources*, World Conservation Monitoring Centre. London: Chapman and Hall.

Goulder, Lawrence H., and Donald Kennedy. 1995. "Valuing Nature." In *Earth Systems: Processes and Issues*. Earth Systems Program, School of Earth Sciences, Stanford University. Draft manuscript.

Grant, Peter. 1986. *Ecology and Evolution of Darwin's Finches*. Princeton: Princeton University Press.

Kant, Immanuel. 1963. "Duties to Animals and Spirits." In *Lectures on Ethics*, translated by Louis Infield. New York: Harper and Row.

Mitchell, Robert Cameron, and Richard T. Carson. 1989. *Using Surveys to Value Public Goods: The Contingent Valuation Method*. Washington, D.C.: Resources for the Future.

Rawls, John. 1973. *A Theory of Justice*. Cambridge, Mass.: Harvard University Press, chapters I–III.

Singer, Peter. 1975. *Animal Liberation*. New York: Random House.

Watson, Richard A. 1983. "A critique of anti-anthropocentric biocentrism." *Environmental Ethics* 5:245–256.

Chapter 4

Valuing Ecosystem Services with Efficiency, Fairness, and Sustainability as Goals

Robert Costanza and Carl Folke

Valuation ultimately refers to the contribution of an item to meeting a specific goal. A baseball player is valuable to the extent he contributes to the goal of the team's winning. In ecology, a gene is valuable to the extent it contributes to the goal of survival of the individuals possessing it and their progeny. In conventional economics, a commodity is valuable to the extent it contributes to the goal of individual welfare as assessed by willingness to pay. The point is that one cannot state a value without stating the goal being served. Conventional economic value is based on the goal of individual utility maximization. But other goals, and thus other values, are possible. For example, if the goal is sustainability, one should assess value based on the contribution to achieving that goal—in addition to value based on the goals of individual utility maximization, social equity, or other goals that may be deemed important. This broadening is particularly important if the goals are potentially in conflict.

There are at least three broad goals that have been identified as important to managing economic systems within the context of the planet's ecological life support system (Daly 1992):

1. assessing and ensuring that the scale of human activities within the biosphere is *ecologically sustainable;*
2. *distributing* resources and property rights *fairly,* both within the current generation of humans and between this and future generations, and also between humans and other species; and

3. *efficiently* allocating resources as constrained and defined by 1 and 2 above, and including both marketed and nonmarketed resources, especially ecosystem services.

Several authors have discussed valuation of ecosystem services with respect to goal 3 above—allocative efficiency based on individual utility maximization (e.g., Mitchell and Carson 1989, Costanza et al. 1989, Dixon and Hufschmidt 1990, Barde and Pearce 1991, Aylward and Barbier 1992, Pearce 1993; see also chapter 3, this volume). In this chapter we explore the implications of extending these concepts to include valuation with respect to the other two goals: (1) ecological sustainability, and (2) distributional fairness. The "Kantian" or intrinsic rights approach discussed by Goulder and Kennedy (chapter 3) is one approach to goal 2, but it is important to recognize that the three goals are not "either-or" alternatives. While they are in some senses independent "multiple criteria" (Arrow and Raynaud 1986), they must all be satisfied in an integrated fashion to allow human life to continue in a desirable way. Similarly, the valuations that flow from these goals are not "either-or" alternatives. Rather than a "utilitarian or intrinsic rights" dichotomy, we must integrate the three goals listed above and their consequent valuations.

Valuations are also the relative weights we give to the various aspects of the individual and social decision problem, and the weights that we give are reflections of the goals and worldviews of the community, society, and culture of which individuals are a part (e.g., Costanza 1991, North 1994, Berkes and Folke 1994). We cannot avoid the valuation issue, because as long as we are forced to make choices we are doing valuation. But we need to be as comprehensive as possible in our valuations and choices about ecosystems and sustainability, recognizing the relationship between goals and values.

This paper is divided into three sections. The first addresses ecosystem valuation in a broader context, in which ecological sustainability and fair distribution are high-priority goals in addition to economic efficiency. The second discusses the assumption of fixed tastes and preferences (which underlies conventional valuation based on individual utility maximization) and looks at the implications of gradually relaxing this assumption for the concept of "consumer sovereignty" and other approaches to social choice. The third section raises the issue of the coevolutionary nature of preference formation, and puts individuals in their dynamic, social, environmental, institutional, and cultural context. As Sen (1995, p. 18) has noted: "Many of the more exacting problems of the contemporary world—ranging from famine prevention to environmental preservation—actually call for *value formation through public discussion*" (our emphasis).

Basing valuation on current individual preferences and utility maximization alone, as in conventional analysis, does not necessarily lead to ecological sustainability or social fairness (Bishop 1993). We advocate a two-tiered approach for combining public discussion and consensus building on sustainability and equity goals with methods for modifying both prices and individual preferences to better reflect these community goals (Rawls 1971, Norton 1995, Costanza et al. 1995). Estimation of ecosystem values based on sustainability goals requires treating preferences as endogenous and coevolving with other ecological, economic, and social variables. Finally, we briefly discuss the possibilities for using integrated ecological economic modeling as a tool for valuation of ecosystem services in this broader context.

Sustainability and Fairness as Goals

Ideally, a framework for economic analysis should contain information about the full implications (economic, social, and ecological) of various alternative policy options relative to existing policy. For every policy option, the various ecological-social-economic linkages should be traced to determine the various consequences for human welfare associated with that option, and where possible the various positive and negative impacts should be quantified and valued (Barbier et al. 1994). Economic analysis is about making choices among alternative uses of scarce resources, and it is in this context that valuation becomes relevant.

When a single goal or criterion is involved, the valuation problem is in principle fairly straightforward. But when multiple goals or criteria are involved, the problem can become much more complicated. A classic example of the multiple criterion problem can be found in the story about the drunkard, the miser, and the health freak (Farquharson 1969, Arrow and Raynaud 1986). All three sit on a committee that has to decide how to spend the money of a foundation earmarked for building a student residence. Three alternatives are determined:

1. no house now (leave the money in the bank to earn interest and build a better house later)
2. a house now without a bar
3. a house now with a bar

Suppose the rankings of the alternatives by the three committee members are:

miser—1, 2, 3

health freak—2, 3, 1

drunkard—3, 2, 1

The winning option depends on the order in which the voting is done and can be manipulated strategically. For example, if the miser were chairman of the committee, he could call a vote first on whether there should be a bar (option 3) or not (options 1 and 2). Since both the miser and the health freak prefer no bar (1 or 2), no bar would be chosen by a two-thirds majority. Then he could call a vote on the remaining two options (now or later), which would yield a two-thirds majority for later (option 1) and an overall ranking of 1, 2, 3. But if the health freak were chairman, he could suggest voting first on the question of whether to build the house now (options 2 or 3) or wait (option 1). The decision to build now would pass by a two-thirds majority. Then he could call a vote on the question of the bar, which would be rejected by another two-thirds majority, yielding an overall ranking of 2, 3, 1. Likewise, if the drunkard were chairman he could propose voting between the option 2 (now without a bar) and options 1 and 3 (either build now or wait). The second grouping would win by a two-thirds majority, since both the drunkard and the miser prefer either option 1 or 3 to option 2. Then a vote between options 1 and 3 would yield a two-thirds majority for option 3 (build now with a bar) and an overall ranking of 3, 1, 2. It can be shown that because of strategic manipulations and other "voting paradoxes" that multi-criteria problems do not have any clear-cut, unambiguous, systematic solutions (Arrow and Raynaud 1986) and it is only a dictatorship of one criterion over the others that could not be manipulated strategically (Satterthwaite 1975).

This result is obtained in an environment of fixed preference orderings and no discussion among committee members (criteria). Social choice theory in general has tended to avoid the issue of the connection between value formation and the decision-making process. As Arrow (1951, p. 7) put it: "we will also assume in the present study that individual values are taken as data and are not capable of being altered by the nature of the decision process itself." One way out of this dilemma is to relax the assumption of fixed preferences and allow the committee members to talk with each other, to convey information, to try to change each other's minds (preference orderings), and possibly to come to a consensus on the rankings, as they would do in a real committee. For example, the drunkard could argue that recent scientific evidence has shown that two glasses of red wine per day actually improves one's health, and this might convince the health freak to change his ordering to 2, 3, 1, or even to 3, 2, 1, especially if some restrictions were put in so that, for example, the bar could serve only beer and wine.

This *value formation through public discussion,* as Sen (1995) suggests, is essential to integrate the three goals of sustainability, fairness, and efficiency and can be seen, in fact, as the essence of democracy. As Buchanan (1954, p. 120) put it: "The definition of democracy as 'government by discussion' implies that individual values can and do change in the process of decision-making." Limiting our valuations and social decision making to the goal of economic efficiency based on fixed preferences prevents the needed democratic discussion of values and options and leaves us with only the "illusion of choice" (Schmookler 1993). What are the implications of all this for the valuation of ecosystem services?

Fixed Tastes and Preferences and Consumer Sovereignty

As discussed above, conventional economic valuation is based on a social decision-making rule sometimes referred to as "consumer sovereignty." By consumer sovereignty is meant that consumer choices are paramount, and that individual consumer preferences, whatever they happen to be and however they are formed, should determine relative value. This rule embodies the assumption that tastes and preferences are fixed and that the economic problem consists of optimally satisfying those preferences. If tastes and preferences are fixed and given, then we do not have to know or care why consumers want what they want; we just have to satisfy their preferences as efficiently as possible. As long as economic efficiency is the only goal, this approach works reasonably well. But as soon as we introduce the goals of social fairness and ecological sustainability, we run into the multi-criterion decision problem (as discussed above), which has no systematic or "procedural" solution. One way out of this predicament is to relax the assumption of fixed tastes and preferences and allow some democratic discussion and modification of values. In addition, tastes and preferences do, in fact, change anyway, especially in the longer term (North 1994). They are shaped by the institutional framework under the influence of education, advertising, changing cultural assumptions, etc. (North 1990). For both of these reasons we need other criteria for what is "optimal" in addition to economic efficiency and more decision rules as well as consumer sovereignty.

Questioning consumer sovereignty raises legitimate concerns regarding the possible manipulation of preferences. If tastes and preferences can change, then who is going to decide how to change them? There is a real danger that a "totalitarian" government might be employed to manipulate preferences to conform to the desires of a select elite rather than the society as a whole. Two points need to be kept in mind, however: (1) preferences are already being manipulated every day; and (2) we can apply open democratic principles to the task of deciding how to manipulate preferences just as easily as we can apply hidden or totalitarian principles. So the question becomes: Do we want preferences to be manipulated outside of democratic discussion and control, either by a dictatorial government or by big business acting through advertising? Or do we want to explore and shape them consciously, based on democratic social dialogue and consensus, with the additional goals of long-term sustainability and social fairness in mind? Either way, this is an issue that can no longer be avoided and one that we believe can best be handled using the principle of "democracy as discussion."

Four Degrees of Consumer Sovereignty

The "consumer sovereignty" principle of social choice is not quite as monolithic as we have portrayed it. There are actually quite a range of opinions and interpretations. Costanza et al. (1995) define four versions of the consumer sovereignty principle as positions on a continuum of degrees of preference endogeneity. These four degrees are labeled: (1) unchanging preferences, (2) preferences as given, (3) commitment to democracy, and (4) democratic preference change.

"Unchanging preferences" implies that preferences are both given and fixed. To say that preferences are given is to say that stated and revealed preferences of individuals will be accepted, at face value, as indicative of the individual's actual welfare. To say that preferences are fixed is to claim that preferences do not change through time. According to this view, preferences are locked in, at least in the sense that they are impossible to change through rational considerations (Stigler and Becker 1977).

A majority of economists adopt a somewhat weaker version of consumer sovereignty, according to which preferences are assumed to be given and fixed only in the methodological sense. Preferences are aggregated from "snapshots," not considered as dynamic processes. If preferences are given and fixed for the duration of the analysis, then they are not influenced by changes in other people's behavior and can be aggregated. But this represents a conscious tradeoff of reality for mathematical precision and explanatory power.

A third degree of consumer sovereignty takes given-ness as a purely methodological decision, admits that preferences change, but makes no attempt to change them in an explicit or systematic manner. This view argues that if we set out to change preferences, we have taken a giant step down the road toward paternalism, expertism, and perhaps even totalitarianism (Randall 1995). Preferences are highly individual, and nobody—not politicians, not philosophers, not social scientists, and certainly not environmental activists—is justified in telling individuals what their preferences should be, according to this view.

The fourth degree is labeled "democratic preference change." If a democratic process, including safeguards for individual rights of present people, is in place, then it makes sense to inject into the debate moral concerns about the well-being of future generations, even if these arguments require questioning and criticizing individuals' sincerely felt current preferences. As in the miser/drunk/health freak example above, discussion and criticism of particular preference orderings may be in the form of rational suasion, of pointing out to people the consequences of their desires, of showing alternative paths to personal satisfaction that have less severe impacts on the future of society, and of modifying valuation procedures to reflect more closely the preference sets that are more likely to lead to ecologically sustainable and socially fair decisions. For short-run problems, it may seem reasonable to assume that preferences are given, but it is less reasonable for long-run problems, and in particular not for problems related to ecological sustainability and social fairness.

There is a huge literature on how preferences change, which we can only touch on here, with relevant research from psychology and economics, in particular recent research on preference reversals (Tversky and Kahneman 1986), revealed preferences, constructed preferences (Gregory et al. 1993), and decision making under uncertainty (Heiner 1983); social psychology and sociology, in particular research on social traps (Platt 1973, Cross and Guyer 1980); anthropology, especially research on coevolutionary adaptation of cultures and ecosystems, and ecological anthropology (Harris 1979); and animal ecology, especially research on animal feeding and foraging preferences.

Coevolving Preferences, Goals, and Values

There are certainly several historical examples of societies that managed to integrate the three goals of ecological sustainability, social fairness, and allocative efficiency. Some of their adaptations still survive (Gadgil et al. 1993, Norgaard 1994). In these societies a pattern of coevolutionary adaptation between social systems and natural systems must have been the norm, with the adaptations in many cases driven by crises, learning, and redesign (Holling et al. 1995a). Individual preferences acted in a cultural setting that promoted sustainability of the combined and coevolving social-ecological system, simply because behaving in a sustainable fashion was a necessity for survival and we only observe the societies that have survived.

Some of the most sophisticated coevolved institutions are common-property arrangements. Examples include Spanish *huertas* for irrigation, Swiss grazing commons (Ostrom 1990), and marine resource tenure systems in Oceania (Johannes 1978). In other areas, such institutions have evolved over a short period of time (on the order of one decade) in response to a management crisis. An example is the Turkish Mediterranean coastal fishery in Alanya (Berkes 1992). There are social mechanisms in place that respond to ecological feedbacks and direct societies' adaptation toward sustainability. The coevolutionary character reflects the fact that ecological and social systems can change qualitatively to generate and implement innovations that are truly creative, in the sense of opportunities for novel cooperation and feedback management (Holling et al. 1995a).

Of course, such social mechanisms for adaptations cannot be captured in a conventional cost-benefit analysis, which only reflects what an aggregate of current individuals prefer, without discussion. The results of a benefit-cost study are not sufficient to address the question of which policy is best relative to all three goals mentioned above, since efficiency in a cost-benefit context does not guarantee sustainability or fairness (Bishop 1993, Perrings 1994).

Thus, we can distinguish at least three types of value that are relevant to the problem of valuing ecosystem services. These are laid out in table 4.1, according to their corresponding goal or value basis. Efficiency-based value (E-value) is described in detail in several recent publications (e.g., Mitchell and Carson 1989, Costanza et al. 1989, Dixon and Hufschmidt 1990, Barde and Pearce 1991, Aylward and Barbier 1992, Pearce 1993; chapter 3, this volume). It is based on a model of human behavior sometimes referred to as *Homo economius,* which suggests that humans act rationally and in their own self-interest. Value in this context (E-value) is based on current individual preferences that are fixed or given (level 1, 2, or 3 of consumer sovereignty, as described above). Little discussion or scientific input is required to form these preferences, and value is simply people's revealed willingness to pay for the good or service in question.

Fairness-based value (F-value) would require that individuals vote for their preferences as a member of the community, not as individuals. This species *(Homo communicus)* would engage in much discussion with other members of the community and come to consensus on the values that would be fair to all members of the current and future community (including non-human species), incorporating scientific information about possible future consequences as necessary. One method to implement this might be Rawls's (1971) "veil of ignorance," by which everyone votes as if they were operating with no knowledge of their own status in current or future society.

Sustainability-based value (S-value) would require an assessment of the contribution to ecological sustainability of the item in question. The S-value of ecosystem services is connected to their physical, chemical, and biological role in the long-term functioning of the global system. Scientific information about the functioning of the global system is thus critical in assessing S-value, and some discussion and consensus building is also necessary.

Table 4.1. Valuation of ecosystem services based on the three primary goals of efficiency, fairness, and sustainability

Goal or Value Basis	Who Votes	Preference Basis	Level of Discussion Required	Level of Scientific Input Required	Specific Methods
Efficiency	*Homo economius*	Current individual preferences	low	low	willingness to pay
Fairness	*Homo communicus*	Community preferences	high	medium	veil of ignorance
Sustainability	*Homo naturalis*	Whole system preferences	medium	high	modeling with precaution

If it is accepted that all species, no matter how seemingly uninteresting or lacking in immediate utility, have a role to play in natural ecosystems (Naeem et al. 1994, Tilman and Downing 1994, Holling et al. 1995b), estimates of ecosystem services may be derived from scientific studies of the role of ecosystems and their biota in the overall system, without direct reference to current human preferences. Humans operate as *Homo naturalis* in this context, expressing preferences as if they were representatives of the whole system. Instead of being merely an expression of current individual preferences, S-value becomes a system characteristic related to the item's evolutionary contribution to the survival of the linked ecological economic system. Using this perspective we may be able to better estimate the values contributed by, say, maintenance of water and atmospheric quality to long-term human well-being, including protecting the opportunities of choice for future generations (Golley 1994, Perrings 1994). One way to get at these values would be to employ systems-simulation models that incorporate the major linkages in the system at the appropriate time and space scales (Bockstael et al. 1995). To account for the large uncertainties involved, these models would have to be used in a precautionary way, looking for the range of possible values and erring on the side of caution.

A Two-Tiered Decision Structure

How does one integrate these three goals and their related forms of value in a social-choice structure that preserves democracy? We advocate a two-tiered conceptual model (Page 1991, Norton 1994, Costanza et al. 1995) that makes value formation and reformation an endogenous element in the search for a rational policy for managing human economic activities. More like the decision-making process going on in the real world and less like most models for evaluating environmental policies, this conceptual model embeds both economic models and ecological models in a larger social process. The first step in that process, however, is political, not scientific. It is necessary for the various elements of a community or society, perhaps through representatives of the stakeholder groups, to propose and discuss various visions that they would set as positive outcomes of a process of economic development over generations. An important part of this will be the ranking of risks and attempts to set some kind of priorities in addressing risk problems. But comparative risk processes are not as important as public discussions of the positive, long-term aspirations of the stakeholders for their region. It may be possible to begin by attempting to agree on some possible management goals and some projects (to be undertaken in willing local communities), to experiment with pilot projects and to evaluate them scientifically in pursuit of shared, if tentative, goals. The implementation of Agenda 21 of the U.N. Conference on Environment and Development in 1992 is one example of such a process, based on a shared vision of a sustainable society formulated by the global community.

The model is hierarchical in the sense that economic models represent large subsystems that are embedded in larger-scale ecological, biogeochemical, and hydrological systems (figure 4.1). We model economic behaviors and activity on a shorter frame of time (several years), while modeling the relationship of the economy to the larger physical systems that form its management context on longer scales of time (decades to centuries). A two-tier system of analysis sorts possible environmental problems and risks according to the likely temporal and spatial scale of their impacts, and applies an appropriate action criterion—such as a cost-benefit criterion or a Safe Minimum Standard criterion—given the scope and scale of possible risks of a policy. The model is an action-based model that includes economic models and ecological models in a larger system that sets goals, engages in experiments and pilot projects in search of those goals, monitors progress toward those goals scientifically, and then factors scientific results into an ongoing public process of revising goals and the policies designed to achieve them (Costanza et al. 1995). It is this learning or "adaptive management" (Walters 1986) that submits policies to rigorous re-examination both with regard to progress toward the stated goals, and also with regard to the "appropriateness" of current individual preferences under various models.

In this context, actively seeking to influence preferences is consistent with a democratic society. In order to operationalize real democracy at least a

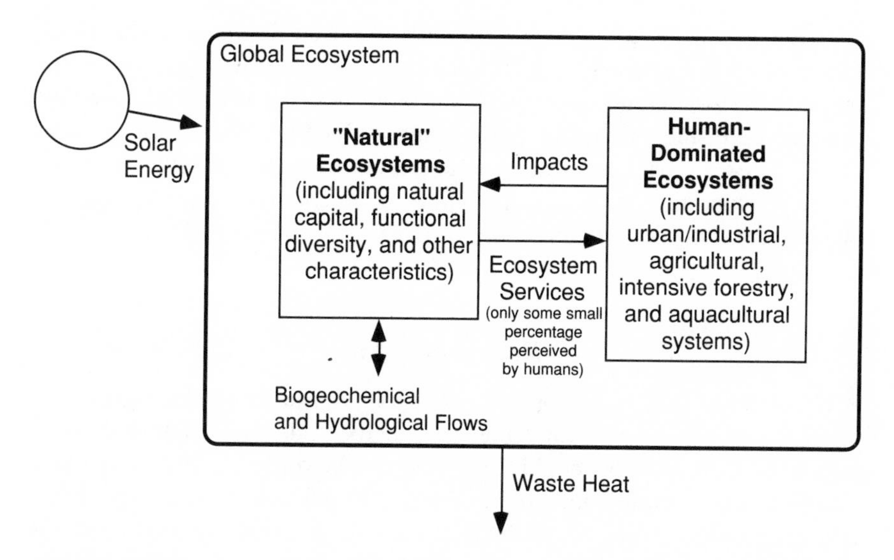

Figure 4.1. Human-dominated ecosystems are parts of the overall global system. Ecosystem services are essential for the development and well-being of human society, but only a fraction of this work is covered by market prices or perceived by humans.

two-tiered decision structure, or, better, a multilayered set of institutions (Ostrom 1990, Hanna 1997), ought to be used. This is necessary in order to eliminate "preference inconsistencies" between the short term and the long term and between local and global goals, a phenomenon described in the social psychology literature as a "social trap" (Platt 1973, Cross and Guyer 1980). There must first be general, democratic consensus on the broad, long-term goals of society. At this level "individual sovereignty" holds in the sense that the rights and goals of all individuals in society must be taken into account, but in the context of a shared dialogue and discussion aimed at achieving the broadest consensus possible. Once the broad goals are democratically arrived at, they can be used to limit and direct preferences at lower levels. For example, once there is general consensus on the goal of sustainability, then society is justified in taking action to change local behaviors that are inconsistent with this goal. It may be justified, for example, to attempt to change either people's preferences for driving automobiles or the price of doing so (or both) in order to change behavior to be more consistent with longer-term sustainability goals. In this way we are utilizing the foresight that we do possess in order to modify short-term cultural evolutionary forces toward achieving our shared long-term goals.

Such a process is going on continuously at various levels in society. From the level of the household to international agreements, the institutional framework (formal and informal norms and rules) constrain and shape the preferences of individuals. Institutions are defined as the humanly devised constraints that structure incentives in human exchange, whether political, social, or economic, and that shape human interactions and the way societies evolve through time (North 1990).

Integrated Ecological-Economic Modeling and Assessment

Addressing the goal of ecological sustainability requires a large measure of scientific assessment and modeling (Faucheux et al. 1996). The process of integrated ecological-economic modeling can help to build mutual understanding, solicit input from a broad range of stakeholder groups, and maintain a substantive dialogue between members of these groups. In the process of adaptive management, integrated modeling and consensus building are essential components (Gunderson et al. 1995). A recent Scientific Committee on Problems of the Environment (SCOPE) project on Integrated Ecological Economic Assessment (IA for short) developed the following basic framework (Costanza and Tognetti 1996). The framework is seen as a creative and learning process rather than a purely technical tool—within which a well-rounded decision can be achieved through the consensus of stakeholders. The process consists of twelve steps and assumes feedback loops from later steps to earlier steps:

1. *Define the focus of attention.* This would likely result from a proposed development opportunity and/or an ecological concern.
2. *Identify stakeholders.* These typically would include the government, business, landowners, nongovernmental organizations, funding agencies, community-based organizations, researchers, etc.
3. *Establish techniques to bring stakeholders together (e.g., roundtable).* This step presupposes that one or more of the stakeholders has sufficient interest to draw the remaining stakeholders to a meeting. It may be that specific stakeholders need to be persuaded that it is in their best interest to convene in such a roundtable. Other stakeholders may need to convince them of the value of developing a participatory approach.
4. *Seek agreement on an acceptable facilitator.* Ideally such a person should be as neutral and unbiased as possible and without a stake in the outcome of the process. The facilitator should nevertheless be

committed to the process and be able to balance the differing powers of the stakeholders.

5. *Define stakeholder interests.* Before the roundtable meeting, stakeholder groupings should be encouraged to meet and discuss their own interests.
6. *Hold roundtable.* The roundtable should ideally be convened jointly by several stakeholders. The agenda should include opportunities for:
   - sharing individual visions
   - identifying complementarity and conflicts
   - agreeing that a process is necessary to address conflicts
   - seeing that integrated assessment is a way forward with the potential to develop consensus and arrive at a "win-win" situation
   - establishing a structure for ongoing dialogue including a stakeholder committee to oversee the process and feedback opportunities to the stakeholder groups and to all stakeholders collectively.
7. *Undertake a scoping exercise.* This process is necessary to identify the key issues, questions, data/information availability, land-use patterns, proposed developments, existing institutional frameworks, timing and spatial consideration, etc. It provides a means to determine whether a specific action will have significant effects on expressed values and to link the model with those values. This scoping exercise is also seen as building trust among the stakeholders, as well as an acceptance of the process. The stakeholders build upon knowledge and capacity.
8. *Build and run a scoping model.* A scoping model provides a relatively quick process of identifying and building in the key components in order to:
   - generate alternative scenarios
   - identify critical information gaps
   - understand the sensitivity of the scenarios to uncertainty
   - identify and agree on additional work to be undertaken by one or more methods of detailed modeling.

Stakeholders participate in the development of the scoping model.

9. *Commission detailed modeling.* Additional information is gathered and the chosen model(s) are modified, extended, and run.
10. *Present models.* Also present results of model scenarios and discuss findings among stakeholders.
11. *Build consensus recommendations.*

12. *Proceed with, and monitor the development of, the preferred scenario.* Learn from the results and iterate the IA process as necessary. Perceptions change as things actually happen, thus the process must permit changing values to influence decisions at each stage. As iterations occur, the scenario conception changes, leading to new issues for resolution among groups.

Several examples of applying this process are discussed in Costanza and Ruth (1996). One example worth noting is in the Patuxent River drainage basin in Maryland, where integrated ecological-economic modeling and analysis are being applied in order to improve understanding of regional systems, assess potential future impacts of various land-use, development, and agricultural policy options, and better assess the value of ecological systems (Bockstael et al. 1995, Reyes et al. 1996). The integrated model will allow stakeholders to evaluate the indirect effects over long-time horizons of current policy options. These effects are almost always ignored in partial analyses, although they may be very significant and may reverse many long-held assumptions and policy predictions. It will also allow us to directly address the functional value of ecosystem services by looking at the long-term, spatial, and dynamic linkages between ecosystems and economic systems (figure 4.2).

While integrated models aimed at realism and precision are large, complex, and loaded with uncertainties of various kinds (Costanza et al. 1990, Bockstael et al. 1995), our abilities to understand, communicate, and deal with these uncertainties are rapidly improving. It is also important to remember that while increasing the resolution and complexity of models increases the amount we can say about a system, it also limits how accurately we can say it. Model predictability tends to fall with increasing resolution due to compounding uncertainties as described above (Costanza and Maxwell 1994). What we are after are models that optimize their "effectiveness" (Costanza and Sklar 1985) by choosing an intermediate resolution where the product of predictability and resolution (effectiveness) is maximized

It is also necessary to place the modeling process within the larger framework of adaptive management (Holling 1978) if it is to be effective. We need to view the implementation of policy prescriptions in a different, more adaptive way, which acknowledges the uncertainty embedded in our models and allows participation by all the various stakeholder groups. "Adaptive management" views regional development policy and management as "experiments," where interventions at several levels are made to achieve understanding and to identify and test policy options (Holling 1978, Walters 1986, Lee 1993, Gunderson et al. 1995). This means that models, and policies

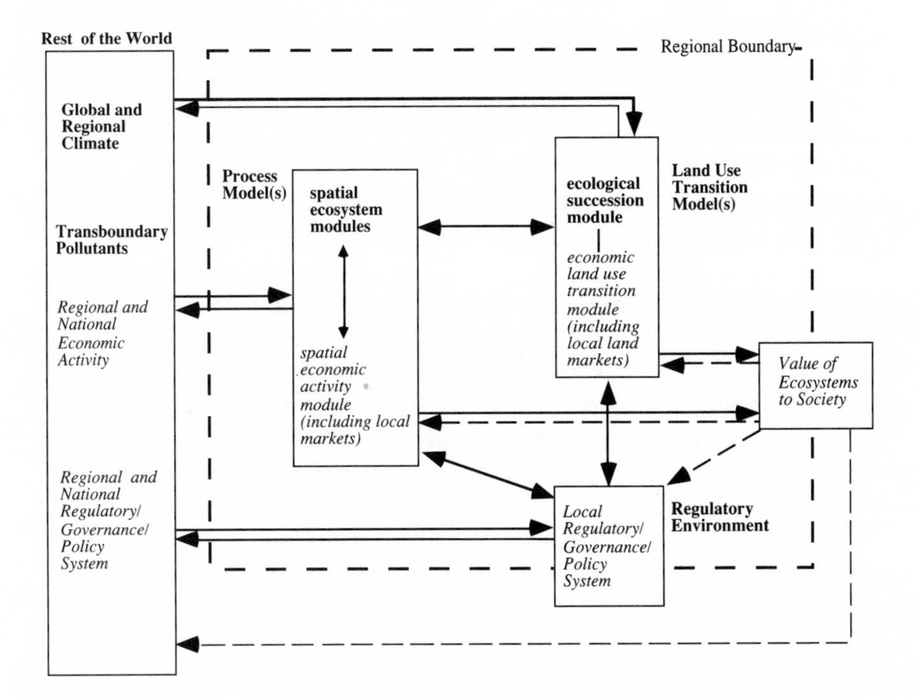

Figure 4.2. Integrated ecological economic modeling and valuation framework.

based on them, are not taken as the ultimate answers, but rather as guiding an adaptive experimentation process with the regional system. More emphasis is placed on monitoring and feedback to check and improve the model, rather than using the model to obfuscate and defend a policy that is not corresponding to reality. Continuing stakeholder involvement is essential in adaptive management.

Conclusions

If economics and other social sciences are to adequately address problems of sustainability, it will be necessary to develop evolutionary models that make preference formation and reformation an endogenous part of the analysis, and to develop mechanisms to modify short-term cultural evolutionary forces in the direction of long-term sustainability and social fairness goals. Society has begun to do this with the recent growing consensus that sustainability is an appropriate long-run, global goal (WCED 1987), but there is still a long way to go in developing explicit, shared visions of a sustainable and desirable society (Meadows 1996).

We believe that society can make better choices about ecosystems if the valuation issue is made as explicit as possible. This means taking advantage of the best information we can muster about ecosystem services and being aware of the different goals of society and their attendant values. In this paper we have discussed the goals of ecological sustainability, social fairness, and economic efficiency as a basis for valuation in an integrated way. Methods for valuation relative to the efficiency goal are well developed. Methods relative to the other two goals need much further development. For valuation relative to fairness we may need to operate behind a "veil of ignorance" as to our status and position in current and future society (Rawls 1971). For valuation relative to sustainability we need to develop truly integrated assessments and models of the quality, quantity, and spatial and temporal dynamics of ecosystem services and the various aspects of their connection to human well-being in the long run. In all cases it also means acknowledging and communicating the huge uncertainties associated with this endeavor, and developing new and better ways to make decisions that achieve our goals in the face of these uncertainties.

Acknowledgments

Ron Trosper, Charles Perrings, Gretchen Daily, and Susan Hanna provided helpful comments on earlier versions of this paper. The Pew Charitable Trusts and the Beijer International Institute for Ecological Economics provided support during the preparation of this manuscript.

References

Arrow, K. J. 1951. *Social Choice and Individual Values*. New York: John Wiley.

Arrow, K. J., and H. Raynaud. 1986. *Social Choice and Multicriterion Decision-Making*. Cambridge: MIT Press.

Aylward, B. A., and E. B. Barbier. 1992. "Valuing environmental functions in developing countries." *Biodiversity and Conservation* 1:34–50.

Barbier, E.B., J. C. Burgess, and C. Folke. 1994. *Paradise Lost? The Ecological Economics of Biodiversity*. London: Earthscan.

Barde, J-P., and D. W. Pearce. 1991. *Valuing the Environment: Six Case Studies*. London: Earthscan.

Berkes, F. 1992. "Success and Failure in Marine Coastal Fisheries of Turkey." In

Making the Commons Work. D.W. Bromley, ed., pp. 161–182. San Francisco: Institute for Contemporary Studies.

Berkes, F., and C. Folke. 1994. "Investing in Cultural Capital for a Sustainable Use of Natural Capital." In *Investing in Natural Capital: The Ecological Economics Approach to Sustainability*, A.M. Jansson, M. Hammer, C. Folke, and R. Costanza, eds., pp. 128–149. Washington, D.C.: Island Press.

Bishop, R.C. 1993. "Economic efficiency, sustainability, and biodiversity." *Ambio* 22:69–73.

Bockstael, N., R. Costanza, I. Strand, W. Boynton, K. Bell, and L. Wainger. 1995. "Ecological economic modeling and valuation of ecosystems." *Ecological Economics* 14:143–159.

Buchanan, J. M. 1954. "Social choice, democracy, and free markets." *Journal of Political Economy* 62:114–123.

Costanza, R., ed. 1991. *Ecological Economics: The Science and Management of Sustainability.* New York: Columbia University Press.

Costanza, R., S. C. Farber, and J. Maxwell. 1989. "The valuation and management of wetland ecosystems." *Ecological Economics* 1:335–361.

Costanza, R., and T. Maxwell. 1994. "Resolution and predictability: An approach to the scaling problem." *Landscape Ecology* 9:47–57.

Costanza, R., B. Norton, and R.C. Bishop. 1995. "The Evolution of Preferences: Why 'Sovereign' Preferences May Not Lead to Sustainable Policies and What to Do About It." Paper presented at the SCASSS (Swedish Collegium for Advanced Study in the Social Sciences) workshop Economics, Ethics, and the Environment, Upsalla, Sweden, August 25–27.

Costanza, R., and M. Ruth. 1996. "Dynamic Systems Modeling for Scoping and Consensus Building." Paper presented at the inaugural conference of the European Chapter of the International Society for Ecological Economics: Ecology, Society, Economy, Université de Versailles, Paris, France, May 23–25.

Costanza, R., and F.H. Sklar. 1985. "Articulation, accuracy, and effectiveness of mathematical models: A review of freshwater wetland applications." *Ecological Modelling* 27:45–68.

Costanza, R., F.H. Sklar, and M.L. White. 1990. "Modeling coastal landscape dynamics." *BioScience* 43:545–555.

Costanza, R., and S. Tognetti, eds. 1996. *Integrated Adaptive Ecological and Economic Modeling and Assessment—A Basis for the Design and Evaluation of Sustainable Development Programs.* Draft Synthesis paper, Scientific Committee on Problems of the Environment (SCOPE), 51 Bld de Montmorency, 75016 Paris, France.

Cross, J.G., and M.J. Guyer. 1980. *Social Traps.* Ann Arbor: University of Michigan Press.

Daly, H. E. 1992. "Allocation, distribution, and scale: Towards an economics that is efficient, just, and sustainable." *Ecological Economics* 6:185–193.

DeBellevue, E.B., T. Maxwell, R. Costanza, and M. Jacobsen. 1993. "Development of a Landscape Model for the Patuxent River Watershed." Discussion Paper No.

10, Maryland International Institute for Ecological Economics, Solomons, Maryland.

Dixon, J. A., and M. M. Hufschmidt. 1990. *Economic Valuation Techniques for the Environment: A Case Study Workbook*. Baltimore: Johns Hopkins University Press.

Farquharson, R. 1969. *Theory of Voting*. New Haven: Yale University Press.

Faucheux, S., D. Pearce, and J. Proops. 1996. *Models of Sustainable Development*. Cheltenham, England: Edward Elgar Press.

Gadgil, M., F. Berkes, and C. Folke. 1993. "Indigenous knowledge for biodiversity conservation." *Ambio* 22:151–156.

Golley, F.B. 1994. "Rebuilding a Humane and Ethical Decision System for Investing in Natural Capital." In *Investing in Natural Capital: The Ecological Economics Approach to Sustainability,* A.M. Jansson, M. Hammer, C. Folke, and R. Costanza, eds., pp. 169–178. Washington, D.C.: Island Press.

Gunderson, L., C.S. Holling, and S. Light, eds. 1995. *Barriers and Bridges to the Renewal of Ecosystems and Institutions*. New York: Columbia University Press.

Gregory, R., S. Lichtenstein, and P. Slovic. 1993. "Valuing environmental resources: A constructive approach." *Journal of Risk and Uncertainty* 7:177–197.

Hanna, S. 1997. "Managing for Human and Ecological Context in the Maine Soft Shell Clam Fishery." In *Linking Social and Ecological Systems: Institutional Learning for Resilience,* F. Berkes and C. Folke, eds. Cambridge, England: Cambridge University Press.

Harris, M. 1979. *Cultural Materialism: The Struggle for a Science of Culture*. New York: Random House.

Heiner, R. 1983. "The origin of predictable behavior." *American Economic Review* 73:560–595.

Holling, C.S., ed. 1978. *Adaptive Environmental Assessment and Management*. London: John Wiley.

Holling, C.S., F. Berkes, and C. Folke. 1995a. "Science, Sustainability and Resource Management." *Beijer Discussion Papers Series,* no. 68. Stockholm, Sweden: Beijer International Institute of Ecological Economics.

Holling, C.S., D.W. Schindler, B.W. Walker, and J. Roughgarden. 1995b. "Biodiversity in the Functioning of Ecosystems: An Ecological Synthesis." In *Biodiversity Loss: Economic and Ecological Issues,* C. Perrings, K.-G. Mäler, C. Folke, C.S. Holling, and B.-O. Jansson, eds., pp. 44–83. New York: Cambridge University Press.

Johannes, R. E. 1978. "Traditional marine conservation methods in Oceania and their demise." *Annual Review of Ecology and Systematics* 9:349–364.

Lee, K.N. 1993. *Compass and Gyroscope: Integrating Science and Politics for the Environment*. Washington, D.C.: Island Press.

Meadows, D. 1996. "Envisioning a Sustainable World." In *Getting Down to Earth: Practical Applications of Ecological Economics,* R. Costanza, O. Segura, and J. Martinez-Alier, eds., pp. 117–126. Washington, D.C.: Island Press.

Mitchell, R. C., and R.T. Carson. 1989. *Using Surveys to Value Public Goods: The Contingent Valuation Method*. Washington, D.C.: Resources for the Future.

Naeem, S., L.J. Thompson, S.P. Lawler, J.H. Lawton, and R.M. Woodfin. 1994. "Declining biodiversity can alter the performance of ecosystems." *Nature* 368:734–737.

Norgaard, R.B. 1994. *Development Betrayed: The End of Progress and a CoEvolutionary Revisioning of the Future*. London: Routledge.

North, D.C. 1990. *Institutions, Institutional Change and Economic Performance*. Cambridge, England: Cambridge University Press.

North, D.C. 1994. "Economic performance through time." *American Economic Review* 84:359–368.

Norton, B.G. 1994. "Economists' preferences and the preferences of economists." *Environmental Values* 3:311–332.

Norton, B.G. 1995. "Ecological integrity and social values: At what scale?" *Ecosystem Health* 1:228–241.

Ostrom, E. 1990. *Governing the Commons: The Evolution of Institutions for Collective Actions*. Cambridge, England: Cambridge University Press.

Page, T. 1991. "Sustainability and the Problem of Valuation." In *Ecological Economics: The Science and Management of Sustainability,* R. Costanza, ed., pp. 88–101. New York: Columbia University Press.

Pearce, D. 1993. *Economic Values and the Natural World*. London: Earthscan.

Perrings, C.A. 1994. "Biotic Diversity, Sustainable Development, and Natural Capital." In *Investing in Natural Capital: The Ecological Economics Approach to Sustainability,* A.M. Jansson, M. Hammer, C. Folke, and R. Costanza, eds., pp. 92–112. Washington, D.C.: Island Press.

Platt, J. 1973. "Social traps." *American Psychologist* 28:642–651.

Randall, A. 1995. "Valuation and Beyond." Paper presented at AAAS (American Association for the Advancement of Science) symposium Ecosystems and Landscapes: Describing and Valuing Whole Ecosystems, Atlanta, Georgia, February 18.

Rawls, J. 1971. *A Theory of Justice*. Oxford: Oxford University Press.

Reyes, E., R. Costanza, L. Wainger, E. DeBellevue, and N. Bockstael. 1996. "Integrated Ecological Economics Regional Modelling for Sustainable Development." In *Models of Sustainable Development*. S. Faucheux, D. Pearce, and J. Proops, eds., pp. 253–277. Cheltenham, England: Edward Elgar Press.

Satterthwaite, M. A. 1975. "Strategy-proofness and Arrow's conditions: Existence and correspondence theorems for voting procedures and social welfare functions." *Journal of Economic Theory* 10:187–217.

Schmookler, A. B. 1993. *The Illusion of Choice: How the Market Economy Shapes Our Destiny*. Albany: State University of New York Press.

Sen, A. 1995. "Rationality and social choice." *American Economic Review* 85:1–24.

Stigler, G.J., and G.S. Becker. 1977. "De gustibus non est disputandum." *American Economic Review* 67:76–90.

Tilman, D., and J.A. Downing. 1994. "Biodiversity and stability in grasslands." *Nature* 367:363–365.

Tversky, A., and D. Kahneman. 1986. "Rational Choice and the Framing of Decisions." In *Rational Choice: The Contrast Between Economics and Psychology,* R.M. Hogarth and M.W. Reder, eds., pp. 67–94. Chicago: University of Chicago Press.

Walters, C. J. 1986. *Adaptive Management of Renewable Resources.* New York: McGraw-Hill.

WCED. 1987. *Our Common Future: Report of the World Commission on Environment and Development.* Oxford: Oxford University Press.

Part II

Overarching Services

Chapter 5

The Interaction of Climate and Life

Susan E. Alexander, Stephen H. Schneider, and Kalen Lagerquist

Natural ecosystems provide humanity with a wide variety of vital public services whose degradation may seriously threaten civilization. One of the services that ecosystems provide is a major influence on the atmospheric composition. Over billions of years the composition of the atmosphere has changed considerably. Through eons of build-up, photosynthesis in bacteria, algae, and (later) plants has provided us with the oxygen in the atmosphere that animals depend on. In addition, oxygen in the stratosphere (the upper atmosphere) generates the protective ozone layer. The abundance of oxygen in the atmosphere, surface waters, and soils also contributes to the self-cleansing ability of the atmosphere through oxidation processes. The concentrations of a variety of oxidizing agents such as ozone (O_3), hydroxyl radicals (OH), and nitrogen dioxide (NO_2), determine the rate at which reduced compounds (e.g., carbon monoxide, or CO) are converted to oxidized ones (e.g., carbon dioxide, or CO_2) that can be more easily removed from the air. While the level of atmospheric oxygen (O_2) is not expected to change appreciably from human activity, the oxidative capacity of the atmosphere, linked to several of the major biogeochemical cycles, may be altered as the steady-state concentrations of OH and other oxidizing agents change.

Natural ecosystems also help to stabilize the climate. The interaction of climate and life is seen through the strength of the atmospheric greenhouse effect as a driving force in global climate change. The natural greenhouse effect operating through clouds, water vapor, carbon dioxide, and other trace gases in the atmosphere keeps the earth's surface habitable. The surface temperature is about thirty-three degrees Celsius higher on average than if these gases or cloud particles were not present. Life can have both positive and negative feedbacks on climate by influencing the relative and absolute amounts of trace gases. Over tens of millions of years, life may have potentially helped to stabilize climate by removing CO_2 as the sun grew brighter, whereas life appears to have destabilized climate during the interglacial–ice age transitions by decreasing CO_2 and methane (CH_4) in cold times relative to warmer eras. These feedbacks, both positive and negative, suggested an analogy to the biological process of coevolution, in which the close association of two interacting species can lead to evolutionary paths that are different because of their co-presence (Ehrlich and Raven 1964). Climate and life have likewise coevolved, influencing the evolutionary paths of each other in ways that would not have occurred had they not been in each other's presence (Schneider and Londer 1984). The goal of this chapter is to identify a wide range of ecosystem services associated with the atmosphere and climate, and introduce some initial attempts to value those services.

Life and Biogeochemical Cycles

Life on earth is inextricably linked to climate through a variety of interacting cycles and feedback loops. In recent years there has been a growing awareness of the extent to which human activities, such as deforestation and fossil fuel burning, have directly or indirectly modified the biogeochemical and physical processes involved in determining the earth's climate. These changes in atmospheric processes can disturb a variety of the ecosystem services that humanity depends on. In addition to helping to maintain relative climate stability and a self-cleansing, oxidizing environment, these services include protection from most of the sun's harmful ultraviolet rays, mediation of runoff and evapotranspiration (which affects the quantity and quality of fresh water supplies and helps control floods and droughts), and regulation of nutrient cycling, among others.

Before further discussion of these services, it is important to review briefly how life and climate interact. The transport and transformation of substances in the environment, through life, air, sea, land, and ice, are known collectively as biogeochemical cycles. These global cycles include the circulation of certain elements, or nutrients, on which life and the earth's climate depend. One way that climate influences life is by regulating the flow of substances through these biogeochemical cycles, in part through atmospheric circulation. Water vapor is one such substance. It is critical for the survival and health of human beings and ecological systems and is part of the climatic state. When water vapor condenses to form clouds, more of the sun's rays are reflected out of the atmosphere into space, usually cooling the climate. Conversely, water vapor is also an important greenhouse gas in the atmosphere, trapping heat in the infrared part of the spectrum in the lower atmosphere. The water or hydrologic cycle intersects with most of the other element cycles, including the cycles of carbon, nitrogen, sulfur, and phosphorus, as well as the sedimentary cycle. The processes involving each one of these elements may be strongly coupled with that of other elements, and ultimately, with important regional- and global-scale climatic or ecological processes.

Managing and finding solutions to many of the important environmental problems facing humanity begin with understanding and integrating biogeochemical cycles and the scales at which they operate. Examples of these links include world climate and the potential threat of global climate change; agricultural productivity and its strong reliance on climatic factors, including temperature and precipitation, and the availability of nutrients; the cleansing of toxics from soils and streams through precipitation and runoff; acid precipitation and the perturbation of ecosystem processes; the depletion of stratospheric ozone and its potential threat to human health and the food chain; and the often destructive interaction with natural cycles of other humanmade compounds such as pesticides and synthetic hormones.

The Hydrologic and Sedimentary Cycles

While the total amount of water found on earth may seem huge, the amount of precipitating freshwater available to people is a tiny fraction of this total. Earth's renewable supply of water is continually distilled and distributed through the hydrologic cycle. It falls from the sky as precipitation, collects in lakes, rivers, and oceans, or seeps into the ground and eventually evaporates or transpires, accumulating as water vapor in clouds, ready to begin the sun-powered cycle again. Water is transferred to the air from the leaves of plants primarily by a process called transpiration. This, combined with evaporation from bodies of water and the soil, is known as evapotranspiration. Evaporation of ocean water is about six times as much globally as evapotranspiration on land, although in the centers of continents evapotranspiration may be the main local source of water vapor. Changes in the global climate may cause changes in the hydrologic cycle. Increases in temperature and evaporation are expected to cause increases in precipitation, which may further affect runoff and soil moisture, and eventually influence vegetation patterns and world agriculture.

The sedimentary cycle is tied to the hydrologic cycle through precipitation. Water carries materials from the land to the oceans, where they can be deposited as sediments. On a shorter time scale, the sedimentary cycle includes the processes of physical or chemical erosion, nutrient transport, and sediment formation, for which water flows are mostly responsible. On a geologically longer time scale, the processes of sedimentation, chemical transformation, uplift, sea floor spread, and continental drift operate. Both the hydrologic and sedimentary cycles are intertwined with the distribution of the amounts and flows of six important elements: hydrogen, carbon, oxygen, nitrogen, phosphorus, and sulfur. These elements, or macronutrients, combine in various ways to make up more than 95 percent of all living things. Appropriate quantities of them in proper balance and in the right places are required to sustain life. Although great stocks of all of these nutrients exist in the earth's crust in different (but not always accessible) forms, at any one time the natural supply of these vital elements is limited. Therefore, they must be recycled for life to regenerate continuously. We describe three of these cycles critical to important ecosystem services in the following sections.

The Nitrogen Cycle

Nitrogen exists in a variety of forms in natural systems, and its compounds are involved in numerous biological and abiotic processes. In its gaseous form of N_2, nitrogen makes up almost 80 percent of the atmosphere. This constitutes the major storage pool in the complex cycle of nitrogen through ecosystems. Some of this gas is converted in the soils and waters to ammonia (NH_3), ammonium (NH_4^+), or many other nitrogen compounds. The process is known as nitrogen fixation, and, in the absence of industrial fertilizers, it is the primary source of nitrogen to all living things. Biological nitrogen fixation is mediated by special nitrogen-fixing bacteria and algae. On the land, these bacteria often live on nodules on the roots of legumes, where they use energy from plants to do their work. In freshwater and, possibly, in marine systems, cyanobacteria fix nitrogen. Once nitrogen has been fixed in the soil or an aquatic system, it can follow two different pathways. It can be oxidized for energy in a process called nitrification or assimilated by an organism into its biomass in a process called ammonia assimilation.

Plants incorporate the appropriate forms of fixed nitrogen into their tissues through their root systems. The plants then use the nitrogen to manufacture amino acids and convert the nitrogen into proteins. Fixed as proteins in the bodies of living organisms, nitrogen eventually returns via the nitrogen cycle to its original form of nitrogen gas in the air. The process of denitrification starts when plants containing the fixed nitrogen are either eaten or die. Fixed nitrogen products in dead plants, animal bodies, and animal excreta encounter denitrifying bacteria that undo the work done by the nitrogen-fixing bacteria. Generally, N_2 is the end product of denitrification, but nitrous oxide (N_2O) is also produced in much smaller quantities (up to 10 percent).

The disruption of the nitrogen cycle by human activity plays an important role in a wide range of environmental problems, from the production of tropospheric (lower-atmosphere) smog to the perturbation of stratospheric ozone and the contamination of groundwater. Nitrous oxide, for example, is a greenhouse gas like carbon dioxide and water vapor that can trap heat near the earth's surface. It also destroys stratospheric ozone. Eventually, nitrous oxide in the stratosphere is broken down by ultraviolet light into nitrogen dioxide (NO_2) and nitric oxide (NO), which can catalytically reduce ozone. Nitrogen oxides are chemically transformed back to either N_2 or nitrate or nitrite compounds, which may later get used by plants after they are washed by the rain back to the earth's surface. Nitrate rain is acidic and can cause ecological problems as well as serve as a fertilizer to vegetation.

The Sulfur Cycle

Another example of a major biogeochemical cycle of significance to climate and life is the sulfur cycle. Living things require certain safe, low levels of this nutrient. The sulfur cycle can be thought of as beginning with the gas sulfur dioxide (SO_2) or the particles of sulfate (SO_4^{2-}) compounds in the air. These compounds either fall out or are rained out of the atmosphere. Plants take up some forms of these compounds and incorporate them into their tissues. Then, as with nitrogen, these organic sulfur compounds are returned to the land or water after the plants die or are consumed by animals. Bacteria are important here as well, since they can transform the organic sulfur to hydrogen sulfide gas (H_2S). In the oceans, certain phytoplankton can produce a chemical that transforms to SO_2 that resides in the atmosphere. These gases can re-enter the atmosphere, water, and soil, and continue the cycle.

In its reduced oxidation state, the nutrient sulfur plays an important part in the structure and function of proteins. In its fully oxidized state, sulfur exists as sulfate and is the major cause of enhanced acidity in both natural and polluted rainwater. This link to acidity makes sulfur important to geochemical, atmospheric, and biological processes such as the natural weathering of rocks, acid precipitation, and rates of denitrification. Sulfur is also one of the main elemental cycles most heavily perturbed by human activity. Estimates suggest that emissions of sulfur to the atmosphere from human activity are at least equal to or probably larger in magnitude than those from natural processes. Like nitrogen, sulfur can exist in many forms: as gases or sulfuric acid particles. Sulfuric acid particles contribute to the polluting smog that engulfs some industrial centers and cities where many sulfur-containing fuels are burned. Such particles floating in air (known as sulfate aerosols) can cause respiratory diseases or cool the climate by reflecting some extra sunlight to space.

The lifetime of most sulfur compounds in the air is relatively short (days). Superimposed on these fast cycles of sulfur are the extremely slow sedimentary-cycle processes of erosion, sedimentation, and uplift of rocks containing sulfur. In addition, sulfur compounds from volcanoes are intermittently injected into the atmosphere, and a continual stream of these compounds is produced from industrial activities. These compounds mix with water vapor and form sulfuric acid smog. In addition to contributing to acid rain, the sulfuric acid droplets of smog form a haze layer that reflects solar radiation and can cause a cooling of the earth's surface. While many questions remain concerning specifics, human modification of the sulfur cycle is generating major physical, biological, and social problems, including acid rain and smog.

The Carbon Cycle

Carbon, the key element of all life on earth, has a complicated biogeochemical cycle of great importance to global climate change. The carbon cycle includes four main reservoirs of stored carbon: as CO_2 in the atmosphere; as organic compounds in living or recently dead organisms; as dissolved carbon dioxide in the oceans and other bodies of water; and as calcium carbonate in limestone and in buried organic matter (e.g., natural gas, peat, coal, and petroleum). Ultimately, the cycling of carbon through each of these reservoirs is tightly tied to living organisms.

Plants continuously extract carbon from the atmosphere and use it to form carbohydrates and sugars to build up their tissues through the process of photosynthesis. Animals consume plants and use these organic compounds in their metabolism. When plants and animals die, CO_2 is formed again as the organic compounds combine with oxygen during decay. Not all of the compounds are oxidized, however, and a small fraction is transported and redeposited as sediment and trapped where it can form deposits of peat, coal, and petroleum. Carbon dioxide from the atmosphere also dissolves in oceans and other bodies of water. Aquatic plants use it for photosynthesis, and many aquatic animals use it to make shells of calcium carbonate ($CaCO_3$). The shells of dead organisms (e.g., phytoplankton or coral reefs) accumulate on the sea floor and can form limestone that is part of the sedimentary cycle. The relevant time scales for these different processes vary over many orders of magnitude, from millions of years for the rock cycle and plate tectonics to days and even seconds for processes like photosynthesis and air-sea exchange.

CO_2 is a trace gas in the earth's atmosphere that has a substantial effect on earth's heat balance by absorbing infrared radiation. This gas, like water vapor (H_2O), CH_4, and N_2O, has a strong greenhouse effect. Life can alter the global concentration of CO_2 over very short time periods. During the growing season, CO_2 decreases in the atmosphere of the temperate latitudes due to the increasing sunlight and temperatures, which help plants to increase their rate of carbon uptake and growth. During the winter dormant period, more CO_2 enters the atmosphere than is removed by plants, and the concentration rises because plant respiration and the decay of dying vegetation and animals occurs faster than photosynthesis. The land mass in the northern hemisphere is greater than that in the southern hemisphere, thus the global concentration of CO_2 tracks the seasonality of terrestrial vegetation in the northern hemisphere more than that of the southern.

Human Modifications of Climate Services

Human activities are significantly perturbing all of these biogeochemical cycles as well as other earth system processes, both directly through industrial processes and indirectly through changing distributions and abundance of life. The atmosphere is of particular importance to the perturbations due to its crucial role in mediating all energy that enters and leaves earth. Overall, the atmosphere is the component that controls the dominant energy flow in the earth's climate system, and solar radiation from the sun provides the energy to make the weather machine work. Embedded in this process are the biogeochemical cycles we have described that operate on a variety of time and space scales and help to regulate flows of energy and materials throughout the earth system (figure 5.1). Yet, while we understand much about the functioning of separate parts of this system, there is still a great deal to be discovered about the feedbacks and linkages that allow these interconnected parts to function as a whole and, in turn, how they will respond to human modification.

Human Disturbance

Life influences the amount of CO_2 in the atmosphere through photosynthesis, respiration, and oceanic absorption. As ecosystems are altered, the balance of these processes will be altered. Human activities are upsetting this

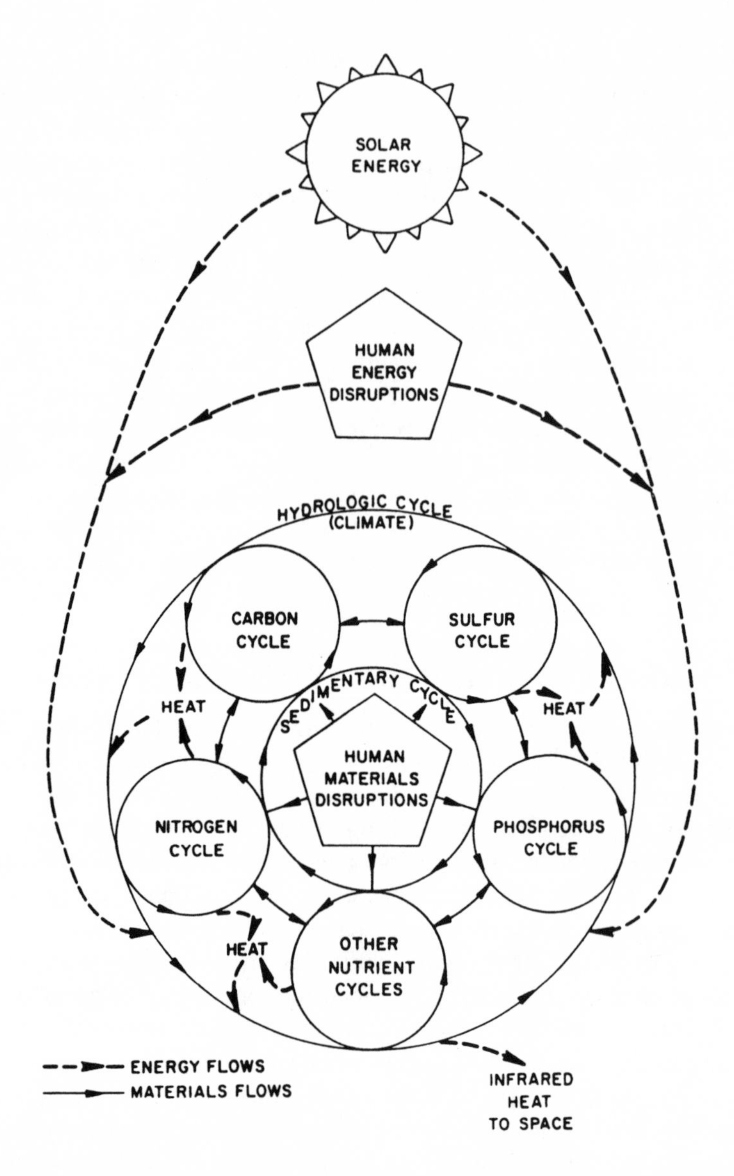

Figure 5.1. Climate and life are linked by a complex web of interconnected cycles. Life on earth depends on the cycling of nutrients through air, water, soil, and living things. The climate mediates the flow of materials through these global cycles. Solar energy degrades to heat at each stage of the cycling process and is eventually returned to space as infrared radiation. The composition of the earth's atmosphere regulates the radiative balance on earth between absorbed solar energy and emitted infrared energy, which, in turn, controls the climate.
Source: Schneider and Morton 1981.

balance and increasing CO_2 in the atmosphere through the burning of fossil fuels and clearing of forests. A significant increase in CO_2 could have dramatic consequences. Mathematical models of the climate suggest that when CO_2 (or its heat-trapping equivalent in other greenhouse gases) doubles (sometime in the middle of the next century if population, economic, and technology trends continue as typically projected), the world will warm up somewhere between one and five degrees Celsius by A.D. 2100 unless other factors counteract or amplify the CO_2–induced change (IPCC 1996a).

Even the lower end of that range is a projected warming at the rate of one degree per hundred years, a factor of ten faster than the one degree per thousand years that has been the typical average rate of natural sustained global temperature change from the end of the ice ages to warmer interglacial times. Should the higher end of the one-to-five degree Celsius range occur, then we could see rates of climate change some fifty times faster than sustained, natural average conditions. Climate largely determines the types of ecosystems that occupy an area. Global climate change at such a rapid rate would force many species to shift their ranges in an attempt to keep up with changing climatic conditions, as occurred during the ice age–interglacial transition ten to fifteen thousand years ago. Migrations of species such as slow-growing trees with large seeds would have to occur much faster than they did in the past to keep up with rapidly shifting climates. Other species could move more easily, raising the likelihood that communities of species could be disassembled (e.g., Root and Schneider 1993). Estimating the rates of global warming in the next century, however, is very controversial because of the uncertainties involved with multiple interacting feedback mechanisms (IPCC 1996a).

Humanity can control climate in ways other than changing greenhouse gas concentrations. Consider the amount of moisture released to the atmosphere through transpiration in the tropical rainforests. The dense vegetation in areas such as the Amazon basin typically recycles the precipitation that falls on it many times over, helping to form heavy cloud cover in the region. The clouds, in turn, reflect sunlight and produce more rain, directly influencing regional climate as well as indirectly perturbing global climate through altering large-scale circulation patterns over the tropics. As humanity deforests regions like the Amazon, not only is CO_2 released into the atmosphere, but changes in the hydrologic cycle will almost certainly affect regional climate and possibly even global climatic patterns. In deforested areas of northeastern Brazil, the cutting of the tropical forests has led to desertification, changing both surface reflectivity and the rate of transpiration. This change in ecosystem character can lead to a destabilizing positive feedback, which may cause an even further reduction in precipitation.

In a recent study on the possible climatic impacts of tropical deforestation, researchers suggest that conversion of forest into cropland or pastures would cause significant changes in the local microclimate (Salati and Nobre 1991). Expected changes include reduction in soil moisture, larger diurnal fluctuation of surface temperature and humidity deficit, and increased surface runoff during the rainy season and decreased runoff during the dry season. Results from general circulation model simulations of large-scale deforestation and conversion to grassy vegetation in the Amazon basin indicate an increase in surface temperature, decrease in evapotranspiration, and significant reduction in precipitation (Lean and Warrilow 1989; Shukla et al. 1990). Depending on the scale of the disturbed areas, local climate changes can lead to regional climate changes, which, in turn, may cause alterations in the global climate through atmospheric connections between tropical circulation and large-scale circulation patterns outside of the tropics. The effect on the ecological systems through changes in the hydrologic cycle, an increase in the dry season, and the disruption of plant-animal interactions may make it difficult for the rainforests to reestablish themselves if they are destroyed. Climate change aside, the implications of this scenario for the conservation of biodiversity are serious.

The provision of fresh water and regulation of its flows through precipitation, evaporation, transpiration, and runoff is mediated by all ecosystems. Forests and other vegetation types are critical components of this ecosystem service, providing free flood and drought relief, among other things. The loss of these services, through land-use change, can exacerbate disasters like spring floods in the Midwest and Southeast resulting from large expanses of land cleared for agriculture, as well as the drainage of wetlands and swamps, which otherwise might have acted as reservoirs for holding excess water or filtering toxic wastes.

Climate Change Uncertainty

The combination of potentially very rapid rates of human-induced climate change at the same time natural habitat has been fragmented for agriculture and development activities and assaulted with a host of chemical agents is unprecedented. It is for these reasons that it is essential to understand not only how much climate change is likely, but just as important, how to characterize and analyze the value of the ecosystem services that might be disrupted. How the biosphere will respond to human-induced climate change is fraught with uncertainty. One thing that is clear is that life, biogeochemical cycles, and climate are linked components of a highly interactive system. An illustration of this linked behavior can be seen in the simultaneous variation of CO_2, CH_4, temperature, and SO_4^{2-} found over time in Antarctic ice cores (see Charlson et al 1992). Temperature, CO_2, and CH_4 are positively correlated with one another, while each is negatively correlated with SO_4^{2-}. More recent data of N_2O, CH_4, and CO_2 over the past three hundred years show an increase in these trace gases that matches the magnitude of the changes in composition that occurred between the ice age and interglacial periods. This change in composition causes more heat to be trapped near the earth's surface. Since the Industrial Revolution the build-up of these and other greenhouse gases has increased the flow of energy to earth's surface by an average of roughly two watts per square meter. Climatologists also generally agree that the global air temperature at the surface has warmed up on average approximately 0.5 ± 0.2 degrees Celsius in the past century. It is this rate of change that appears very large compared to the sustained temperature changes from the ice ages to the interglacials in recent earth history.

Uncertainties become more significant when projections of climatic impacts are considered. The combination of increasing population and increasing energy consumption per capita is expected to contribute to increasing CO_2 and sulfate emissions over the next century, but projections of the extent of the increase are very uncertain. Central estimates of emissions suggest a doubling of current CO_2 concentrations by the middle of the twenty-first century, leading to typical projected warming ranging, as mentioned earlier, from one degree to more than five degrees by the second half of the twenty-first century. Warming at the low end of this uncertainty range could still have significant implications for species adaptations, whereas warming of five degrees or more could have catastrophic effects on natural and managed ecosystems, produce serious coastal flooding, and involve other impacts on natural and human systems. The overall cost of these impacts in "market sectors" of the economy could easily run into many tens of billions of dollars annually (Smith and Tirpak 1988, IPCC 1996b). Although fossil fuel use contributes substantially to the cause of the impacts, associated costs are not included in the price of conventional fuels; they are externalized. Internalizing these environmental externalities into economic benefit-cost analyses (see Goulder and Kennedy, chapter 3, this volume) is a principal goal of international climate policy advocates.

Economic Analyses

We now turn to analyzing a few of the specific ecosystem services that link climate and life, and use the subjective probabilities of potential climate change impacts to provide a crude metric for assigning dollar values to certain aspects of these services.

Valuing Climate Extremes

Over the past several years climate in the United States has become a much-talked-about media topic. Extreme events have become more conspicuous recently. During the summer of 1988 the Midwest experienced a record-breaking heat wave and associated drought that led to a 30 percent reduction in crop production that year. In 1993 record rains led to summer catastrophic flooding of the Mississippi River and its tributaries. The winters of 1994 and 1995 brought severe cold spells across the country. Trends in the U.S. climate since the beginning of this century include a 5 percent increase in precipitation since 1970 over that of the previous seventy years; a severe moisture surplus in more than 30 percent of the country in each of three different years during this time period (the Mississippi flooding is an example of this type of extreme event); and an average daily temperature increase of 0.3 degrees Celsius since the turn of the century (Karl et al. 1995). A Climate Extremes Index produced by the National Climate Center supports the belief that the United States has experienced more climate extremes in recent decades, and a Greenhouse Climate Response Index shows an increase in anticipated U.S. greenhouse climate response indicators (figure 5.2). These indices combine data on weather extremes such as droughts, wet winters, severe rainstorms, and other events. While qualitatively consistent with generalized predictions for global greenhouse warming, the magnitude and persistence of these trends cannot yet be considered conclusive evidence of linkage. At the same time, however, normal variation in weather patterns may not be able to explain the increase in weather extremes since the mid-1970s, except perhaps as chance events with less than a 10 percent probability (e.g., Karl et al. 1995). As part of our evaluation, we can anticipate costs associated with global change and place a preliminary value on some of the ecosystem services that could be affected.

Catastrophic floods and droughts are cautiously projected to increase in both frequency and intensity with a warmer climate and the influence of human activities such as urbanization, deforestation, depletion of aquifers, contamination of groundwater, and poor irrigation practices (IPCC 1996a). Humanity remains vulnerable to extreme weather events. For example, consider that between 1965 and 1985 in the United States floods claimed 1,767 lives and caused more than $1.7 billion in property damage (Dracup and Kendall 1990). This estimate is based on federal expenditures because information on private insurance losses and costs is unavailable. Ultimately,

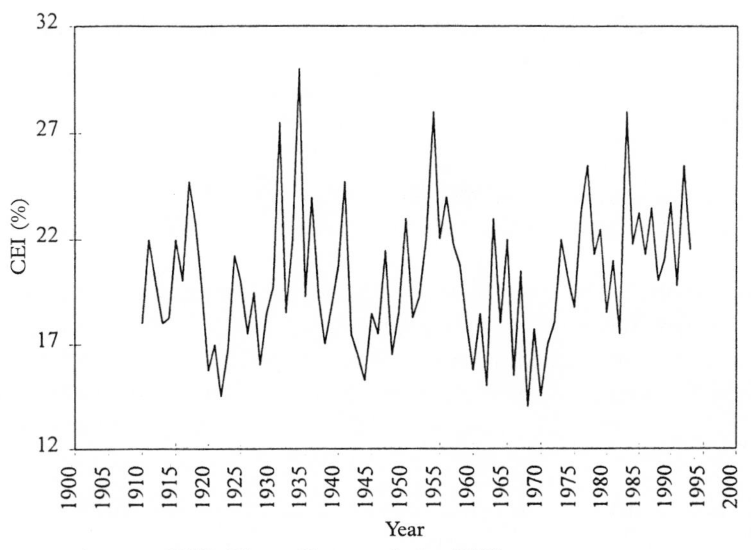

An annual U.S. Climate Extremes Index (CEI).

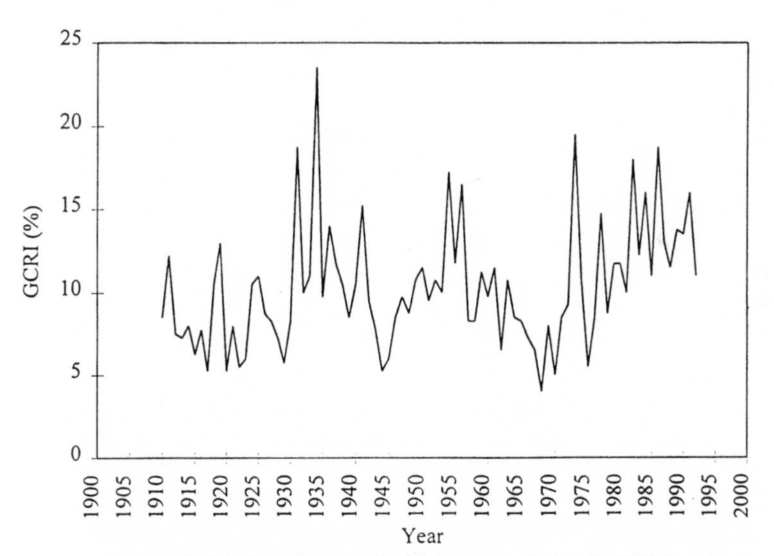

An annual U.S. Greenhouse Climate Response Index (GCRI) based on greenhouse climate response indicators.

Figure 5.2. Variations in an annual U.S. Climate Extremes Index (expected value 20 percent) and an annual Greenhouse Climate Response Index (expected value 10 percent) based on greenhouse climate response indicators over the past century.
Source: Karl et al. 1995.

the effects of these floods are felt across a wide range of economic sectors, as can be seen with the overall cost evaluation of the Midwest flood of 1993 (table 5.1).

In the 1993 Midwest flood, 9 states and 525 counties declared disasters. The estimated federal response and recovery costs include $4.2 billion in direct federal expenditures, $1.3 billion in payments from federal insurance programs, and more than $621 million in federal loans to individuals, businesses, and communities. In the upper Mississippi Valley states of Minnesota, Nebraska, North Dakota, and South Dakota, as well as Wisconsin and northern Iowa, losses were primarily agricultural. In Illinois, central Iowa, and Missouri, major losses occurred in agriculture as a result of bottomland flooding, but urban areas also sustained damages. Numerous im-

Table 5.1. Summary of federal expenditures for the Midwest flood of 1993 (in millions of dollars)

	Missouri	Iowa	Minnesota	Illinois	Other States[a]	Total
USDA	141.6	376.2	446.2	63.3	512.2	1,699.9
FEMA	291.5	189.8	62.9	197.5	290.9	1,098.0
HUD	152.1	107.7	29.8	94.9	75.1	500.0
Commerce	51.9	48.5	7.9	8.4	23.8	201.3
USACE	128.7	9.7	0.3	70.3	12.0	253.1
HHS	19.3	22.8	4.0	7.4	15.2	75.0
Education	4.5	11.1	0.8	1.4	2.2	100.0
Labor	15.0	15.0	5.0	10.0	19.6	64.6
National Community	1.0	1.2	0.7	0.4	0.7	4.0
DOT	73.5	22.1	7.3	33.3	36.9	146.7
EPA	7.6	4.6	2.2	5.3	12.4	34.0
DOI	5.1	2.1	6.0	11.8	8.3	41.2
TOTAL	891.8	810.8	573.1	504.0	1,009.3	4,217.8

[a]Denotes costs combined, including those for the states of Kansas, Nebraska, North Dakota, South Dakota, and Wisconsin.

Abbreviations: USDA, United States Department of Agriculture; FEMA, Federal Emergency Management Agency; HUD, Housing and Urban Development; USACE, United States Army Corps of Engineers; HHS, Department of Health and Human Services; DOT, Department of Transportation; EPA, Environmental Protection Agency; DOI, Department of the Interior.

Source: Interagency Floodplain Management Review Committee report to the Administrative Floodplain Management Task Force, 1994.

pacts of the flooding are still largely unknown, including cumulative effects of releases of hazardous material such as pesticides, herbicides, and other toxics; effects on groundwater hydrology and groundwater quality; distribution of contaminated river sediments; and alteration of forest canopy and subcanopy structure. In addition, the loss of tax revenue has not been quantified for the Midwest flood. It is important to note that while not all costs of the 1993 flood can be calculated in monetary terms, both quantifiable and nonquantifiable costs were significant in magnitude and importance. While we are not claiming that this event was directly caused by anthropogenic climate change, it does allow a rough estimate of the magnitude of costs should such changes cause increased extremes, a cautiously anticipated assessment by groups such as the Intergovernmental Panel on Climate Change (IPCC 1996a).

Like floods, severe droughts of the twentieth century have affected both the biophysical and socioeconomic systems of many regions. Drought analyses indicate that even reasonably small changes in annual streamflows due to climatic change can have dramatic impacts on drought severity and duration. For example, changes in the mean annual streamflow of a region of only ± 10 percent can cause changes in drought severity of 30 to 115 percent (Dracup and Kendall 1990). Damage estimates from the 1988 drought in the midwestern United States show a reduction in agricultural output by approximately one-third, as well as billions of dollars in property damage.

Hurricanes can also cause devastation in the tens of billions of dollars. Warmer surface waters in the oceans currently produce stronger hurricanes (that is, they are warm-season phenomena). Other meteorological factors are involved, though, that may act to increase or decrease the intensity of hurricanes. An increase in intensity of hurricanes with warmer waters is plausible, yet speculative given the number of factors involved. There is little doubt, however, of the heightened damage that would be due to more intense hurricanes.

Damage assessment is one possible way in which we can relate the cost of more inland and coastal floods, droughts, and hurricanes to the value of preventing the disruption of climate stability. In the 1993 Midwest flood example, we delineated the costs of a single event. We now turn to an example of a more integrated analysis: the cost assessment of future sea level rise along U.S. coasts associated with possible ice-cap melting or with ocean warming and the resulting thermal expansion of the waters. In a probability distribution of future sea level rise by 2100, changes range from slightly negative values to a meter or more rise, with the midpoint of the distribution being approximately half a meter (Titus and Narayanan 1996). A number of studies have assessed the potential economic costs of sea level rise along the developed coastline of the United States. For a 50 cm rise in sea level by the year 2100, estimates of potential costs range from $20.4 billion (Yohe et al. 1996) to $138 billion (Yohe 1989) in lost property. How do the costs of prevention compare to the losses potentially sustained by increasing floods and droughts or by future sea level rises? The next sections explore this topic of placing a value on climate changes and abatement.

Valuing Carbon

There is already a historic background on the evaluation of carbon that includes climate change policies such as the introduction of carbon taxes that reduce greenhouse gas emissions through increasing prices of carbon-based fuels proportional to the amount of carbon they emit (Nordhaus 1992). Another mechanism for reducing greenhouse gas emissions is through an international tradable emissions permit system intended to limit emissions of certain pollutants (Grubb et al. 1994). These policies and others constitute ways of balancing the economic costs of emissions with some assumed benefit of averting the loss of ecosystem services (called "climate damage"). For example, William Nordhaus's imposed carbon taxes range from a few dollars per ton to hundreds of dollars per ton in computer model scenarios. He has shown, in the context of this economic model and its assumptions, that this carbon tax would cost the world economy anywhere from less than 1 percent in gross national product to a several percent loss by the year 2100. Even a 1 percent loss in GDP, based on his assumed baseline of 460 percent growth in personal income between 1965 and 2100, amounts to trillions of dollars per year by 2100. This cost of preventing a degrading of the climatic environment, however, should be compared to estimates of the societal value of the climate services.

Any comprehensive attempt to evaluate the societal value of climate change should include such things as loss of species diversity, loss of coastline from increasing sea level, environmental displacement of persons, and agricultural losses. Nordhaus (1992) first estimated the climate damage at 1 percent reduction in GNP based on market sector losses for a central estimate of climate change. This was criticized (e.g., Oppenheimer 1993, Schneider 1993) as too narrow a view of climate as a type of public good since it reflected neither nonmarket values (e.g., species loss) nor climate "surprise" scenarios (e.g., see Schneider and Root 1995). In response, Nordhaus (1994) conducted a survey of conventional economists, environmental economists, atmospheric scientists, and ecologists. Their estimates of loss of gross world product (GWP) resulting from a three-degree Celsius warming by 2090 varied between a loss of 0 and 21 percent of GNP with a mean of 1.9 percent (Nordhaus 1994). Even a 2 percent loss of GWP, however, represents climate damage of hundreds of billions of dollars annually. For a six-degree Celsius warming scenario, the respondents predicted a loss of the world economy ranging from 0.8 to 62 percent with a mean estimate of 5.5 percent. A striking difference was noted between respondents from different academic disciplines, with natural scientists' estimates of economic impact twenty to thirty times higher than conventional economists'.

While it is impossible to estimate credibly a numerical value on all of the ecosystem services provided through the maintenance of the carbon cycle at its present state, it may be useful to look at land-use change and loss of biomass, mostly through deforestation, as a source of atmospheric CO_2. In a very simplistic and preliminary evaluation, we can use the rates of net deforestation to calculate a value for carbon. For example, global loss of aboveground biomass from deforestation in the tropics is approximately 1–3 gigatons/year over the past ten years (Food and Agriculture Organization 1993). This amounts to 2–5 gigatons of carbon in carbon dioxide released into the atmosphere each year from deforestation and forest degradation (this does not include the 6-gigaton carbon emissions from the burning of fossil fuels). Much of the carbon from biosphere emissions is taken up immediately by vegetation, however, leaving approximately 1–2.5 gigatons of net carbon added to the atmosphere each year. We can apply the concept of carbon taxation for emissions to an ecosystem service valuation of retaining the carbon in the forests. Using a range of carbon taxes from typical macroeconomic models (e.g., Gaskins and Weyant 1993) between $1 per ton and $100 per ton of carbon, the net value of carbon lost each year amounts to between $1 and $250 billion/year. However, use of optimizing economic models to estimate climate damage is highly unsatisfying, since these studies use very limited and often ad hoc assumptions that both over- and underestimate the likely damages to various market and nonmarket sectors.

Methods of Valuation

The need for alternative methods of evaluation of these climate-related ecosystem services is quite clear when examining preliminary public opinion responses of global warming. In a controversial method called contingent valuation (see chapter 3), respondents are surveyed to determine how much they would be willing to pay to prevent a given global climate change scenario from happening or how much money they would require to permit a given amount of change. The difficulties with this type of valuing of environmental goods and processes are immense, especially since much of the evaluation is subjective. Public opinion depends, in part, on people's exposure to the issues and the level of education and information on these issues they have received.

In a Southern California study, the contingent valuation technique was applied to determine the influence of potential changes in temperature and precipitation resulting from global warming on respondents' willingness to pay (Berk and Schulman 1995). Factorial survey methods were used to present a variety of hypothetical climate scenarios to a sample of six hundred Southern California residents. Respondents were provided with a baseline microclimate for the region before future climate scenarios were evaluated. For example, for residents living in coastal communities, the baseline climate over the past ten years was described as having a summer average high temperature of 75 degrees Fahrenheit, with daily highs ranging between 70 and 80 degrees, and an average of thirteen inches per year of rain. One possible future scenario over the next ten years included a summer average high temperature of 100 degrees Fahrenheit, with daily highs ranging from 80 to 120 degrees (the latter typical of Death Valley, California), and an average of twenty inches per year of rain. With these and other scenarios, predicted probabilities were determined from the respondents' willingness to pay for the abatement of different mean high temperatures. In this scenario, respondents were willing to pay an average of $140 to offset a mean high temperature of 100 degrees, while a mean high temperature of 80 degrees was worth approximately $100 (figure 5.3). This represents a 40 percent increment in willingness to pay for a 20-degree rise in temperature and other scenario characteristics. Note, however, that unlike the Nordhaus 1994 Survey of Experts (all of whom assigned accelerating damage costs to climate change scenarios as they became larger), the Southern California residents reached a plateau (see figure 5.3) in their willingness to pay to prevent 120-degree Fahrenheit mean temperatures as compared to 110-degree Fahrenheit mean temperatures.

However, the actual damages to the L.A. basin residents of mean high temperatures of 110 or more degrees Fahrenheit (which would imply occasional extreme heat waves similar in temperature to Death Valley mean highs) would be orders of magnitude more costly, we believe, than a 100-degree Fahrenheit mean high temperature, as such extreme heat would decimate most existing vegetation and threaten the lives of tens of thousands of elderly and other persons vulnerable to heat stroke. For just such reasons, Berk and Schulman (1995) strongly caution against taking the dollar values from the survey literally or using them in cost-benefit analyses, as they confound several sources of value including stewardship and altruism. In addition, some of the climate increases were well above the range of current scientific estimates of greenhouse warming (IPCC 1996a). The survey was not done in conjunction with atmospheric scientists and climatologists who

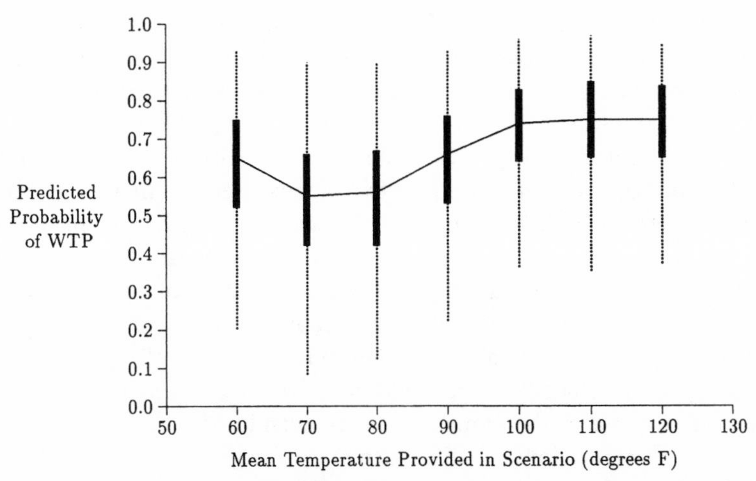

Figure 5.3. Predicted probabilities of Los Angeles survey respondents' willingness to pay (WTP) for the abatement of different mean daily high temperatures.
Source: Berk and Schulman 1995.

could provide more realistic climate scenarios or ecologists, public health officials, or others who could help the respondents realize what such warming might mean for trees, birds, or people. Contingent valuation of the hypothetical good is possible when people believe the survey scenario. We present this type of evaluation study to highlight how difficult it is to find acceptable methods to place values on the climatic components of ecosystem services. In this survey case, the background of the respondents as well as their (limited) prior knowledge of the impacts of greenhouse warming played a large role in the survey outcomes. At the same time, however, contingent valuation points out that people are willing to pay to preserve ecosystem services as well as the tremendous need for education to help citizens more realistically value climate and other environmental services.

Conclusions

The ongoing disturbances of the atmosphere that affect the biogeochemical and physical processes that determine the climate may influence human and natural systems in profound ways. We have attempted to outline a few of the major ecosystem services that are associated with climate and the atmosphere, as well as introduce the challenging task of quantifying, and ultimately monetizing, these services. Current monetized estimates of climate damage by the middle of the twenty-first century from typical climate change scenarios range from slight economic benefit to a trillion or more dollars lost annually, with most macroeconomic assessments assuming a 1–2 percent annual loss to GWP from "best guess" projected climate change. Moreover, the interacting processes and biogeochemical cycles occurring across a wide spectrum of scales lead to synergistic effects that are not usually considered and sometimes not even known (i.e., surprises) when we attempt to disaggregate and value ecosystem services (e.g., Schneider and Turner 1995). The deforestation of the Amazon basin is one example of interacting scales where land-use change affecting local and regional climate may also produce a net global residual. Even if the mosaic of regional effects averages itself out globally, there could be residual effects arising from heterogeneous forcing of the climate in areas outside of the tropics (i.e., regional high concentrations of sulfate aerosols or tropospheric ozone).

Ecosystems both mediate and respond to the climate system through a variety of physical, biological, and chemical feedback cycles. The uncertainty of resulting synergisms and potential global effects, as exemplified in the Amazon basin, points to the important challenge of defining and understanding the processes that link species and ecosystems with climate. With increasing knowledge, we can better anticipate ecological responses under changing climate scenarios. Meanwhile, humanity continues to perform this potentially trillion dollar unnatural experiment on "Laboratory Earth" (Schneider 1997).

Acknowledgments

The authors sincerely thank Gretchen Daily for advice on manuscript development, organization, and analysis. Joseph Coughlan, Paul Ehrlich, and three anonymous reviewers provided very helpful comments on various drafts of the manuscript.

References

Berk, R. A., and Schulman, D. 1995. "Public perceptions of global warming." *Climatic Change* 29: 1–33.

Charlson, R. J., Orians, G. H., Wolfe, G. V., and Butcher, S. S. 1992. "Human Modifications of Global Biogeochemical Cycles." In Butcher, S. S., Charlson, R. J., Orians, G. H., and Wolfe, G. V. (eds.), *Global Biogeochemical Cycles*. New York: Academic Press, pp. 353–361.

Dracup, J.A., and Kendall, D.R. 1990. "Floods and Droughts." In Waggoner, P. E.

(ed.), *Climate Change and U.S. Water Resources*. New York: John Wiley, pp. 243–267.

Ehrlich, P. A., and P. H. Raven. 1964. "Butterflies and plants: A study in coevolution." *Evolution* 8: 586–608.

FAO (Food and Agriculture Organization). 1993. *Forest Resources Assessment, 1990: Tropical Countries*. FAO Forestry Paper 112. Rome: U.N. Food and Agriculture Organization.

Gaskins, D., and Weyant, J. 1993. "EMF-12: Model Comparisons of the costs of reducing CO_2 emissions." *American Economic Review* 83: 318–323.

Grubb, M., Duong, M. H., and Chapuis, T. 1994. Optimizing Climate Change Abatement Responses: On Inertia and Induced Technology Development. In Nakicenovic, N., Nordhaus, W. D., Richels, R., and Toth, F.L. (eds.), *Integrative Assessment of Mitigation, Impacts and Adaptation to Climate Change*. Laxenberg, Austria: International Institute of Applied Systems Analysis, pp. 513–534.

Interagency Floodplain Management Review Committee. 1994. *Sharing the Challenge: Floodplain Management into the Twenty-first Century*. Report to the Administration Floodplain Management Task Force. Washington, D.C.: U.S. Government Printing Office.

IPCC (Intergovernmental Panel on Climate Change). 1990. Houghton, J. T., Jenkins, G. J., and Ephraums, J. J. (eds.), *Climate Change: IPCC Scientific Assessment*. Cambridge, England: Cambridge University Press.

IPCC. 1996a. J. J. Houghton, Meira Filho, L. G., Callander, B. A., Harris, N., Kattenberg, A., and Maskell, K., (eds.). *Climate Change 1995. The Science of Climate Change: Contribution of Working Group I to the Second Assessment Report of the Intergovernmental Panel on Climate Change*. Cambridge, England: Cambridge University Press.

IPCC. 1996b. R. T. Watson, Zinyowere, M. C., and Moss, R. H., (eds.). *Climate Change 1995. Impacts, Adaptations, and Mitigation of of Climate Change: Scientific-Technical Analyses: Contribution of Working Group II to the Second Assessment Report of the Intergovernmental Panel on Climate Change*. Cambridge, England: Cambridge University Press.

Karl, T. R., Knight, R..W., Easterling, D. R., Quayle, R. G. 1995. "Trends in U.S. climate during the twentieth-century." *Consequences* 1: 2–12.

Lean, J., and Warrilow, D. A. 1989." Climatic impact of Amazon deforestation." *Nature* 342: 411–413.

Myers, N. 1993 "Environmental refugees in a globally warmed world." *Bioscience* 43: 752–761.

Nordhaus, W. D. 1992. "An optimal transition path for controlling greenhouse gases." *Science* 258: 1315–1319.

Nordhaus, W. D. 1994. *Managing the Global Commons: The Economics of Change*. Cambridge, Mass: MIT Press.

Nordhaus, W. D. 1994. "Expert opinion on climatic change." *American Scientist* 82: 45–51.

Oppenheimer, M. 1993. Letter. *Science* 259: 1382–1383.

Root, T., and Schneider, S. H. 1993. "Can large-scale climatic models be linked with multiscale ecological studies?" *Conservation Biology* 7: 256–270.

Root, T. L., and Schneider, S. H. 1995. "Ecology and climate: Research strategies and implications." *Science* 269: 334–341.

Salati, E., and Nobre, C. A. 1991. "Possible climatic impacts of tropical deforestation." *Climatic Change* 19: 177–196.

Schneider, S. H. 1993. Letter. *Science* 259: 1381.

Schneider, S. H., and Londer, R. 1984. *The Coevolution of Climate and Life*. San Francisco: Sierra Club Books.

Schneider, S. H., and Morton, L. 1981. *The Primordial Bond: Exploring Connections between Man and Nature through the Humanities and Sciences*. New York: Plenum Press.

Schneider, S. H. 1997. *Laboratory Earth: The Planetary Gamble We Can't Afford to Lose*. New York: Basic Books.

Schneider, S. H., and Root, T. L. 1996. "Impacts of Climate Change on Biological Resources." In M. Mac, P. Opler, and P. D. Doran, (eds.), *Status and Trends Report*. Washington, D.C.: National Biological Service.

Schneider, S. H., and Turner, B. L. 1995. "Anticipating Global Change Surprise," In S. J. Hassol and J. Katzenberger (eds.), *Elements of Change 1994*. Aspen Global Change Institute, Aspen, Colorado, pp. 130–145.

Shukla, J., Nobre, C. A., and Sellers, P. 1990. "Amazonia deforestation and climate change. *Science* 247: 1322–1325.

Smith, J. B., and Tirpak, D. A., eds. 1988. *The Potential Effects of Global Climate Change on the United States: Draft Report to Congress*, vols. 1 and 2. U.S. Environmental Protection Agency, Office of Policy Planning and Evaluation, Office of Research and Development. U.S. Washington, D.C.: Government Printing Office.

Titus, J. G., and Narayanan, V. 1996. "The risk of sea level rise." *Climate Change* 33(2): 151–212.

Yohe, G. 1989. "The cost of not holding back the sea—economic vulnerability." *Ocean and Shoreline Management* 15: 233–255.

Yohe, G., Neumann, J., Marshall, P., and Ameden, H. 1996. "The economic cost of greenhouse induced sea level rise for developed property in the United States. *Climatic Change*.

Chapter 6

Biodiversity and Ecosystem Functioning

David Tilman

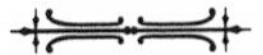

This chapter addresses how biodiversity may influence the supply of ecosystem goods and services. Biodiversity, or biological diversity, is the variety of life at all levels of organization, from the level of genetic variation within and among species to the level of variation within and among ecosystems and biomes. For convenience or necessity, biologists have tended to focus studies of biodiversity on the number of species in an ecosystem, which is called species diversity or species richness. The rapid expansion of human activities across the earth, and the subsequent modification of natural ecosystems into systems managed for human benefit, has led both to dramatic increases in species extinctions and to much lower biodiversity within managed ecosystems. This has raised numerous concerns, including the possibility that the functioning and stability of earth's ecosystems might be threatened by this loss of biological diversity (Ehrlich and Ehrlich 1981, Schulze and Mooney 1993). The goods and services provided by ecosystems depend on ecosystem functioning and on the susceptibility of this functioning to drought, floods, invasions by exotic organisms, and other disturbances. Thus, there may be a link between biodiversity and the ability of ecosystems to provide goods and services to humanity. This chapter reviews the literature on this subject to address three major questions: (1) Does the productivity of ecosystems depend on their biodiversity? (2) Does ecosystem stability depend on biodiversity, i.e., are more diverse ecosystems more resistant to and more able to recover from disturbances? (3) Does the long-term sustainability of ecosystem functioning depend on ecosystem biodiversity? This chapter does not address the economic valuation of biodiversity, but rather lays out ecological principles relevant to calculations of valuation in the chapters that follow.

The idea that biodiversity influences ecosystem functioning is venerable, apparently first suggested by Darwin (1872), who noted that ecosystem productivity depended on biodiversity. As quoted in McNaughton (1993), Darwin stated, "The more diversified in habits and structures the descendants . . . become, the more places they will be enabled to occupy. . . . If a plot of ground be sown with one species of grass, and a similar plot be sown with several distinct genera of grasses, a greater number of plants and a greater weight of dry herbage can be raised in the latter than the former case." McNaughton (1977, 1993) and others have expanded on this diversity-productivity hypothesis. Odum (1953), MacArthur (1955), and Elton (1958) noted that the larger the number of species in an ecosystem, the greater would be the number of interspecific interactions linking them and determining the functioning of the ecosystem. Because of this, they hypothesized that ecosystems that are more species rich should be more resistant to perturbations and disturbances because they would contain more alternative pathways for the flow of energy and the internal cycling of nutrients. As this tight internal recycling is interrupted by the loss of biodiversity, ecosystems are thought to become more open and thus lose the nutrient capital on which their sustained productivity had been based (e.g., Vitousek and Hooper 1993). These three hypotheses, which are clearly interrelated, are more fully developed below, as are the various observational and experimental studies that have been used to test them. Although there are many unanswered questions, in total this review shows that the ability of ecosystems to provide a sustainable flow of goods and services to humans is likely to be highly dependent on biodiversity, which, itself, can be sustained only if humans alter their present course of action.

Biodiversity and Ecosystem Productivity

Humans depend on living plants for the production of food, forest products, and many other goods essential for human life. The total of all materials produced by the growth of plants in a period of time (most often a year) is called ecosystem primary production, or, more simply, ecosystem productivity. Ecosystem productivity has been hypothesized to be higher when more plant species are present because differences among species in methods of resource capture should allow more diverse communities to more fully utilize their limiting resources. For instance, some plant species have physiologies and morphologies that allow them to grow best during cooler and wetter weather, whereas others grow better during hotter and drier weather. If a cool-season and a warm-season species were to grow together, these complementary features might lead to greater total primary productivity across the full growing season than possible when either species grew alone. Similarly, plant species differ in the depth in the soil profile at which they are rooted, again potentially allowing a fuller exploitation of soil resources, and thus greater ecosystem productivity, in more species-rich communities. Indeed, the physiologies, morphologies, and life histories of plant species differ one from the other in a multitude of ways (e.g., Chapin 1980, Chabot and Mooney 1985, Grime 1979) that might allow mixtures of several species to more fully utilize limiting resources than would a monoculture of any one species. Such considerations have led to the general expectation that, all else being equal, plant primary productivity should be an increasing function of the number of plant species in a community.

Three general relations between diversity and ecosystem functioning have been proposed (figure 6.1; Vitousek and Hooper 1993). The linear relationship of curve 1 implies that each species added to or removed from an ecosystem would have the same impact on ecosystem processes as that of any other species. As Vitousek and Hooper noted, this seems unlikely. The flat relationship of curve 3 means that, after one, two, or some small number of species are present, additional species would have no effect on an ecosystem process. Such an abrupt and low limit to the effects of diversity also seems unlikely. Vitousek and Hooper hypothesized that the most likely response would be saturating or asymptotic, as in curve 2, because this means that each added species shares an increasingly great proportion of its

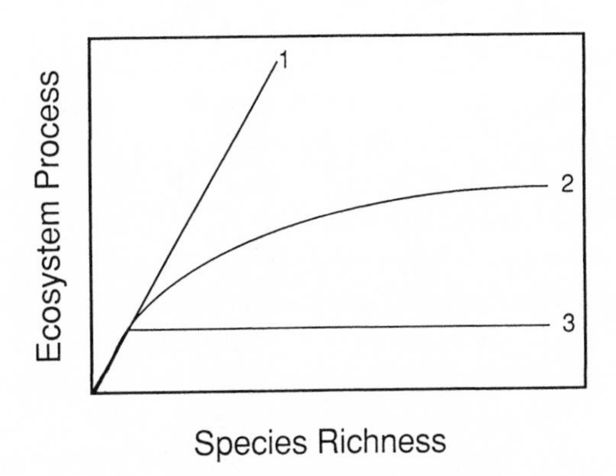

Figure 6.1. Three qualitatively different potential relationships between an ecosystem process (e.g., productivity, resistance to disturbance, resilience, etc.) and biodiversity. Based on Vitousek and Hooper (1993).

traits with existing species and thus does less to diversify the functioning of the ecosystem than the initial species.

There are two major reasons biodiversity may influence ecosystem productivity. The first may be called the sampling-competition effect. In any given habitat, a particular plant species will be the most productive, a different species will be the next most productive species, etc. Because plant species compete, the most productive species often will come to dominate a habitat in which it is present. All else being equal, the probability of having a highly productive species present will increase with plant diversity. This causes total community productivity to increase, on average, with plant diversity, with the resulting curve looking much like curve 2 of figure 6.1 (Tilman et al. 1997 in review). This illustrates a major effect of diversity. The more diverse an ecosystem is, the more likely it is to contain one or more superior species that will come to strongly influence the functioning of that ecosystem. Greater diversity allows a greater sampling of the full potential of biodiversity, and competitive interactions magnify the differences among species, causing productivity or some other aspect of ecosystem functioning to increase with diversity.

The second major reason biodiversity may influence ecosystem productivity is complementary resource use by different species. Consider a terrestrial ecosystem in which two factors—soil water and soil nitrogen—constrain productivity. All plants require both water and nitrogen for survival and growth. There will be some levels of soil water and soil nitrogen that are so low that no plant species are able to survive. These levels are indicated by the unshaded region of figure 6.2A. The higher levels of water and nitrogen in the shaded region of this figure allow one or more species to survive and grow. Any given species is only able to survive and grow in a portion of the shaded region. Because water and nitrogen are nutritionally essential resources for plants, the shape of this region is rectangular (Tilman 1982). The range of water and nitrogen levels at which a species can survive and grow is indicated by what is called its resource-dependent growth isocline. The isocline for species A is shown in figure 6.2B. Species A can survive and grow in the shaded region. Other species, which differ one from the other in their physiologies, morphologies, and life histories, are able to occupy other regions (figure 6.2C–F). Note that the portion of the water-nitrogen plane shown in these figures is such that it spans the spatial and temporal ranges of soil water and soil nitrogen in the ecosystem of interest. Note, also, that the five species shown differ in their requirements for nitrogen and water, and that each species has the right angle corner of its isocline touching the line that separates the region of survival from the region of death. This line is an interspecific tradeoff curve (Tilman 1988). The proportion of the region of growth that is covered by the shaded isocline of one or more species

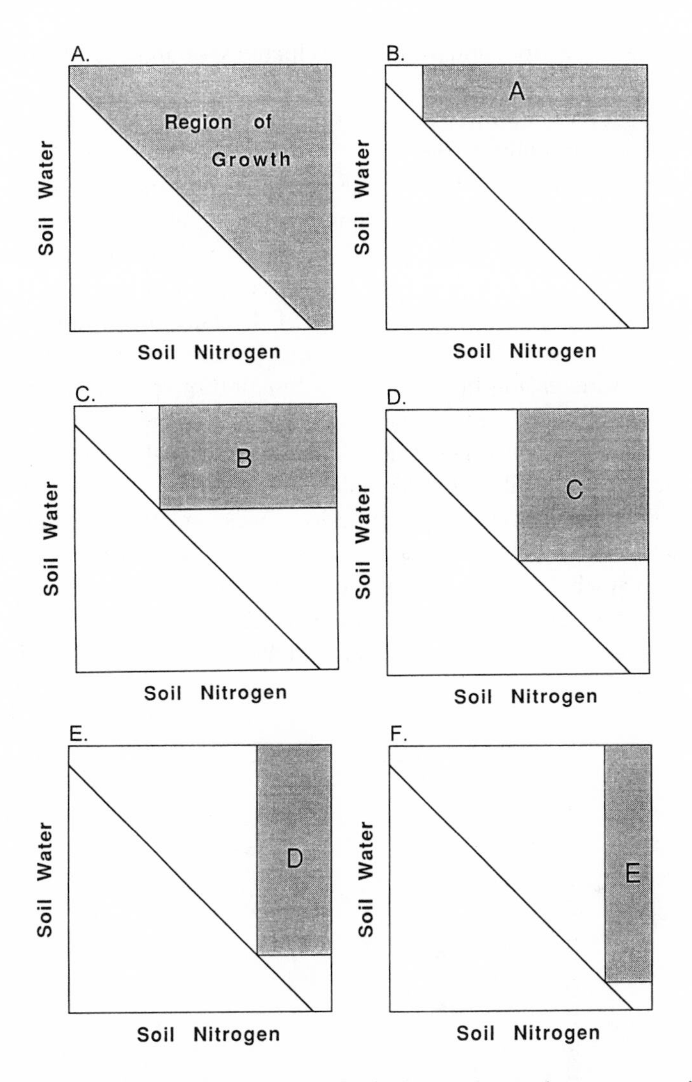

Figure 6.2. A. The shaded portion of this figure shows the concentrations of soil nitrogen and soil water that can sustain the growth of at least one species of plant in a habitat. This livable habitat itself is heterogeneous, with the various points in the shaded region of the graph representing different localities within the habitat. **B–F.** These shaded regions with right angle corners are resource-dependent growth isoclines for species A through E (see Tilman 1982). For instance, species A can just survive for soil nitrogen and water concentrations on the inside edge of its isocline, and will grow for concentrations in the shaded region. These five species have an interspecific tradeoff in their requirements for nitrogen and water, with species A having a low requirement for nitrogen but a high requirement for water. Species E, in contrast, has a high requirement for nitrogen and a low requirement for water. The species were chosen to be evenly spaced along the interspecific tradeoff curve.

is approximately proportional to the productivity of an ecosystem containing those species.

On average, monocultures of these species cover about 45 percent of the potential available habitat (figure 6.3). Mixtures of these species provide greater coverage (figure 6.4), with the best coverage, and thus greatest primary productivity, provided by the combination of all five species living in the same habitat. The results for all possible combinations of these five species, taken 0, 1, 2, 3, 4, or 5 at a time, are shown in figure 6.3. This illustrates two basic principles of the effects of diversity on productivity and, perhaps, on other aspects of ecosystem functioning. First, the resulting relationship is an increasing but asymptotic function of species richness. This occurs because, on average, each additional species is increasingly similar to the collection of pre-existing species, i.e., each additional species contributes less to the resource utilization capability of the total community. The relationship is increasing because each added species gives better coverage of the physical conditions that constrain productivity. Second, the resulting relationship is stochastic, i.e., is predicted to have variance in productivity that depends on the actual identities of the species that co-occur at any given level of diversity. This variance is lower at higher diversity because higher di-

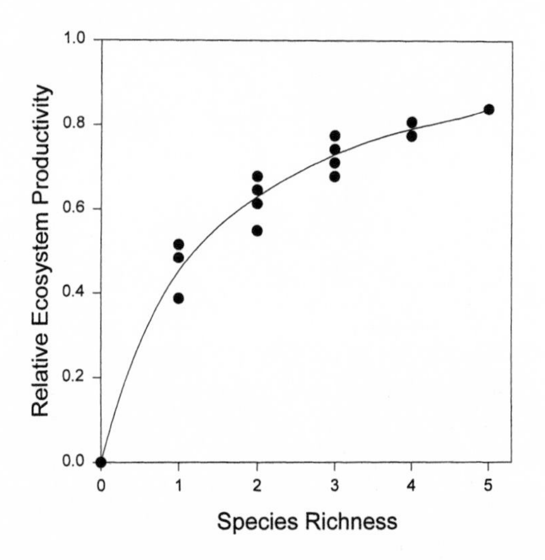

Figure 6.3. The isoclines illustrated in figure 6.2 can be used to determine the amount of the livable habitat of figure 6.2A that is "covered" by various combinations of 1, 2, 3, 4, or all 5 species, using the species of figure 6.2B–F. The total amount of livable habitat that is covered by all species present in a habitat is a measure of the relative productivitiy of that plant community. This was determined for all possible combinations of these species taken 0, 1, 2, 3, 4, and 5 at a time, with results shown above.

versity represents a fuller sampling of the potential total community coverage, but this effect is clearer for cases with more species than used in figures 6.3 and 6.4. This variance, and the clear effects of particular species combinations, highlights the important point that the effects of biodiversity on ecosystem functioning come from differences among species and from interactions among species. It might be possible to find a group of functionally similar species for which changes in diversity have minimal effects on

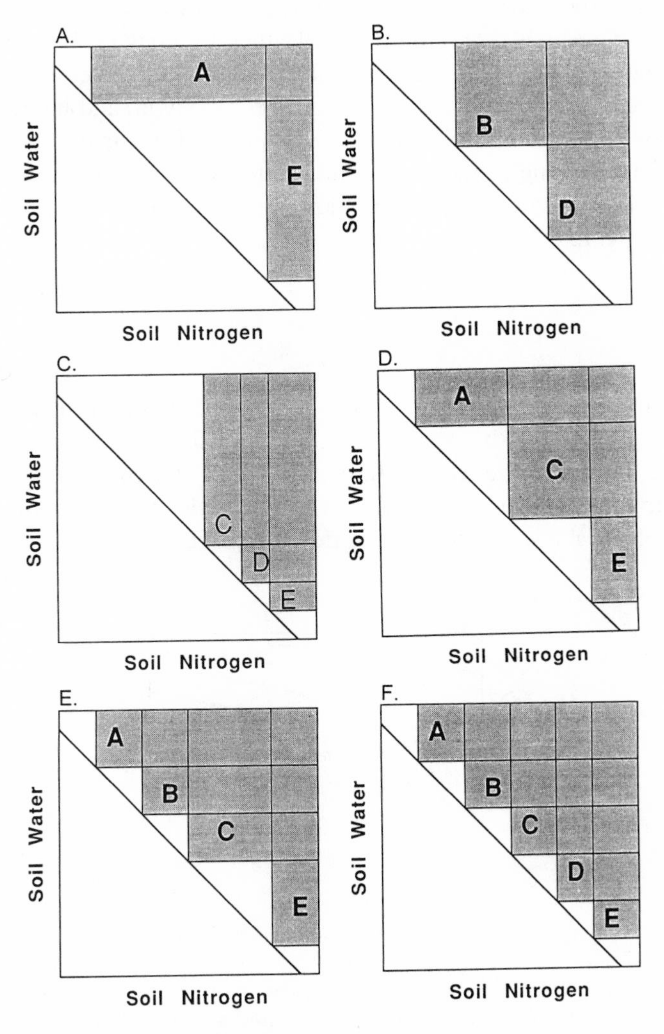

Figure 6.4. Examples of habitat coverage (i.e., amount of livable portion of graph that is covered by all species present) for two combinations of 2 species (parts A and B), two combinations of 3 species (parts C and D), and for 4 and 5 species (parts E and F).

productivity. Diversity, per se, is no guarantee of greater productivity, unless the more diverse communities contain in them species with a greater range of relevant traits. The effects of diversity will be distinguishable only in comparisons of communities whose compositions are an unbiased sample of the total species pool. Thus, it is important in testing for the effects of biodiversity on ecosystem functioning to assure that experimental communities of different diversity levels are random and independent subsets of the total species pool.

Although productivity saturates at a low level of diversity in the simple example of figures 6.2–4, real-world complexities are likely to shift this curve over as shown in figure 6.5. The more limiting factors in an ecosystem, and the more complex the ecosystem (e.g., heterogeneity in, and limitation by, water, soil nitrogen, soil phosphorus, calcium, pH, plant diseases, insect herbivores, changing climate, etc.), the greater will be the range of species traits required to "cover" these conditions and to lead to near maximal productivity (figure 6.5).

Tests

As Darwin asserted, it has long been known by agriculturalists that increases in diversity, at least in the range of one to four or five plant species, leads to increased primary productivity. Although it was written from a different perspective, Harper's (1977) synthesis and analysis of competition among pasture plants shows just this. Using deWit replacement diagrams of com-

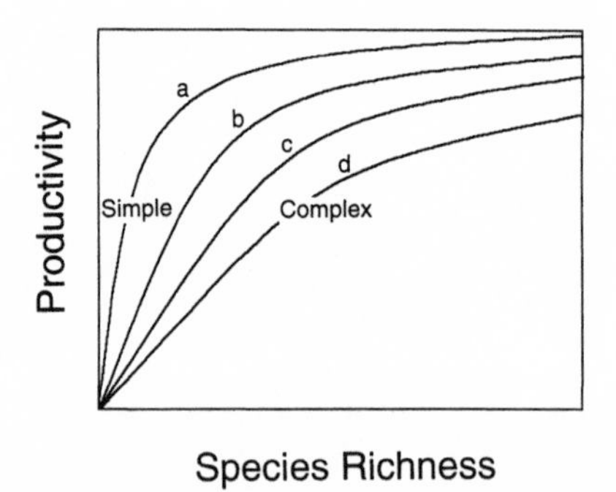

Figure 6.5. The hypothesized relationship between productivity (and perhaps other aspects of ecosystem functioning) and species richness for ecosystems that differ in their complexity. The relationship may saturate at lower diversity in simple habitats but at higher richness in more complex habitats, a possibility suggested by Chris Field.

petitive effects, Harper found that the total "yield" (productivity) of two-species plots was greater than that of either species alone when two species stably coexisted. This increase in productivity measured the effect of increasing species richness from one to two. Of the cases Harper reviewed, the greatest increases in productivity came in mixtures of markedly different species, such as a grass and a forb and especially a grass and a legume.

Trenbath (1974) performed a major analysis of agricultural experiments in which two or more crop species were grown together. Trenbath's analysis of 572 different intercrop systems showed that, on average, intercrops have about 10 percent higher yields than more traditional single-species monocultures. Numerous additional studies performed since Trenbath were reviewed by Swift and Anderson (1993), who concluded, "The evidence seems unequivocal that, given a number of qualifications, intercrops can be designed which will outyield sole crops" (p. 24). A cause of some of these increased yields was the presence of a legume, which helped overcome one of the most common limitations of agricultural yield, nitrogen availability. There have been many other studies designed to understand other causes for such effects, and these have provided a diversity of answers (Swift and Anderson 1993). For instance, greater diversity in plant species can increase the diversity of decomposer species and thus influence nutrient cycling and productivity. Greater plant diversity can also provide habitat for predators and parasites that attack herbivores or other crop pests. And, greater crop diversity can allow more efficient utilization, spatially or seasonally, of limiting resources.

There have been fewer studies of the diversity-productivity hypothesis in natural ecosystems. Comparative studies of the effects of diversity on productivity are potentially confounded because it is also known that productivity affects diversity. For example, experimental manipulations of productivity lead to changes in diversity (e.g., Lawes and Gilbert 1880a, 1880b; Tilman 1987, 1994; Goldberg and Miller 1990), with increased productivity most often leading to decreased diversity. In contrast, comparative studies in natural ecosystems have often shown a unimodal relationship between productivity and diversity (Grime 1973; Tilman 1980, 1982; Huston 1980, 1994; Tilman and Pacala 1993). This suggests that causality may occur in both directions, i.e., diversity may influence productivity and productivity may influence diversity. As such, comparative studies in natural ecosystems may be illustrating the long-term total effects of both directions of causality and may not give insight into either process alone. Experimental studies are needed that control one process and observe its effects on the other.

I know of three such studies. In a greenhouse study, various numbers of species were randomly drawn from a pool of 16 species to establish communities that contained from 1 to 16 plant species, with 64 monocultures,

20, 30, and 40 replicates, respectively, of the pots containing 2, 4, and 8 species, and 10 replicates of the pots containing all 16 species (Naeem et al. 1995). This study showed the expected effect of plant diversity on productivity, with average productivity increasing with diversity (figure 6.6A). The pot-to-pot variance (standard deviation) in productivity was also lower at higher species richness (figure 6.6B). In total, this study by Naeem et al. supports both of the major predictions of theory. It also shows that results observed in agricultural experiments on two-species mixtures apply to mixtures of uncultivated plants containing many species. In a laboratory experiment, species diversity on several trophic levels (plants, decomposers, herbivores) was directly and simultaneously manipulated such that there were low-, moderate- and high-diversity ecosystems (Naeem et al. 1994). The highest diversity ecosystems had the greatest productivity. The third study established 147 field plots, each 3 m × 3 m, that were planted with 1, 2, 4, 6, 8, 12, or 24 prairie species (Tilman et al. 1996b). Each plot contained species chosen by a separate random draw of the appropriate number of species from a pool of 24 prairie species. In the second year of growth, productivity, as measured by the total cover of all plant species in a plot, was a significantly increasing function of species richness (figure 6.6C), remarkably similar to the relationships predicted for the sampling-competition effect and the resource-complementarity effect (figure 6.3).

An important application of the diversity-productivity relationship was proposed by Bolker, Pacala, Bazzaz, Canham, and Levin (1995). Calibrated models of forest growth predicted, for forest stands growing at elevated atmospheric CO_2 levels (i.e., levels of the near future), that forest stands containing many spatially intermingled species would be at least 30 percent more productive than stands planted to a single species. This means that biodiversity may lead to a 30 percent increase in the amount of atmospheric carbon dioxide removed by forests and stored in the forest ecosystem. Such added carbon storage that may result from biodiversity could play a crucial role in allowing global reforestation to ameliorate effects of high rates of CO_2 production and thus to moderate global climate change.

In total, these studies provided broad and general support for the diversity-productivity hypothesis. They illustrate that the effects of diversity are not direct but are based on the greater range in species traits associated with unbiased increases in diversity. They also show that each unique combination of species may have a different effect, and thus that variation in productivity is expected within communities of equal diversity but different compositions. This variance is reduced by higher diversity. Stated differently, the theoretical basis for the effects of diversity on primary productivity predicts that both species composition and species diversity are major determinants of the primary productivity of plant communities. Finally, the results suggest that genetic diversity within a single species may lead to

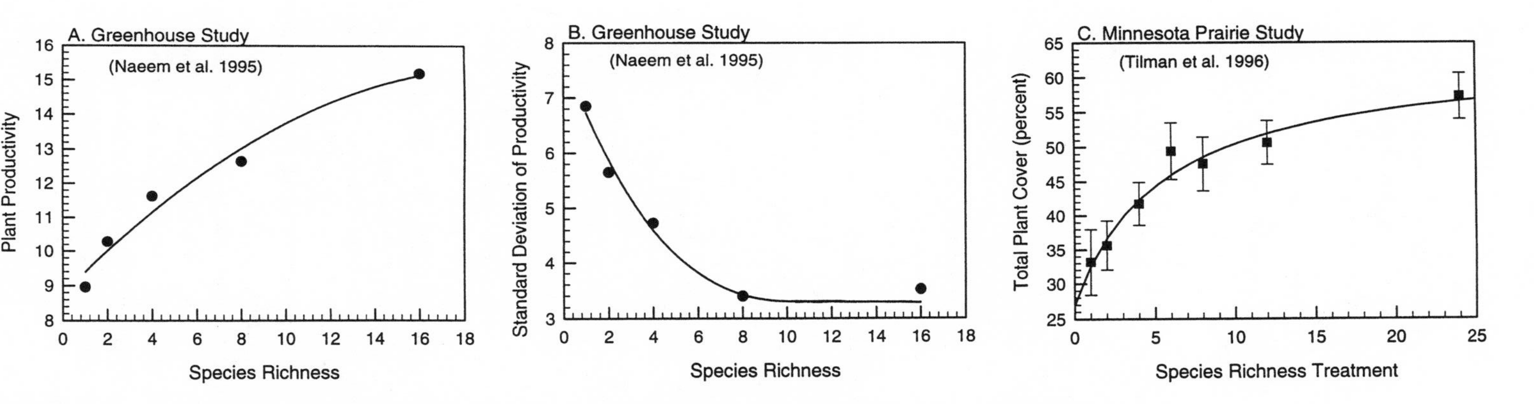

Figure 6.6. **A.** Average aboveground plant productivity (g/pot) in a greenhouse experiment by Naeem et al. (1995) in which random draws were used to select 2, 4, or 8-species subsets from a pool of 16 British grassland plant species. There were 64 monocultures (4 replicates of each of the 16 species), 20 replicates of 2-species pots, 30 replicates of 4-species pots, 40 replicates of 8 species pots, and 10 replicates of all 16 species. **B.** There was marked pot-to-pot variation in productivity at each of the levels of species diversity. This variability was measured as the standard deviation of productivity at each level of species diversity. As shown here, this variability was much greater in monocultures, and declined with increased species richness. Data for A and B are from Naeem et al. (1995) and were kindly provided by Shahid Naeem so that they could be accurately regraphed here. **C.** The dependence of productivity, as estimated by total plant cover, on plant species richness in a field experiment in which from 1 to 24 prairie plant species were added to plots. Mean and standard errors are shown for each level of experimentally imposed species richness.
Source: Modified from Tilman et al. (1996b).

greater productivity than for a genetically uniform species, which has been found in barley and wheat (Allard and Adams 1969). This also is supported, in a somewhat oblique way, by the greater productivity associated with hybrid crop varieties. Hybrid varieties are plants that have been bred to have great heterozygosity at loci, i.e., to have great genetic variation within each individual. The work reviewed above suggests that spatially intermingled mixtures of different hybrid varieties could lead to even greater productivity than obtained from a single hybrid variety.

Stability and Biodiversity

All forests, croplands, grasslands, and other ecosystems experience natural variations and disturbances such as droughts, heavy rains, unusually hot or cool growing seasons, hail, and outbreaks of various pests, diseases, and pathogens. These disturbances can greatly decrease the abilities of these ecosystems to provide ecosystem goods and services. Many modern agricultural practices, such as choice of genetic varieties, irrigation, and use of pesticides, are necessitated because of the high susceptibility of agricultural ecosystems, especially those planted to a single species, to such disturbances. Different ecosystems respond differently to disturbance. Some ecosystems are fairly stable, which means that they are not greatly impacted by disturbance, whereas others are less stable and can have great losses of productivity following disturbances. Odum (1953), MacArthur (1955), and Elton (1958) proposed that ecosystem stability depended on diversity. These early ideas were perhaps most clearly articulated in Elton's (1958) book, in which he asserted that more diverse communities were less oscillatory in response to environmental variation and less subject to invasion by novel species. He cited several lines of evidence supporting this conclusion, including the still well-supported observations that islands, which are species poor, are more readily invaded by alien species than comparable mainland areas, and that simplified agricultural ecosystems are more subject to oscillation and pest invasion than diverse natural communities.

This hypothesis was explored in greater detail mathematically by Gardner and Ashby (1970) and May (1973). Gardner and Ashby showed that randomly assembled groups of species interacting with randomly chosen strengths were markedly less stable as the number of interacting species was increased, and that stability depended on the degree of connectance among species. May explored this hypothesis by developing a highly structured (i.e., nonrandom) multispecies competitive community in which he could vary the number of competing species. He found that the dynamics of the species became increasingly less stable as species diversity increased, and concluded that complexity led, in general, to lower stability. A period of great debate followed, during which ecology gained greater mathematical sophistication, and concepts and definitions related to diversity were clarified (e.g., Pimm 1979, 1984, 1993; King and Pimm 1983). This debate was resolved by the realization that increased diversity stabilizes the functioning of the total ecosystem, but that diversity can destabilize the dynamics of individual species (King and Pimm 1983, Tilman 1996). This resolution was foreshadowed by May (1974), who mused, "If we concentrate on any one particular species our impression will be one of flux and hazard, but if we concentrate on total community properties (such as biomass in a given trophic level) our impression will be of pattern and steadiness."

Pimm made an important distinction between two different components of stability. The first he called resistance. The resistance to disturbance of an ecosystem is a measure of the magnitude of change in the ecosystem in response to a particular intensity of disturbance. Pimm hypothesized that more diverse ecosystems might be more resistant to perturbation, i.e., change less in response to perturbation. The second component of stability was resilience, which measures the rate of recovery from perturbation. Resilience is best measured as the specific rate of recovery, i.e., the rate of recovery divided by the magnitude of the original deviation. Again, Pimm suggested that resilience might be greater at greater diversity, but Lockwood and Pimm (1994) questioned if this was as likely to depend on diversity as was resistance.

To understand why ecosystem stability may depend on diversity, consider two hypothetical species that compete with each other. One species is totally resistant to a given disturbance, and the second is totally susceptible. When the disturbance occurs, monocultures of the first species maintain their original biomass, whereas monocultures of the second species have their biomass fall to zero. Assuming that the two species had approximately equal abundances before the disturbance, the average biomass across many such areas containing just one or the other species would have fallen to half of the average before the disturbance. What would happen in areas containing both species? The immediate effect would be that susceptible species would be driven to zero, and the other species would not be harmed, again leading to an average biomass of half of pre-disturbance levels (assuming approximately equal abundances of the competing species). However, the disturbance-resistant species now would be living in areas from which its competitors had been removed. In these areas it should be able to increase in abundance up to its carrying capacity. This would cause biomass in all these areas to increase approximately twofold. This compensatory increase in biomass, with two species areas having double the average productivity of area containing one species, is a measure of the maximal potential stability conferred on these two-species areas by their diversity. Clearly, there is a temporal component to this, and the actual magnitude of the stabilizing effect of diversity would depend on the time available for recovery and on the growth rates of the resistant species.

Tests

A test of these ideas was provided by the severe drought that struck the midwestern United States beginning in 1987 and peaking in 1988 (Tilman and El Haddi 1992). On average, across 207 permanent Minnesotan grassland plots that differed in plant diversity and that had been annually sampled for ecosystem productivity, plant diversity, and many other variables starting in 1982, drought caused plant productivity to fall to less than ½ of its pre-drought average. However, during the drought, the most species-poor plots had their productivity fall to about 1/12 of their pre-drought level, whereas the most species-rich plots had productivity fall to only about half (figure 6.7, Tilman and Downing 1994). The ratio of productivity during the drought to that before the drought is a direct measure of resistance to drought. There was clear dependence of drought resistance on diversity. Although there were several other factors that potentially confounded the relationship in figure 6.7, thorough analyses showed that none of these changed the highly significant dependence of drought resistance on species diversity shown there (Tilman and Downing 1994, Tilman 1996).

Further analyses showed that compensatory increases in drought-resistant species were a major cause of greater drought resistance in more diverse plots (Tilman 1996). Moreover, more diverse plots had less year-to-year variation in productivity during nondrought years (Tilman 1996). Analyses using Pimm's rigorous definition of resilience showed that resilience tended to increase with diversity, but this was not as consistent or strong an effect as for drought resistance (Tilman 1996). The responses of individual plant species were also analyzed. Unlike the response for total community plant biomass (i.e., productivity), the abundances of individual species were slightly, but statistically significantly, more variable year-to-year at higher diversity, thus supporting May's (1973) prediction. Thus, this evidence supports the hypothesis that diversity increases the resistance of ecosystem productivity to perturbation, but that diversity decreases the resistance of individual species abundances to perturbation (Tilman 1996).

Many other studies of diversity-stability relationships, though less well replicated than that just discussed, have yielded amazingly similar results. In a study of Serengeti grasslands, McNaughton (1977) found that grasslands with higher plant diversity recovered more rapidly after grazing by the African buffalo. He also found that green plant biomass, both in four neigh-

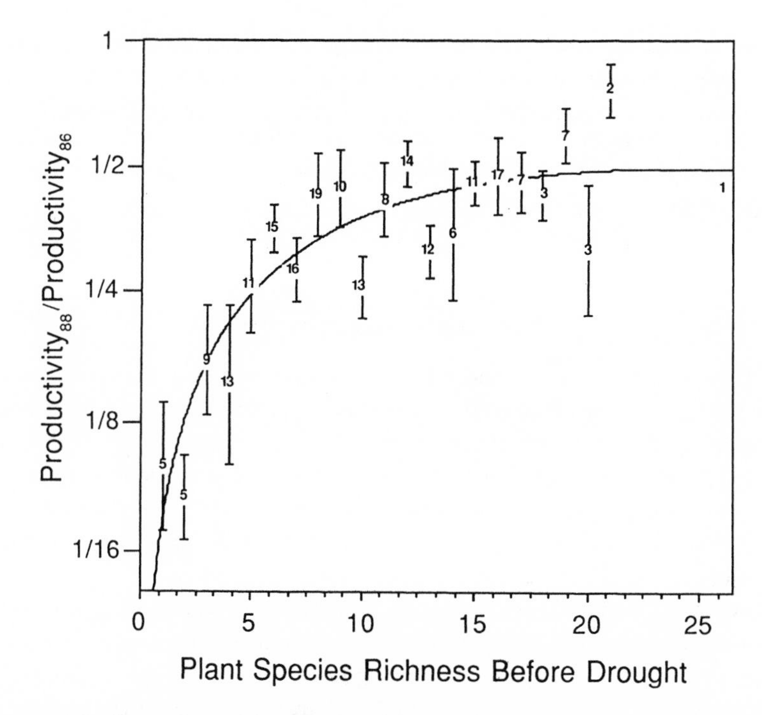

Figure 6.7. Drought resistance of prairie grassland vegetation and its dependence on species richness. Drought resistance was measured as the ratio of productivity (plant biomass) during the 1988 drought divided by that in 1986 before the drought began. This ratio was calculated for each of 207 grassland plots that contained from 1 species to 26 species. The numbers indicate how many plots had a given level of plant species richness in 1986 (before the drought), and each number is placed at the average productivity ratio for that level of species richness. The vertical bars show the variation (standard error) about each mean.
Source: Redrawn from Tilman and Downing 1994.

boring grassland stands and in twenty-eight stands across a rainfall gradient, was more resistant to climatic variations in the more diverse stands. Frank and McNaughton (1991) found a significant effect of plant diversity on the resistance of Yellowstone National Park grasslands to drought. Lepŝ et al. (1982) found that drought led to a lower decrease in standing crop in a species-rich grassland field than in a species-poor one. Several other such studies are reviewed in McNaughton (1993). These studies, in total, demonstrate a consistent effect of biodiversity on the resistance of ecosystem processes, especially productivity, to perturbation.

All of these examples illustrate that biodiversity has an "insurance value" in that it helps minimize the costs of various unpredictable events. The pattern for ecosystem resilience is less clear, with some studies showing a significant effect and others not revealing any significant relationship between ecosystem resilience and diversity. Finally, both the results in Minnesota grasslands and theory suggest that species diversity does not have a stabilizing effect on the population fluctuations of individual species. However, it seems plausible that intraspecific genetic diversity might increase resistance to disturbance for the abundance of an individual species.

Ecosystem Sustainability

Farming, forestry, grazing, and other human activities that harvest products from ecosystems may lead to progressively lower ecosystem productivity, or alternatively may be done in such a way that the productive capacity of the ecosystem is sustained. Because degradation implies a reduction in the natural underpinnings of productivity and poses a threat to maintaining ecosystem goods and services, there is growing interest in assuring ecosystem sustainability.

An often articulated idea is that morphological, physiological, and other differences among species would allow an ecosystem containing more species to more fully exploit its limiting resources (e.g., Odum 1969; McNaughton 1977, 1993). This increased resource exploitation is a major reason it is thought that diversity may increase productivity, as discussed above. However, there is another ramification. The more efficient utilization of limiting resources should decrease their availabilities in the environment. For soil nutrients this would minimize the amounts that could be leached through the soil and into the groundwater. This would help conserve these nutrients within the ecosystem (e.g., Ewel et al. 1991, Vitousek and Hooper 1993). Moreover, the litter produced by different species differs in its effects on nutrient cycling (Vitousek et al. 1987, Wedin and Tilman 1990, Pastor et al. 1984). It may be that more diverse plant communities would support a more diverse decomposer community that would be better at retaining nutrients within the ecosystem (Vitousek and Hooper 1993, Swift and Anderson 1993). The best field experiment that tests these ideas is a study of tropical succession that Ewel et al. (1991) performed in Costa Rica. The researchers established plots that differed in the number of species of successional plants and then followed the nutrient budgets of these plots. They found that plots with low plant diversity had greater leaching losses of soil nutrients, showing that diversity can play a significant role in the maintenance of soil fertility and thus productivity (Ewel et al. 1991, Vitousek and Hooper 1993). Similarly, Tilman et al. (1996b) observed significantly lower leaching loss of soil nitrogen in more diverse plots in their experimental study using prairie species. This could have a significant long-term impact on the nutrient capital of an ecosystem and thus on the sustainability of its productivity, species composition, soil organic matter, etc. Ecosystems attain a sustainable level of functioning when, on average, over periods long enough to include their disturbance cycles, rates of loss and gain of organic matter and nutrients are in balance. If decreased biodiversity were to lead to greater losses of nutrients and organic matter, the long-term effect would be lower average amounts of organic matter and limiting nutrients, which would lead to lower fertility and productivity and likely to changes in species abundances and community composition.

Tests

I know of no rigorous long-term field tests of the dependence of ecosystem sustainability on biodiversity. The available evidence showing the dependence both of productivity and of nutrient retention on biodiversity argues heavily for sustainability depending on biodiversity. We need long-term field experiments in which biodiversity is controlled and effects on sustainable ecosystem functioning are observed.

Conclusions

This review has found that many aspects of the stability, functioning, and sustainability of ecosystems depend on biodiversity. This dependence is not some direct or magical effect of biodiversity, but rather reflects the increased functional roles that are possible in ecosystems that contain more species. The current evidence shows strong dependence on biodiversity of the resistance of ecosystem functioning to disturbance, indicating that more diverse ecosystems are more stable. However, it is less clear from current studies if ecosystem resilience similarly depends on biodiversity. In both simple agricultural ecosystems and in natural ecosystems, the primary productivity of communities increases, probably in a saturating manner, with biodiversity. This suggests that management practices that maintain diverse forest, grassland, and aquatic ecosystems may help assure the sustained production of ecosystem goods and services. However, we still have only rudimentary knowledge of many of these processes. There have been few long-term experiments. Many processes and many ecosystem types have never been explored experimentally. We need better knowledge of the number of species required to assure the sustainability of various ecosystem functions and how this depends on spatial patterning, spatial scale, and time. The answers to these questions will be of great importance for managing ecosystems to achieve sustainable flows of the goods and ecosystem services essential for human life.

Acknowledgments

I thank Clarence Lehman for comments on this manuscript, Nancy Larson for assistance in its preparation, and the National Science Foundation and the Andrew Mellon Foundation for financial support.

References

Allard, R. W., and J. Adams. 1969. "Population studies in predominantly self-pollinating species: XIII. Intergenotypic competition and population structure in barley and wheat." *American Naturalist* 103: 621–645.

Bolker, B. M., S. W. Pacala, F. A. Bazzaz, C. D. Canham, and S. A. Levin. 1995. "Species diversity and ecosystem response to carbon dioxide fertilization: Conclusions from a temperate forest model." In press.

Chabot, B. F., and H. A. Mooney, eds. 1985. *Physiological Ecology of North American Plant Communities.* Chapman and Hall, London.

Chapin, F. S., III. 1980. "The mineral nutrition of wild plants." *Annual Review of Ecology and Systematics* 11:233–260.

Darwin, C. 1872. *The Origin of Species,* 6th London edition. Thompson & Thomas, Chicago.

Ehrlich, P., and A. Ehrlich. 1981. *Extinction.* Ballantine Books, New York.

Elton, C. S. 1958. *The Ecology of Invasions by Animals and Plants.* Methuen & Co., London.

Ewel, J. J., M. J. Mazzarino, and C. W. Berish. 1991. "Tropical soil fertility changes under monocultures and successional communities of different structure." *Ecological Applications* 1:289–302.

Frank, D. A., and S. J. McNaughton. 1991. "Stability increases with diversity in plant communities: Empirical evidence from the 1988 Yellowstone drought." *Oikos* 62:360–362.

Gardner, M. R., and W. R. Ashby. 1970. "Connectance of large dynamic (cybernetic) systems: Critical values for stability." *Nature* 228:784.

Goldberg, D. H., and T. E. Miller. 1990. "Effects of different resource additions on species diversity in an annual plant community." *Ecology* 71:213–225.

Grime, J. P. 1973. "Competitive exclusion in herbaceous vegetation." *Nature* 242:344–347.

Grime, J. P. 1979. *Plant Strategies and Vegetation Processes.* John Wiley, Chichester, England.

Harper, J. L. 1977. *Population Biology of Plants.* Academic Press, New York.

Huston, M. 1980. "Soil nutrients and tree species richness in Costa Rican forests." *Journal of Biogeography* 7:147–157.

Huston, M. A. 1994. *Biological Diversity: The Coexistence of Species on Changing Landscapes.* Cambridge University Press, Cambridge, England.

King, A. W., and S. L. Pimm. 1983. "Complexity, diversity, and stability: A reconciliation of theoretical and empirical results. *American Naturalist.* 122:229–239.

Lawes, J. B., and J. H. Gilbert. 1880. "Agricultural, botanical and chemical results of experiments on the mixed herbage of permanent grassland, conducted for many years in succession on the same land. I. The agricultural results." *Philosophical Transactions of the Royal Society* 171:289–415.

Lawes, J. B., J. H. Gilbert, and M. T. Masters. 1900. "Agricultural, botanical and chemical results of experiments on the mixed herbage of permanent meadow, conducted for more than twenty years in succession on the same land. Part II. The botanical results—Section I. *Philosophical Transactions of the Royal Society* (B) 192:139–210.

Lepŝ, J., J. Osbornová-Kosinová, and M. Rejmánek. 1982. "Community stability, complexity and species life history strategies." *Vegetatio* 50:53–63.

Lockwood, J. L., and S. L. Pimm. 1994. "Species: Would any of them be missed?" *Current Biology* 4:455–457.

MacArthur, R. H. 1955. "Fluctuations of animal populations and a measure of community stability." *Ecology* 36:533–536.

May, R.M. 1973. *Stability and Complexity in Model Ecosystems.* Princeton University Press, Princeton, N.J.

May, R. M. 1974. *Stability and Complexity in Model Ecosystems.* 2nd edition. Princeton University Press, Princeton, N.J.

McNaughton, S. J. 1977. "Diversity and stability of ecological communities: A comment on the role of empiricism in ecology." *American Naturalist* 111:515–525.

McNaughton, S. J. 1993. "Biodiversity and function of grazing ecosystems." In E.-D. Schulze, and H.A. Mooney, eds., *Biodiversity and Ecosystem Function,* pp. 361–383. Springer-Verlag, Berlin, Germany.

Naeem, S., L. J. Thompson, S. P. Lawler, J. H. Lawton, and R. M. Woodfin. 1994. "Declining biodiversity can alter the performance of ecosystems." *Nature* 368:734–737.

Naeem, S., L. J. Thompson, S. P. Lawler, J. H. Lawton, and R. M. Woodfin. 1995. "Empirical evidence that declining species diversity may alter the performance of terrestrial ecosystems." *Philosophical Transactions of the Royal Society of London, Series B* 347:249–262.

Odum, E. P. 1953. *Fundamentals of Ecology.* Saunders, Philadelphia.

Odum, E. P. 1969. "The strategy of ecosystem development." *Science* 164:262–270.

Pastor, J., J. D. Aber, C. A. McClaugherty, and J. M. Melillo. 1984. "Above-ground production and N and P cycling along a nitrogen mineralization gradient on Blackhawk Island, Wisconsin." *Ecology* 65:256–268.

Pimm, S. L. 1979. "Complexity and stability: Another look at MacArthur's original hypothesis." *Oikos* 33:351–357.

Pimm, S. L. 1984. "The complexity and stability of ecosystems." *Nature* 307:321–326.

Pimm, S. L. 1993. "Biodiversity and the balance of nature." In E.-D. Schulze and

H.A. Mooney, eds., *Biodiversity and Ecosystem Function*, pp. 347–359. Springer-Verlag, Berlin, Germany.

Schulze, E. D., and H. A. Mooney, eds. 1993. *Biodiversity and Ecosystem Function*. Springer-Verlag, Berlin, Germany.

Swift, M. J., and J. M. Anderson. 1993. "Biodiversity and ecosystem function in agricultural systems." Pp. 15–41, in E.-D. Schulze, and H.A. Mooney, eds., *Biodiversity and Ecosystem Function*. Springer-Verlag, Berlin, Germany.

Tilman, D. 1980. "Resources: A graphical-mechanistic approach to competition and predation." *American Naturalist* 116:362–393.

Tilman, D. 1982. "Resource competition and community structure." *Monographs in Population Biology*. Princeton University Press, Princeton, N.J.

Tilman, D. 1987. "Secondary succession and the pattern of plant dominance along experimental nitrogen gradients." *Ecological Monographs* 57(3):189–214.

Tilman, D. 1988. Plant Strategies and the Dynamics and Structure of Plant Communities. Princeton University Press, Princeton, N.J.

Tilman, D. 1994. "Competition and biodiversity in spatially structured habitats." *Ecology* 75:2–16.

Tilman, D. 1996. "Biodiversity: Population versus ecosystem stability." *Ecology* 77:350–363.

Tilman, D., and J. A. Downing. 1994. "Biodiversity and stability in grasslands." *Nature* 367:363–365.

Tilman, D., and A. El Haddi. 1992. "Drought and biodiversity in grasslands." *Oecologia* 89:257–264.

Tilman, D., and S. Pacala. 1993. "The maintenance of species richness in plant communities." In R. E. Ricklefs and D. Schluter, eds., *Species Diversity in Ecological Communities*, pp. 13–25, University of Chicago Press, Chicago.

Tilman, D., C. L. Lehman, and C. Yin. 1996. "Habitat destruction and deterministic extinction in competitive communities." *American Naturalist*, in review.

Tilman, D., D. Wedin, and J. Knops. 1996. "Productivity and sustainability influenced by biodiversity in grassland ecosystems." *Nature* 379:718–720.

Trenbath, B. R. 1974. "Biomass productivity of mixtures." *Advances in Agronomy* 26:177–210.

Vitousek, P. M., and D. U. Hooper. 1993. "Biological diversity and terrestrial ecosystem Biogeochemistry." In E. D. Schulze and H. A. Mooney, eds., *Biodiversity and Ecosystem Function*, pp. 3–14. Springer-Verlag, Berlin, Germany.

Vitousek, P. M., L. R. Walker, L. D. Whiteaker, D. Mueller-Dombois, and P. A. Matson. 1987. "Biological invasion by *Myrica faya* alters ecosystem development in Hawaii." *Science* 238:802–804.

Wedin, D., and D. Tilman. 1990. "Species effects on nitrogen cycling: A test with perennial grasses." *Oecologia* 84:433–441.

Chapter 7

Ecosystem Services Supplied by Soil

Gretchen C. Daily, Pamela A. Matson, and Peter M. Vitousek

Like virtually all land-based organisms, humans depend on soil for essential material goods and ecosystem services. Soil represents an important component of a nation's assets, one that takes hundreds to hundreds of thousands of years to build and very few to be wasted away (Oldeman et al. 1990). Civilizations have drawn great strength from productive earth; conversely, the loss of productivity through mismanagement is thought to have ushered once flourishing societies to their ruin (Adams 1981; Hillel 1991). Today, soil productivity remains an important determinant of the economic status of nations and, especially in the case of poor nations, of their prospects for future development (Wali 1992; Ehrlich et al. 1995).

Although a casual glance suggests little more than ground-up rock, soil is actually a complex, dynamic ecosystem, sustaining physical processes and chemical transformations vital to terrestrial life. Like a sponge, soil absorbs precipitation and gradually meters it out to plant roots and into subterranean aquifers and surface streams. Soil shelters seeds and provides physical support and nourishment to plants. It consumes wastes and the remains of dead plants and animals, rendering their potential toxins and human pathogens harmless, while recycling their constituent materials into forms usable by plants. In the process, soil organisms regulate the fluxes of the important greenhouse gases, CO_2, CH_4, and N_2O. Soil plays a critical role in fueling the entire terrestrial food chain and is an important feature of many aquatic systems as well.

Characterization of Soil

First we describe the formation and composition of soil, as a basis for understanding both its capacity to perform each of these services and its vulnerability to human impact.

Soil Genesis and Structure

Soil genesis begins with the physical disintegration and chemical decomposition of rock, earth's solid crust. Temperature changes; erosive forces of water, ice, and wind; and living organisms exert physical stresses that break down rock into unconsolidated material called regolith. Concurrently, the chemical processes of weathering set upon newly exposed rock surfaces, transforming the "primary" minerals (those stable under the temperatures and pressures deep within earth where the rock was formed) into "secondary" forms (minerals stable at the surface of earth), with a release of essential plant nutrients in the process. These transformations are accelerated by the presence of oxygen, by water and its dissolved salts, and by acids derived from the atmosphere and from the microbial breakdown of plant residues. Thus soil formation begins; but much more remains to be done before fertile soil is born.

As weathering continues, soil undergoes horizon differentiation—the separation and diversification of horizontal zones. Plants and animals add organic material to the surface; their residues shape both the physical and the biotic structure of soils, and their functioning. Generally, mature soils consist of a surface layer of mostly organic matter in all stages of decomposition, underlain by a mixture of smaller amounts of organic matter with inorganic material such as sands, silts, and clays, all of which rests on progressively less weathered layers of subsoil, regolith, and bedrock. Five paramount factors determine the ultimate character of soil: climate; living soil organisms and plants; topography; the nature of the parent material; and the soil's age (the time that it has undergone the processes of soil genesis; Jenny 1980). A sixth factor, human activity, has become increasingly significant.

One important structural aspect of soil that greatly influences its function is soil texture, or the size of its constituent mineral particles. By definition, clay refers to particles of 2 micrometers diameter; silt particles range in diameter from 2 up to 20 micrometers; and sand refers to the next largest size class, up to 2 millimeters in diameter. Texture influences the rate of movement of water through soil, the capacity for its storage, and a soil's susceptibility to erosion and waterlogging from poor cultivation and irrigation practices. Coarse soils permit rapid infiltration but do not retain much water and erode easily; at the other extreme, fine-textured clays have slow percolation rates but high storage capacities, making them more susceptible to waterlogging.

Human beings cannot significantly accelerate the gradual process of soil formation. The deep, rich agricultural soils underlying the world's breadbaskets today were born in remote periods of human history; they represent an inheritance of natural capital, upon whose bequest future generations depend. Human activities can dramatically alter the course of soil formation and maturation, however, setting it back abruptly through the exhaustion and dispersion of soil surface layers. For example, 50–100 years of farming the North American Great Plains has reduced nutrient levels by up to 50 percent; similar losses have taken place in fewer than ten years in the tropics (Bolin et al. 1983). Fifty years of farming in part of Minnesota in the early colonial period reduced surface soil carbon and nitrogen levels by two-thirds, leading to farm abandonment. At least 100 years—and perhaps as much as 200 to 250 years—are required for natural processes to restore these resources to their former levels (Tilman 1987, 1990; D. Tilman, personal communication 1996). Deforestation, overgrazing, and poor cultivation practices have deleterious impacts on soil worldwide.

Soil Composition

Beyond weathered rock and dead organic matter, soil is a mixture of living animals and microorganisms, gases, and water. The relative proportion of these soil constituents varies significantly. The organic soils typical of marshes, bogs, and swamps, for example, consist primarily of organic matter (> 50 percent by volume). The much more prevalent mineral soils are characterized typically by about 40–55 percent mineral particles and only 1–10 percent organic matter by volume; the rest is pore space, filled by water and/or air. Soil organisms comprise a relatively tiny fraction of soil, generally no more than around 0.1 percent by mass (Janick et al. 1974). Each of these soil components is crucial to the supply of ecosystem services.

Humus refers to the relatively stable fraction of soil organic matter, made up of the residues of plants, animals, and microorganisms that resist decomposition. Clay and humus together control most physical properties of soil. They help in the formation and maintenance of stable soil aggregates by forming "contact bridges" with their charged surfaces; these aggregates control the movement and retention of water and air in soil, required for the uptake of nutrients by plant roots. Humus has a greater nutrient- and water-retention capacity than clay (on a weight basis), but since clay generally makes up a greater proportion of a soil, it usually contributes at least as much or more to a soil's chemical and physical properties.

Soil organisms carry out the decomposition of organic material and the transformation of organic nutrients into mineral forms usable by plants. They are remarkable in their diversity and abundance. Many thousands of species of soil crustacea, mites, termites, springtails, millipedes, worms, bacteria, actinomycetes, fungi, algae, protozoa, and other taxa are believed to exist, but most have yet to be discovered and described. The area under a square meter of pasture soil in Denmark yielded 45,000 oligochaete worms, 48,000 small insects and mites, and 10 million nematodes (Overgaard-Nielsen 1955). These soil invertebrates act as "mechanical blenders," breaking up and mixing plant material, microbial excrements, and other matter (Jenny 1980). As much as 10 MT/ha-yr of material is passed through the bodies of earthworms, for example, leaving nutrient-rich "casts" that enhance soil stability, aeration, and draining; the nitrogen flux directly through earthworm tissue may exceed 100 kg/ha-yr (Lee 1985).

Soil microorganisms are much more numerous still. A pinch (one gram) of soil may yield a million fungal propagules (i.e., spores, resting stages, hyphal fragments; Chanway 1993) and over a billion bacterial cells (Rouatt and Katznelson 1961) of unknown numbers of species. Microorganisms are fascinating in the extremes of environmental conditions in which they occur (Edwards 1990). For example, a diverse flora of microfungi was found in the sparse soil of Antarctica, sheltered in rock fissures (Baublis et al. 1991), and deep in boreholes in South Carolina (Sinclair and Ghiorse 1989). Wherever they are found, soil microorganisms play crucial roles in the circulation of matter (UNEP 1995, Section 6.2.7).

Yet except for some economically important species, such as known mutualists and pathogens in agroecosystems, soil organisms remain poorly understood in terms of their global diversity, their functions, and their interactions (UNEP 1995, Section 6.2.7). Microbial species are especially difficult to study: The exchange of genetic material among bacteria and from bacteria to other taxa makes basic classification problematic; many microbes are difficult to isolate and culture; and their communities vary tremendously in time and over short spatial distances. For the most part, microbial communities have been viewed as "black boxes"—often compartmentalized by functional, trophic, or size grouping—with the aim of characterizing the box through measures of inputs and outputs, rather than direct exploration of the complex interactions occurring within (although much more sophisticated approaches are being developed; e.g., Holben et al. 1988; Bull 1992; Korner and Laczko 1992; De Leij et al. 1994). The dependence of human society upon soil biotic diversity, as it relates to soil structure and fertility and plant quality, thus remains one of society's most important scientific mysteries. At this point the scientific community cannot assess the magnitude, let alone the societal consequences, of human impacts on soil biodiversity (Anderson 1995).

Ecosystem Services Supplied by Soil

The ecosystem services supplied by soil are so tightly interrelated as to make any discrete classification arbitrary. We have structured our characterization and valuation around six services: (1) buffering and moderation of the hydrological cycle; (2) physical support of plants; (3) retention and delivery of nutrients to plants; (4) disposal of wastes and dead organic matter; (5) renewal of soil fertility; and (6) regulation of major element cycles.

Buffering and Moderation of the Hydrological Cycle

An enormous amount of water, about 119,000 km^3, is precipitated annually over the earth's land surface—enough to cover it to an average depth of 1m (Shiklomanov 1993). This water is soaked up by ("infiltrates") soils and is gradually metered out to plant roots and into subterranean aquifers and surface streams; the rate at which it moves is determined by humus and clay. Without soil, it would rush off the land in flash floods. Plants and plant residues protect this service, shielding the soil from the full, destructive force of raindrops. Rain on denuded landscapes compacts the surface and turns soil rapidly to mud (especially if it has been loosened by tillage), which clogs surface cavities in the soil, reduces infiltration, increases runoff, and further enhances clogging in a positive feedback. Detached soil particles are splashed downslope and carried off by running water (Hillel 1991).

Water infiltration is closely correlated with soil organic matter content. Human disturbance (such as cultivation) reduces soil organic matter levels—typically by 25 percent or more—by simultaneously reducing the annual input of plant residues and increasing decomposition through elevated soil temperature, aeration, and moisture (Harrison et al. 1993). Loss of organic matter makes soils more prone to erosion and reduces their water-holding capacity. Erosion, in turn, is associated with reductions in infiltration (by up to 90 percent) and tends to carry away the soil material richest in organic matter, in another positive feedback (Pimentel et al. 1993). These losses alter stream flow regimes by amplifying seasonal patterns of runoff—increasing the frequency, severity, and unpredictability of high- and low-flow periods. The increased force associated with higher flows erodes stream channels, lowers water quality, and generally degrades aquatic habitat (NRC 1993).

Living vegetation is also crucial to the hydrological service of soils. In addition to protecting soil from erosion, plants transpire water from soil back into the atmosphere. Vegetation clearance disrupts this link in the water cycle and leads to potentially dramatic increases in surface runoff, along with nutrient and soil loss. The experimental clearing of a New Hampshire forest provides a classic example: Vegetation was cut and left on the ground with minimal disturbance to the forest floor; thereafter, herbicide was applied to prevent regrowth for a three-year period. The result was a 40 percent increase in average stream flow; during one four-month period of the experiment, runoff exceeded pre-clearance values by over 400 percent (Bormann et al. 1968). Because water and soil losses in one area are necessarily gained elsewhere, such local impacts may have significant off-site effects (e.g., flooding; deterioration of recreational and navigational waterways; and damage to facilities for water storage, conveyance, and treatment). On a much larger scale, the extensive deforestation in Himalayan highlands appears to have exacerbated recent flooding in Bangladesh, although the relative roles of human and natural forces remain uncertain (Ives and Messerli 1989).

At the same time that vegetation removal in some sites leads to increased runoff and reduced soil water recharge, other sites may experience alteration in subsurface water reservoirs as deep-rooting plants are removed. In the Australian wheatlands, substitution of annual agricultural crops and pasture for deeply rooted, perennial native vegetation led to a dramatic rise in the water table. This drew naturally occurring salts, which rain had leached into lower soil horizons away from plant roots, up to the soil surface, where they are poisoning the land. The direct annual cost of salinization and waterlogging was estimated (in 1989) at over US$120 million. Overall, the loss of ecosystem services resulting from natural vegetation removal—including erosion, soil compaction, and water repellence—amounted to more than US$500 million per year, as measured by lost production (Lefroy et al. 1993). West Australian farmers have solicited the collaboration of ecologists to restore native vegetation to the wheatlands in a scheme designed to both enhance agricultural income and to preserve and foster the many valued services supplied by soil and vegetation (Saunders et al. 1993).

Thus soil interacts with vegetation to play a key regulatory role in the hydrological cycle. Human society could not survive without, nor substitute for, this service on the scale required and, indeed, even relatively small disruptions have proven costly. Complete control of the hydrological cycle would require the construction and maintenance of a system beyond the wildest dreams—or nightmares—of engineers.

Physical Support of Plants

Plants don't build nests or perform parental care. Instead, virtually all terrestrial plants depend on soil for the sheltering of their offspring, in some cases for many years until conditions become favorable for sprouting. As they mature, plants remain dependent on soil for nourishment and physical support. While one occasionally finds an alpine flower or even a tree growing out of a rock crevice, soil is essential to maintaining populations of even the hardiest species, as the barren landscapes on highly eroded soils testify.

Human-engineered hydroponic systems grow plants in the absence of soil and represent a technological substitute whose structural costs provide a lower-bound index to the value of the physical support service. Hydroponic systems use root anchoring substances such as sand and gravel or industrially produced vermiculite and rockwool, rather than soil. In some cases, no solid medium is used at all, but simply plastic containers or troughs through which nutrient solution is pumped; for large plants, however, such as many of those that form the basis of the human diet, other "above-ground" support is then required. Hydroponic culture is often conducted in glass houses, which provide an indirect form of physical support by eliminating the forces of wind and rain.

The cost of simply the physical support trays and stands amounts to about US$55,000 per ha of Nutrient Film Technique systems, a new and popular hydroponic system, in the United Kindom (FAO 1990). These require replacement after perhaps a decade of use, whereas, with proper care, the physical support service of soil lasts indefinitely.

Retention and Delivery of Nutrients to Plants

Tiny soil particles play a critical role in supplying plant nutrition. Those less than 2 microns in diameter, primarily bits of humus and clays, carry a surface electrical charge that is generally (but not invariably) negative. This electrical property gives soil its exchange capacity, that is, its ability to retain and exchange positively charged nutrient cations released in the decay of organic matter or added to the soil as fertilizer. These ions (such as K^+, Na^+, NH_4^+, Ca^{+2}, and Mg^{+2}) can be reversibly exchanged for other ions (primarily H^+ and Al^{+3}) in the soil solution. Soil exchange capacity is measured in terms of the number of negatively charged sites available in a given amount of soil. The small size of humus and clay particles gives them tremendous charged surface area (and number of exchange sites) per unit mass; for example, the external surface area of 1 g of clay is at least a thousand times that of 1 g of coarse sand. Clays vary widely in their exhange capacities, and humus has the greatest capacity of all.

Soil exchange capacity is crucial to regulating soil fertility. It retains nutrients near the surface of the soil, which otherwise would quickly be leached away, out of reach of plant roots. Instead, the negative charges on clay and humus particles hold nutrient cations in proximity to roots, allowing them to be taken up gradually. A typical exchange reaction is:

$$\boxed{\begin{matrix}\text{clay or}\\ \text{humus}\end{matrix}}\ Ca^{+2} + 2\,H^{+} \leftrightarrow \boxed{\begin{matrix}\text{clay or}\\ \text{humus}\end{matrix}}\ \begin{matrix}H^{+} + Ca^{+2}\\ H^{+}\qquad\end{matrix}$$

Hydrogen ions are available chiefly through the dissociation of carbonic acid (H_2CO_3); this acid forms primarily from the dissolution of CO_2 in soil water. CO_2 is made available by root respiration, decomposition, and diffusion from the atmosphere.

Hydroponic systems rely on technology for the nutrient retention and delivery service, offering insight to the feasibility and costs of substituting for that service. In hydroponic systems, water and plant nutrients must be continuously fed to plant roots. The margin for error in nutrient supply is much smaller in soil-less culture relative to cultivation on land. Soil can act as a buffer in the application of fertilizers, holding the fertilizer ions on soil exchange sites until required by plants. In contrast, even small excesses of nutrients applied hydroponically may be lethal.

Regulating conditions inside hydroponic glass houses (including nutrient concentrations, pH, and salinity of the nutrient solution; air and solution temperature; humidity; light; pests; and plant diseases) is a complex, delicate, and often high-tech undertaking. As a United Nations evaluation put it, "Failure of electrical power supply or water supply can mean total loss in a matter of hours if no alternative source is available, while an error in analysis of the solution in sub-irrigation systems can lead to grave trouble. . . . No one should embark upon serious application of these methods without first studying the subject extensively and, if in doubt of his own knowledge of the scientific principles involved, ensuring access to expert advice at short notice. . . . Wherever reasonably fertile soils exist, soilless methods cannot compete with normal outdoor culture, and no logical person would expect them to do so unless labor costs became prohibitive" (FAO 1990, pp. 135–137).

Construction of a modern hydroponics system in the United States costs an estimated US$850,000 per hectare (FAO 1990). Computers and sophisticated equipment for nutrient solution analysis, operated by highly trained personnel are commonly required. This estimate assumes that all necessary infrastructure and services, such as the supply of reliable electricity, sufficiently pure water, and maintenance expertise, already exist. Operating costs are variable but also high; they include energy for running pumps and fans, maintenance of complex equipment, and replacement and disposal of (frequently nonbiodegradable) plant-growing media and support.

No attempt has been made to grow food hydroponically on remotely the scale required to feed a small city, much less all humanity. Because of high capital and operating costs, use of hydroponics is restricted largely to high-priced, specialty crops grown in climatically homogeneous Japan and Northwestern Europe. In the United States, relative climatic heterogeneity (which allows fresh vegetables to be grown domestically year-round) and low-priced, rapid transportation have discouraged hydroponic culture, and many attempts have failed (FAO 1990; Hoffman 1988). Worldwide, the area under hydroponic culture is only a few thousand hectares; individual enterprises typically operate on 1 ha or a few ha (Benoit 1987; FAO 1990; Kobayashi et al. 1988). By contrast, global cropped area is about 1.4 billion ha (USDA 1993).

Growing hydroponically the 1.8 billion MT of grain and 1.3 billion MT of other crops produced annually today (USDA 1993) would be tremendously more expensive than outdoor soil cultivation. This is evident simply considering that irrigation is now limited, largely for economic reasons, to a mere 16 percent of the world's cropland. Hydroponic culture requires much more careful and costly irrigation than does field cultivation.

Dead Organic Matter and Wastes

Uncountable numbers of terrestrial organisms die each year, and their remains are consumed by organisms in soil. In the processes of decomposition, soils render harmless many potential human pathogens in waste and in the remains of dead organisms. Soil organisms produce potent antibiotic compounds, such as penicillin and streptomycin, manufactured by a soil fungus and a soil bacterium, respectively. The discovery of penicillin marked a tremendous advance in medicine, led to the development of a multibillion-dollar pharmaceutical industry (see chapter 14), and has extended and improved innumerable human lives.

While living, herbivores and higher trophic consumers produce excrement, most of which goes unnoticed, being rapidly processed by soil. Exceptions occur where concentrations of humans and their livestock exceed the waste assimilation capacity of waterways and land. Human waste is typically discharged into waterways, but the waste of livestock is generally returned to the soil—an estimated 90 percent thereof in the United States (Bertrand 1983, cited in Sanchez et al. 1989) and probably a similar fraction elsewhere. Livestock represent a nontrivial source of waste, numbering over 20 billion (there are about 4.3 billion cattle, sheep, goats, pigs, equines, buffaloes, and camels and about 17.2 billion chickens; WRI 1994).

How much waste and dead organic matter is produced and processed each year? A rough approximation of this is the total amount of live organic matter produced each year because, in a steady-state situation, the production of new plant material is balanced by the respiration of plants themselves, their consumers, and decomposers (Odum 1969). Thus, net primary production (NPP)—the amount of energy fixed by primary producers (mostly plants) minus what they use to support their own life processes—is a rough index of the quantity of dead organic matter and waste generated and decomposed. The fraction of NPP consumed by animals is actually relatively minor in most terrestrial ecosystems (Schlesinger 1991).

NPP on land amounts to about 132 billion MT (dry-weight) of organic matter per year (Ajtay et al. 1979). Of this, today's 5.7 billion people consume directly just over 1 billion MT of dry organic matter per year as food (plant and animal products). In addition, humanity co-opts 43 billion MT (32 percent) of total NPP in the form of wasted food, forest products, crop and forestry residues, pastures, and so on (Vitousek et al. 1986; note that these latter estimates are conservative, having been made a decade ago, when the human population was at about 5 billion). Were it not decomposed, just the 44 billion MT shunted through the human enterprise would rapidly cover the planet's land surface.

Decomposition is critical to the production of food, fodder, timber, cotton and other fiber, biomass fuels, pharmaceuticals, and other living sources of material well-being. One needs not only to get rid of the dead organic matter and waste associated with the production and consumption of these goods, but to recycle it as well, or else face rapid exhaustion of essential material constituents of life. There is no human technology that can substitute for the functions of soil organisms in doing this. In the space of the period at the end of this sentence, diverse microbial species process the particular compounds whose chemical bonds they can cleave and pass along to other species, in assembly-line fashion, end-products and by-products of their specialized reactions.

In a fantasy world largely without soil, one could assemble microbial communities in huge decomposition plants and hope that they could process, and be sustained by, the waste and dead organic matter fed to them. One would have no idea which of millions of soil species to include; at best, one could inoculate the vats with communities drawn at random from samples of soil. Moreover, the costs of transporting the dead organic matter and waste to, and distributing the nutrients from, the decomposition plants would be enormous. The complexity of the transformation process makes it difficult to envision such a system working at all.

Anyone who has driven by and smelled a sewage treatment plant on a bad day will recognize that even the small-scale, incomplete decomposition of human wastes in such plants cannot be continuously controlled. The lack of adequate sewage treatment (or any at all) in most cities of the world raises further doubts about substituting technology for this service on the much larger scale required to recycle all of the natural by-products of the human enterprise.

Renewal of Soil Fertility

The processing by soil organisms of dead organic matter and waste replenishes the nutrients required for primary production and thus fuels the cycle of life. In undisturbed natural systems, nutrient cycles tend to be closed, with inputs roughly matching outputs. Human disturbance may dramatically disrupt this approximate steady-state, inducing major changes in water and nutrient fluxes across system boundaries. Deforestation, for instance, increases runoff, nutrient loss, and erosion; these effects are magnified and may carry on indefinitely when forested land is converted permanently to agriculture (Vitousek 1983). In agricultural systems, some nutrient-containing organic matter is removed in harvest and, in many situations, more goes up in the smoke of agricultural residue burning; much is also lost due to elevated decomposition rates (which increase with tillage and irrigation). Until the recent intensification of agriculture, replenishment of nutrients occurred naturally (although not always completely) by nitrogen-fixing and other soil microorganisms. Now a basic tenet of farming is to replace the nutrients lost through crop harvest by adding organic or inorganic fertilizers.

Soil supply of plant nutrients is thus partially substituted for, and also augmented, through the industrial production and application of fertilizers. The benefits of fertilization were appreciated many centuries ago and were, until around 1850, realized exclusively through application of natural organic materials. The use of these materials in industrial farming today, however, is negligible, accounting for less than 1 percent of total nitrogen fertilizer consumption in the United States, for example (Tisdale et al. 1985).

With few exceptions, wherever farmers can afford them, inorganic synthetic fertilizers are used instead. Global inorganic fertilizer consumption grew from 30 million MT in 1960 to over 140 million MT in the late 1980s; it has fallen slightly (and most likely temporarily) since then, to just over 120 million MT in 1993/94 (FAO 1994), for a complex of reasons related to macroeconomic and farm policy, as well as to agronomic factors.

The principal fertilizers used worldwide supply—in various forms—nitrogen, phosphorus, and potassium; 1993/94 consumption levels were 73 million MT, 29 million MT, and 19 million MT, respectively (FAO 1994). Plants also require a suite of other nutrients, for whose supply farmers must rely increasingly on synthetic fertilizers; these are often mixed in with the principal fertilizers. They include sulfur, calcium, and magnesium, as well as a variety of "trace elements" that are required in tiny amounts, such as iron, manganese, zinc, copper, boron, molybdenum, cobalt, and chlorine.

Nitrogen

To what extent does, and could, industrial synthesis of these plant nutrients substitute for their supply by soils? In answering this question, we focus on nitrogen because it limits primary productivity over more of the earth than any other element, and it is consumed in tremendous quantities in agriculture. Nitrogen gas, technically dinitrogen or N_2, composes about 80 percent of earth's atmosphere. The element nitrogen (N) is an essential constituent of all forms of life. Nucleic acids, the building blocks of the genetic material essential for reproduction, and proteins are composed of N-containing compounds, as well as many other biological materials. In total, the living organisms on land and in marine and aquatic environments contain ~15 billion MT of N.

Approximately 90 percent of annual N uptake by plants results from internal recycling of nutrients by soils. But ultimately, this N is derived from the atmosphere: An estimated 100 million MT of N are pulled out of the atmosphere by N-fixing, terrestrial organisms each year (Schlesinger 1991). The ability to fix N is restricted to certain soil bacteria and blue-green algae. Bacteria in the genus *Rhizobium* fix N in a symbiotic relationship with leguminous plants, carrying on the reaction in root nodules and delivering ammonia (NH_3) or organic N directly to the plant roots. Some free-living bacteria, such as *Azotobacter* and *Clostridium,* also fix N, but the resultant ammonia is generally assimilated by the bacteria as rapidly as it is produced; it only becomes available after the bacteria die and decompose. Blue-green algae are important N-fixers in paddy rice farming (and in the oceans). Less than 10 million MT of N are fixed by lightning (Vitousek and Matson

1993), for a total annual "natural" flux of about 110 million MT N from the atmosphere into forms available to terrestrial plants.

By how much has this flux been augmented by human activities? Industrial N fixation and leguminous crop production now add, respectively, about 80 million MT and 30 million MT N annually. An additional 25 million MT N is fixed each year through fossil fuel combustion (Vitousek and Matson 1993).

Nitrogen fertilizer production remains very energy intensive, and its cost fluctuates with energy prices; it now costs roughly US$150 per MT (1990 prices; Bacon 1995). Depending in part on management practices, however, N-fertilizer uptake by plants may be quite inefficient, often accounting for well under 50 percent of the added fertilizer (Legg and Meisinger 1982; Robertson 1993; Strong 1995). Assuming an efficiency of 50 percent, it costs approximately US$300 to supply, through inorganic fertilization alone, each MT of N in crops, not including local distribution and application costs.

A lower-bound estimate of the total annual value of natural N-fertilization of crops is US$45 billion [calculated as (NPP on cultivated land) × (the fraction of crop plants made up by N) × (the price of N-fertilizer) / (the efficiency of N-fertilizer uptake) = (15 billion MT/yr) × (0.01g N/g NPP) × (US$150/MT) /(0.50); derived from Vitousek et al. 1986; Schlesinger 1991]. The lower-bound total value of natural N-fixation on land is about US$33 billion per year (US$300/MT × 110 million MT/yr). The lower-bound total annual value of natural N-fertilization on all land (including N-fixation and internal recycling of N) is about US$320 billion per year [calculated as the annual nitrogen requirement of land plants (1.2 billion MT; Schlesinger 1991) minus the amount supplied anthropogenically (~135 million MT, including the flux from clearing tropical lands; Vitousek and Matson 1993) times the price and efficiency of N-fertilizer application and uptake, given above].

These valuations do not consider important external costs of industrial fertilizer production, such as those associated with the energy (mostly fossil fuel) inputs or with the disruption of the N cycle. Moreover, they represent a small, although important and illustrative, part of the picture, given the many elements other than nitrogen required in plant nutrition.

Regulation of Major Element Cycles

Soils are a key factor in regulating the earth's major element cycles (e.g., of carbon, nitrogen, and sulfur). The amount of carbon and nitrogen stored in soils dwarfs that in vegetation, for example, by factors of ~1.8 and ~18, respectively (Schlesinger 1991). The importance to society of maintaining an approximate steady-state in the stocks and fluxes of major elements can be most easily appreciated by considering the consequences of their recent disruption. The consequences are costly, long-term, and in many instances irreversible on a time scale of interest to society.

Alterations in the carbon and nitrogen cycles, in particular, have the greatest potential to drive global changes in the earth's chemistry. Increased fluxes of carbon to the atmosphere, such as occur when land is converted to agriculture (resulting typically in a loss of at least 25 percent of soil organic matter) or when wetlands are drained (resulting in even greater proportional carbon losses), contribute to the build-up of carbon dioxide and methane, important greenhouse gases, in the atmosphere (Schlesinger 1991; see chapter 5). Changes in nitrogen fluxes through anthropogenic N-fixation, biomass burning, and tropical land clearing cause build-ups of nitrous oxide, a potent greenhouse gas that is also involved in the destruction of the stratospheric ozone shield. These and other changes in the nitrogen cycle also cause acid precipitation and pollution of freshwater, estuarine, and coastal marine ecosystems, resulting in eutrophication and contamination of ground and surface drinking water sources by high nitrate-nitrogen levels (see chapter 18).

Marginal Costs of Soil Loss and Degradation

It should now be obvious that human life could not be sustained without soil. Soil not only supplies essential material ingredients of human well-being, but also yields many less tangible values: artistic expression through (and practical value of) ceramics; a diversity of cultures honed to local soils and vegetation; and the aesthetic beauty of complex, breathtaking landscapes that have captured the imagination of artists for centuries. The total value of soil is incalculable, as it includes the existence value of human society and of millions of other species.

While the cost of losing all soil would be infinite, to what extent is soil actually being eroded and degraded today, and what is the marginal cost thereof? Soil degradation results primarily from water and wind erosion (84 percent) but also from chemical and physical deterioration (16 percent; the latter refer to, for example, nutrient depletion, salinization, or acidification; and waterlogging or compaction, respectively; Oldeman et al. 1990). The global extent and severity of soil degradation are poorly documented. Estimates of the net rate of soil loss range from 24 billion MT/year (Brown 1984) to 75 billion MT/year (Pimentel et al. 1995). An estimated ~2 billion

Table 7.1. Global extent of human-induced soil degradation

Severity of Degradation	Total Degraded Area (million ha)	Percentage of Total Degraded Area	Degraded Area as a Percentage of Vegetated Land
Light	750	38.0	6.5
Moderate	910	46.2	7.9
Severe	300	15.2	2.6
Extreme	10	0.6	0.01
TOTAL	1,970	100.0	17.0

Source: Oldeman et al. 1990.

ha—17 percent of earth's vegetated land surface—have undergone human-induced soil degradation since 1945 (table 7.1; Oldeman et al. 1990). Soil productivity on the land classified as lightly degraded could be restored with improved agricultural practices, such as crop rotation and minimum tillage. Rehabilitation of the moderately and severely degraded soils, however, would require financial incentives and technical expertise beyond the present reach of the average farmer and most developing nations. Extremely degraded soil is beyond any human capacity for improvement.

The direct costs of soil erosion, as measured by the cost of replacing lost water and nutrients on agricultural land, amount to an estimated US$250 billion per year globally (Pimentel et al. 1995). An additional cost of about US$150 billion (per annum) is incurred in the form of off-site damages to recreation; human health; private property; navigation; facilities for water storage, conveyance, and treatment; and so on. In the United States, these on- and off-site costs amount to roughly US$44 billion per year. In contrast, control measures would amount to a comparatively small US$8.4 billion annually (Pimentel et al. 1995); effective and economical measures include, for instance, no-till cultivation, crop rotations, terracing and contour planting, and use of windbreaks (Carter 1994). There is considerable uncertainty in the magnitudes estimated here (Crosson 1995). Nonetheless, the bottom line is that soil erosion may be extremely costly even in the short term, and that the benefits of many prevention measures are likely to greatly outweigh the costs.

Benefit-cost analyses of soil conservation at the global level, although very uncertain, also suggest that control measures make economic sense (Chou and Dregne 1993). For example, the United Nations estimates the total direct on-site income foregone as a result of desertification, a form of land degradation (primarily human induced) in arid regions, at US$42.3 billion per year, whereas estimates of the direct annual cost of all preventive and rehabilitational measures range between US$10 billion and US$22.4 billion (UNEP 1991; Dregne et al. 1992). These comparisons assume that control measures would be fully effective. From a biophysical perspective, this assumption may be valid, although so few restoration efforts have been made on a large scale that one cannot be sure. From social, economic, and political perspectives, however, the barriers to the implementation of such measures often appear formidable.

Nonetheless, these estimates are probably conservative because they ignore the costs of human suffering, of loss of intangible values, and of deleterious off-site and longer-term effects. While soil formation rates are variable and poorly documented, the consequences of soil degradation are likely to manifest themselves for years to come. Studies of succession on newly formed habitats, such as on the earth exposed with the retreat of glaciers (reviewed in Tilman 1988, pp. 214–216), suggest that a minimum of 100–200 years may be required to form productive land once topsoil has been lost. In general, an estimated 200–1,000 years are required for the regeneration of 2.5 cm of lost topsoil (Pimentel et al. 1993). Human-induced loss of the capacity of land to supply food, fodder, fuelwood, timber, and other goods and services represents one of the most serious threats facing society today (Daily 1995).

Conclusions

Soil provides an array of ecosystem services that are so fundamental to life that their total value could only be expressed as infinite.

Elucidating the marginal tradeoffs associated with alternative human activities is crucial to informing alternative courses of action in the management of soil resources. Research is needed to better characterize the ecosystem services supplied by soil, especially with regard to the role of soil biodiversity in the functioning of soil. Better understanding is also needed of the interrelationships of different services supplied by soil and other systems, and the consequences of disrupting one service upon the functioning of others. In addition, much better documentation and monitoring of changes in soil conditions globally and the biogeochemical cycles that they regulate would constitute a very worthwhile investment.

These uncertainties notwithstanding, it is clear that the squandering of soil as is occurring worldwide today is increasingly detrimental socially and economically. Human well-being can be maintained and fostered only if earth's soil resources are as well. The implementation of policy to convert agricultural, pastoral, and forestry systems to sustainable enterprises would not only protect services supplied by soil, but would simultaneously greatly alleviate pressure on many other valuable ecosystem services.

Acknowledgments

Helpful comments on an early draft of this chapter were provided by Michael Dalton, Barbara Dean, Sarah Hobbie, David Layton, and Kristy Manning. G. Daily was supported by the generosity of Peter and Helen Bing and the Heinz and Winslow foundations during the writing of this chapter.

References

Adams, R. McC. 1981. *Heartland of Cities: Surveys of Ancient Settlement and Land Use on the Central Floodplain of the Euphrates*. Chicago: University of Chicago Press.

Ajtay, G., P. Ketner, and P. Duvigneaud. 1979. "Terrestrial Primary Production and Phytomass." In *The Global Carbon Cycle*, B. Bolin, E. T. Degens, S. Kempe, and P. Ketner, eds., pp. 129-182. Chichester, England: John Wiley.

Alexandratos, N. 1995. *World Agriculture: Towards 2010*. Chichester, England: John Wiley.

Anderson, J. 1995. "Soil Organisms as Engineers: Microsite Modulation of Macroscale Processes." In *Linking Species and Ecosystems*, C. G. Jones and J. H. Lawton, eds., pp. 94–106. New York: Chapman & Hall.

Bacon, P., ed. 1995. *Nitrogen Fertilization in the Environment*. New York: Marcel Dekker.

Baublis, J., R. Wharton Jr., and P. Volz. 1991. "Diversity of micro-fungi in an Antarctic dry valley." *Journal of Basic Microbiology* 31: 3–12.

Benoit, F. 1987. "High-technology glasshouse vegetable growing in Belgium." *Soilless Culture* 3: 20–29.

Bertrand, A. R. 1983. "The Key to the Agricultural Ecosystem." In *Nutrient Cycling in Agricultural Ecosystems*, R. R. Lowrance, R. L. Todd, L. E. Asmussen, and R. A. Leonard, eds., pp. 3–12. Athens: University of Georgia Agricultural Experiment Station.

Bolin, B., P. Crutzen, P. Vitousek, R. Woodmansee, E. Goldberg, and R. Cook. 1983. "Interactions of Biogeochemical Cycles." In *The Major Biogeochemical Cycles and Their Interactions*, B. Bolin and R. B. Cook, eds., pp. 1–39. Chichester, England: John Wiley.

Bormann, F., G. Likens, D. Fisher, and R. Pierce. 1968. "Nutrient loss accelerated by clear-cutting of a forest ecosystem." *Science* 159: 882–884.

Brown, L. R. 1984. "Conserving Soils." In *State of the World 1984*, L. R. Brown, ed., pp. 26–51. New York: Norton.

Bull, A., ed. 1992. "Special Issue: Biodiversity amongst microorganisms and its relevance." *Biodiversity and Conservation* 4(1): 219–347.

Carter, M. R., ed. 1994. *Conservation Tillage in Temperate Agroecosystems*. Boca Raton, Fla.: Lewis Publishers.

Chanway, C. 1993. "Biodiversity at Risk: Soil Microflora." In *Our Living Legacy: Proceedings of a Symposium on Biological Diversity*, M. A. Fenger, E. H. Miller, J. A.

Johnson, and E. J. R. Williams, eds., pp. 229–238. Victoria: Royal British Columbia Museum.

Chou, N.-T., and H. Dregne. 1993. "Desertification control: Cost/benefit analysis." *Desertification Control Bulletin* 22: 20–26.

Crosson, P. 1995. "Soil erosion estimates and costs." *Science* 269: 461–464.

Daily, G.C. 1995. "Restoring value to the world's degraded lands." *Science* 269: 350–354.

De Leij, F., J. Whipps, and J. Lynch. 1994. "The use of colony development for the characterization of bacterial communities in soil and on roots." *Microbial Ecology* 27: 81–97.

Dregne, H., M. Kassas, and B. Rozanov. 1992. "A new assessment of the world status of desertification." *Desertification Control Bulletin*, 20: 6–18.

Edwards, C. 1990. *Microbiology of Extreme Environments*. New York: McGraw-Hill.

Ehrlich, P., A. Ehrlich, and G. Daily. 1995. *The Stork and the Plow: The Equity Answer to the Human Dilemma*. New York: Putnam Press.

FAO (United Nations Food and Agriculture Organization). 1990. *Soilless Culture for Horticultural Crop Production*. Rome: FAO.

FAO (United Nations Food and Agriculture Organization). 1994. AGROSTAT - PC. Domain X: "Fertilizers." Rome: UNFAO.

Harrison, K., W. Broecker, and G. Bonani. 1993. "The effect of changing land use on soil radiocarbon." *Science* 262: 725–726.

Hillel, D. 1991. *Out of the Earth: Civilization and the Life of the Soil*. New York: Free Press.

Hoffman, C. 1988. "Biotechnology: The future is now." *Appalachia* (Spring): 7–13.

Holben, W., J. Jansson, B. Chelm, and J. Tiedje. 1988. "DNA probe method for the detection of specific microorganisms in the soil bacterial community." *Applied and Environmental Microbiology* 54(3):703–711.

Ives, J. and B. Messerli. 1989. *The Himalayan Dilemma: Reconciling Development and Conservation*. London: Routledge.

Janick, J., W. Schery, F. W. Woods, and V. W. Ruttan. 1974. *Plant Science*, 2nd. edition. San Francisco: W.H. Freeman.

Jenny, H. 1980. *The Soil Resource*. New York: Springer-Verlag.

Judson, S. 1981. "What's Happening to Our Continents?" In *Use and Misuse of Earth's Surface*, B. Skinner, ed., pp. 12–139. Los Altos, Calif.: Kaufman.

Kobayashi, K., Y. Monma, S. Keino, and M. Yamada. 1988. "Financial results of hydroponic farmings of vegetables in central Japan." *Acta Horticulturae* 230: 337–341.

Korner, J., and E. Laczko. 1992. "A new method for assessing soil microorganism diversity and evidence of vitamin deficiency in low diversity communities." *Biology and Fertility of Soils* 13: 58–60.

Lee, K. 1985. *Earthworms: Their Ecology and Relationships with Soils and Land Use*. New York: Academic Press.

Lefroy, E., R. Hobbs, and M. Scheltema. 1993. "Reconciling Agriculture and Nature Conservation: Toward a Restoration Strategy for the Western Australian Wheatbelt." In *Nature Conservation 3: Reconstruction of Fragmented Ecosystems*, D. A. Saunders, R. J. Hobbs, and P.R. Ehrlich, eds., pp. 243–57. Perth: Surrey Beatty & Sons.

Legg, J., and J. Meisinger. 1982. "Soil nitrogen budgets." *Agronomy* 22: 503–566.

Likens, G. E., F. H. Bormann, R. S. Pierce, and W. A. Reiners. 1978. "Recovery of a deforested ecosystem." *Science* 199: 492–496.

NRC (National Research Council). 1993. *Soil and Water Quality*. Washington, D.C.: National Academy Press.

Odum, E. 1969. "The strategy of ecosystem development." *Science* 164: 262-270.

Oldeman, L., V. van Engelen, and J. Pulles. 1990. "The Extent of Human-Induced Soil Degradation, Annex 5." In *World Map of the Status of Human-Induced Soil Degradation: An Explanatory Note*, rev. 2nd edition, L. R. Oldeman, R. T. A. Hakkeling, and W. G. Sombroek, eds. Wageningen: International Soil Reference and Information Centre.

Overgaard-Nielsen, C. 1955. "Studies on enchytraeidae 2: Field studies." *Natura Jutlandica*, 4: 5–58.

Pierce, F., W. Larson, R. Dowdy, and W. Graham. 1983. "Productivity of soils: Assessing long-term changes due to erosion." *Journal of Soil and Water Conservation* 38: 39–44.

Pimentel, D., J. Allen, A. Beers, L. Guinand, A. Hawkins, R. Linder, P. McLaughlin, B. Meer, D. Musonda, D. Perdue, S. Poisson, R. Salazar, S. Siebert, and K. Stoner. 1993. "Soil Erosion and Agricultural Productivity." In *World Soil Erosion and Conservation*, D. Pimentel, ed., pp. 277–292. Cambridge, England: Cambridge University Press.

Pimentel, D., C. Harvey, P. Resosudarmo, K. Sinclair, D. Kurz, M. McNair, S. Crist, L. Shpritz, L. Fitton, R. Saffouri, and R. Blair. 1995. "Environmental and economic costs of soil erosion and conservation benefits." *Science* 267: 1117–1123.

Robertson, G. 1993. "Fluxes of Nitrous Oxide and other Nitrogen Trace Gases from Intensively Managed Landscapes: A Global Perspective." In *Agricultural Ecosystem Effects on Trace Gases and Global Climate Change*, L. Harper, A. Mosier, J. Duxbury, and D. Rolston, eds., pp. 95–108. Madison, Wisc.: American Society of Agronomy.

Rouatt, J., and H. Katznelson. 1961. "A study of bacteria on the root surface and in the rhizosphere soil of crop plants." *Journal Applied Bacteriology* 24: 164–171.

Sanchez, P., C. Palm, L. Szott, E. Cuevas, and R. Lal. 1989. "Organic Input Management in Tropical Agroecosystems." In *Dynamics of Soil Organic Matter in Tropical Ecosystems*, D. Coleman, J. M. Oades, and G. Uehara, eds., pp. 125–152. Honolulu: University of Hawaii Press.

Saunders, D., R. Hobbs, and P. Ehrlich, eds. 1993. *Nature Conservation 3: Reconstruction of Fragmented Ecosystems*. Perth: Surrey Beatty & Sons.

Schlesinger, W. H., 1991. *Biogeochemistry: An Analysis of Global Change*. San Diego, Academic Press.

Shiklomanov, I. 1993. "World Fresh Water Resources." In *Water in Crisis: A Guide to the World's Fresh Water Resources*, P. Gleick, ed., pp. 13–24. New York: Oxford University Press.

Sinclair, J. and W. Ghiorse. "Distribution of aerobic bacteria, protozoa, algae, and fungi in deep subsurface sediments." *Geomicrobiology Journal* 7: 15–32.

Strong, W. 1995. "Nitrogen Fertilization of Upland Crops." In *Nitrogen Fertilization in the Environment*, P. Bacon, ed., pp. 29–169. New York: Marcel Dekker.

Tilman, D. 1988. *Dynamics and Structure of Plant Communities*. Princeton: Princeton University Press.

Tilman, D. 1987. "Secondary succession and the pattern of plant dominance along experimental nitrogen gradients." *Ecological Monographs* 57: 189–214.

Tilman, D. 1990. "Constraints and tradeoffs: Toward a predictive theory of competition and succession." *Oikos* 58: 3–15.

Tisdale, S., W. Nelson, and J. Beaton. 1985. *Soil Fertility and Fertilizers*. New York: Macmillan.

UNEP (United Nations Environment Programme). 1991. *Status of Desertification and Implementation of the United Nations Plan of Action to Combat Desertification*. Nairobi: UNEP.

UNEP (United Nations Environment Programme). 1995. *Global Biodiversity Assessment*. New York: UNEP.

USDA (United States Department of Agriculture). 1993. *World Agriculture: Trends and Indicators, 1970–91*. Washington, DC: USDA.

Vitousek, P. 1983. "The Effects of Deforestation on Air, Soil, and Water." In *The Major Biogeochemical Cycles and Their Interactions*, B. Bolin and R. B. Cook, eds., pp. 223–245. Chichester, England: John Wiley.

Vitousek, P., P. Ehrlich, A. Ehrlich, and P. Matson. 1986. "Human appropriation of the products of photosynthesis." *BioScience* 36: 368–373.

Vitousek, P., and P. Matson. 1993. "Agriculture, the Global Nitrogen Cycle, and Trace Gas Flux." In *Biogeochemistry of Global Change: Radiatively Active Trace Gases*, R. Oremland, ed. New York: Chapman & Hall.

Wali, M. K., ed. 1992. *Ecosystem Rehabilitation*, vol. 2. The Hague: SPB Academic Publishers.

WRI (World Resources Institute). 1994. *World Resources 1994–95*. New York: Oxford University Press.

Chapter 8

Services Provided by Pollinators

Gary Paul Nabhan and Stephen L. Buchmann

Over the last two decades, there have been many attempts to promote the conservation of biodiversity through demonstrating how many different plants are relied upon by humanity for "its daily bread." Seldom, however, has faunal diversity—and its inherent requirement of wildlands habitat—been adequately linked to global food productivity or stability. One of these linkages is the pollination services provided to cultivated food crops by wild and managed animals that require foraging and nesting habitats adjacent to croplands.

In this chapter, we argue that provision of these pollination services involves far more wild species and far more habitat types than have been considered in most discussions of agriculture's dependence on biodiversity. Through the 1970s, for example, publications from the U.S. National Academy of Sciences (1975) argued that a meager 20 crop plant species keep humankind from starvation, but that there were many other "underexploited crops" that could be further commercialized to buffer global food production from the perils of monoculture. In an American Association for the Advancement of Sciences symposium volume, Felger and Nabhan (1976: p. 144) emphasized that "presently only a handful of [plant] species support civilization" as staple crops, but that pre-industrial hunter-gatherers drew upon another 30,000 species for food. Of those 30,000 species, an estimated 3,000 wild plants that played major roles in prehistoric subsistence were probably suitable for "agronomic development to substantially increase agricultural diversity."

In a more definitive answer to the question, "How many plants feed the world?", Prescott-Allen and Prescott-Allen (1990: p. 371) determined that 103-108 species contribute 90 percent of the national per capita supplies of food plants for the 146 countries in which the FAO collects data, suggesting that "plant species diversity remains a significant factor for [stabilizing the] world food supply." Similar arguments have been used by Harlan (1975) and by the Precott-Allens (1986) to direct attention toward the conservation of wild species in the gene pools of crops; they establish how many congeneric species have already contributed wild genes that are "hidden" in our food supplies. Robert and Christine Prescott-Allen (1990: p. 372) have provided a thoughtful commentary on how these estimates of biodiversity in the world food supply can be used in arguments for conservation:

> If it were true that very few plants fed the world, a strong case could be made for focusing conservation efforts not on saving species that might have potential as new food crops but on maintaining the full range of genetic variability within the major food species and their wild relatives. If, on the other hand, people continue to rely for their nutrition on a variety of plant species, a more convincing argument can be made for conserving both this wider array of species and the diversity of genetic variants that comprise each species. . . . The results suggest that . . . a conservation priority is to maintain both this wider array . . . and the diversity of genetic variants . . .

However, these conservation arguments are somewhat based on the largely unfulfilled potential for formerly utilized wild species to re-enter the world food supply (National Academy of Sciences 1975; Felger and Nabhan 1978), or for more "wild genes" to be used in crop improvement (Harlan 1975). The last two decades of new crop development have resulted in the re-entry of only a few formerly underexploited plants such as grain amaranths into the American food economy. Likewise, at the present time, only a small proportion of all plant breeders routinely use wild species that were formerly neglected in their crop improvement programs, and there remains considerable resistance among breeders to investing more of their time in exploiting such untried sources of genetic diversity.

In short, these arguments are somewhat like those of "rainforest activists" who claim that tropical forests should be saved for their as-yet-undiscovered cures for cancer and AIDS and, by doing so, inadvertently sanction the elimination of "worthless" plant species, which get screened for anti-tumor or anti-virus agents but show no value in treating diseases and are thus deemed to be of no value (Ehrenfeld 1976; Nabhan 1995). As Ehrenfeld (1981, p. 192) has argued,

> The various reasons advanced to demonstrate that these non-resources really are useful or potentially valuable are not likely to be convincing even when they are truthful or correct. . . . Thus the conservation dilemma is exposed: humanists will not normally be interested in saving any non-resource, any fragment of Nature that is not manifestly useful to humankind . . .

Ehrenfeld (1978, p. 205) has therefore called for two changes in the way we assess the value of biological diversity:

1. by reducing the exaggerations and distortions in conservationists' appraisals of the direct value of biodiversity to the world food supply and environmental stability; and
2. by identifying the non-economic values inherent in all natural communities and species, according them "an importance at least equal to that of indirect economic values."

We wish to accept Ehrenfeld's challenges by assessing both the economic and noneconomic values of the world's pollinator diversity, a component of biodiversity that clearly depends on pesticide-free wildlands. We will draw on recent surveys of the diversity of animals involved in pollinating the world's diversity of wild plants. We will then assess the value of pollinators involved in modern commercial agriculture in the United States, while noting where there are data gaps regarding the economic value of pollination services to global agriculture. Specifically, we define agricultural pollination services as the contributions made to crop seed, fruit, or fiber yields owing to the effective pollination of flowers by animal vectors.

The economic value of agricultural pollination services can be directly measured by comparing the yield (loss) of the crop in the absence of these animals with the yield in the presence of the pollinators in question; then by assessing any increases in crop production costs to produce the same yield without the help of the pollinators (Southwick and Southwick 1992). In particular, we will contrast the diminishing services of managed and feral honey bees (*Apis mellifera*) in crop production with the often underestimated services provided by a variety of other wild and semi-managed pollinators. Finally, we will consider the economic and noneconomic values lost if the recently emerged pollination crisis continues for very long.

Faunal Diversity of Pollinators

While recently collaborating on a book entitled *The Forgotten Pollinators*, we realized that there has never been a definitive global (or even hemispheric) estimate of the total number of animal species involved in pollination (Buchmann and Nabhan 1996). In order to approximate a first assessment of the faunal diversity of pollinators, we reviewed the number of species comprising the invertebrate and vertebrate genera, families, and orders in which there are more known effective pollinators than there are pollen or nectar cheaters, robbers, or avoiders. The survey results—included in appendix 5 of Buchmann and Nabhan (1996) in order to elicit additional refinement from ecologists and taxonomists—are summarized here in table 8.1. Our preliminary estimate is that there are more than 1,200 vertebrate species involved in pollination; for invertebrates, the number is greater than the 100,000 pollinators estimated to be present in the tropics by Roubik (1995) and considerably less than the 220,000 species falling within the taxonomic groups dominated by pollinators of flowering plants (Wilson 1992; Buchmann and Nabhan 1996).

In turn, Renner (1995) has recently attempted to estimate how many plants depend on animals to ensure effective pollination at levels necessary for reproduction and survival. Of the estimated 240,000 species of flowering plants for which one or more pollen vectors have been recorded, 219,850 are pollinated by animals; 20,000 are pollinated by wind or are self-fertile; and 150 have water-dispersed pollen. Floral rewardlessness—the lack of nectar or pollen available to animals—has evolved independently in twenty-three plant families (Renner 1995). Nevertheless, the vast majority of angiosperms invest considerable proportions of their total energy budgets

Table 8.1. Pollinator classes for the world's wild flowering plants (angiosperms[a])

Pollination Categories	Estimated Pollinator Taxa
Wind (abiotic)	20,000
Water	150
All insects	289,166
Bees	40,000
Hymenoptera (bees and wasps)	43,295
Butterflies/moths	19,310
Flies	14,126
Beetles	211,935
Thrips	500
All vertebrates	1,221
Birds	923
Bats	165
Mammals other than bats	133

[a]*N* = 240,000 flowering plant species.

Sources: Data assembled from various sources, including Buchmann and Nabhan (unpublished).

in producing secondary chemicals and elaborate structures that either attract or reward animal pollinators (Renner 1995). Plants invest so much in animal attractants and rewards because a very high percentage of their flowers do not receive frequent and effective enough visits from pollinators to ensure adequate levels of seed set.

In a critical survey of controlled field pollination experiments accomplished for 186 flowering plant species, Burd (1994) determined that at least 46 percent of them were pollinator-limited in their seed set rather than being resource-limited by lack of adequate moisture or photosynthates to ensure reproduction.

There are profound implications of Burd's (1994) assessment that under conditions unaffected by anthropogenic effects, the reproductive success of nearly half of the world's plants (studied to date) may be more limited by pollinator scarcity than by the vagaries of weather, soil fertility, or floral browsers and seed parasites. If the 186 studies included in the survey truly reflect natural conditions, as Martin Burd suggests that they do (pers. comm., 1995), then far fewer of these plants may achieve adequate seed set under disrupted ecological conditions. Recent field studies have documented reduced pollinator visitation, pollination of fruits and/or seed set in plant populations disrupted by human settlement patterns (Hendrix 1994; Jennersten 1988); by introduced pests or competitors (Aizen and Feinsinger 1993; Roubik 1978); by agrichemicals (Kevan 1977; Suzan et al. 1984; Tepedino 1979); and habitat degradation and fragmentation (Suzan et al. 1995; Washitani et al. 1994; Rathcke and Jules 1994). Despite the facts that only a small percentage of plants depend solely on one animal pollinator and that such a great variety of animals are involved in pollination, the studies cited above suggest that many plant populations stranded in human-disrupted habitats may no longer be producing enough seeds to regenerate themselves.

In extreme cases, physical fragmentation of plant and animal populations may reduce plant fertility near zero; this so-called Allee effect is most pronounced in remnant populations covering less than 200 m, although the smaller the number of individuals per plant population, the lower the seed production per plant will be (Lamont et al. 1993). To be sure, there are not enough definitive multiyear comparisons of plant reproductive success between "natural" habitats and "disrupted" habitats to predict the magnitude of aggravated pollinator limitation. Nevertheless, adequate availability of pollinators—at the right place and time, in sufficient numbers—can no longer be considered a given for all or even most plants. Considering that 165 genera of vertebrate pollinators (including 186 species) are now considered of conservation concern to IUCN (Nabhan 1996), it is clear that the diversity of pollinators available to both wild and domesticated plants is diminishing. While the long-term effects of diminished pollinator diversity on natural vegetation are difficult if not impossible to predict, the following sections provide an appraisal of pollinator diversity and abundance on agricultural production.

Faunal Diversity of Agricultural Pollinators

Until recently, surveys of crop pollinators relied heavily upon data collected primarily to underscore the importance of honey bees (*Apis* spp.) in the world economy. When McGregor (1976) compiled data from 130 crops in the United States and 400 crops worldwide, he determined that 15–30 percent of all human food is derived from honey bee–pollinated crops. This estimate, as well as those in Crane and Walker (1984) tend to underestimate the role of other, wild invertebrate pollinators. Recently, Roubik (1995) attempted to compile an exhaustive survey of the pollinators of some 1,509 cultivated plant species, and we have analyzed his data on the confirmed pollinators of 1,330 crop plants. Bees were confirmed pollinators for 72.7 percent of the crop species, and as possible pollinators of another 10.2 percent; *Apis* honey bees were specifically mentioned for only 15.5 percent of these listed crops. Other major taxonomic groups of pollinators included flies (18.8 percent), bats (6.5 percent), wasps (5.2 percent), beetles (5.1 percent), birds (4.1 percent), moths (2.9 percent), butterflies (1.5 percent), and thrips (1.3 percent). Thirty-seven genera of invertebrates and seven genera of vertebrates were specifically mentioned in Roubik's (1995) compendium (table 8.2).

The importance of the animal vectors of crop pollen is underscored by the fact that at least 72 percent of 1,330 crop species inventoried by Roubik (1995) are represented by one or more cultivars that require pollen movement by some vector; in other words, they are not capable of self-pollination. Less than 2 percent of Roubik's crops relied exclusively upon wind-pollination (table 8.2). Roughly two-thirds of the crops in the world appear to be obligately outcrossing cultivars and land races that require visits by animal pollinators to set fruit and seed.

For a moment, let us focus exclusively on the 103–108 crop species that Prescott-Allen and Prescott-Allen (1990) claim to be those which feed the world by providing 90 percent of national per capita food supplies for 146 countries. Of those 100-plus crops, the 71 species that are bee-pollinated are visited not only by *Apis* (19), but by the following other bee genera: *Amegilla* (1), *Ancyloscelis* (1), *Bombus* (3), *Chalicodoma* (1), *Exomalopsis* (1), *Lipotriches* (1), *Megachile* (3), *Melipona* (3), *Peponapis* (4), *Xenoglossa* (4), and *Xylocopa* (1). In addition, the following numbers of major crops have their flowers visited by thrips (4); wasps (2); flies (5); beetles (6); moths (1); and other insects (7). Vertebrate pollinators of crops include unspecified birds, as well as bats in the following genera: *Eonycteris; Macroglossus;* and

Table 8.2. Pollinator classes for cultivated food plants of the world[a]

Pollen Vectors	Number of Floral Host Species	Percentage of Total (N = 1,509 species)	Corrected Percentage (without 549 unknown vectors) (N = 960 species)
Wind (abiotic)	47	3.10	4.90
All vertebrates	155	10.27	16.15
Birds	52	3.45	5.42
Bats	103	6.83	10.73
Thrips	12	0.80	1.25
Butterflies and moths	35	2.32	3.65
Flies	179	11.86	18.65
Beetles	48	3.18	5.00
All bees	918	60.83	95.63
Non-*Apis* bees	796	52.75	82.92
Honey bees	122	8.08	12.71
Wasps	46	3.05	4.79
Other insects	66	4.37	6.88
Unknown vectors	549	36.38	57.19

[a]N = 1,509 species comprising the majority of world crop or medicinal plants.

Sources: Pollinator estimates are from Roubik (1995) and personal observations (Buchmann and Nabhan, unpublished).

Pteropus. Conservatively, we estimate that at least twenty genera of animals other than honey bees provide pollination services to the world's hundred most important crops, and they collectively pollinate at least as many crop species as managed *Apis* colonies pollinate. As we shall discuss shortly, managed honey bee colonies are in decline in many parts of the world. This global trend begs the question of whether wild pollinators will play an increasingly important role in pollinating crops that historically have been pollinated by honey bees. The outcome depends on how much society values all pollination services, particularly those of wild animals, and how much we value wild habitats required by pollinators.

Economic Valuation of Pollination Services

Until recently, the only published economic assessment of pollinator services on a national or global basis have been those for managed European honey bee colonies; even state-level estimates of a single wild pollinator's value are rarities (Menke 1952). Crane and Walker (1984), for example, estimated that the global value of crops pollinated by honey bees exceeds the value of the annual commercial honey crop by fifty times. The work of Robinson et al. (1989 a and b) exemplified early estimates of economic pollination parameters by building a model based on the increased production of the forty largest U.S. crops attributed to exclusive pollination by honey bees. They estimated that in the United States in 1985, the increased seed (and fruit) yield and enhanced produce quality achieved through pollination by honey bees alone could be valued at $9.7 billion (Robinson et al. 1989b). This value is more than sixty times greater than the combined sum of all pollination service charges paid to beekeepers by farmers ($60.9 million per year) and all federal subsidies provided at that time ($80.8 million per year). Robinson et al. (1989b, p. 152) concluded that "total monetary returns to farmers, as measured in added crop value, are many times higher than the current costs of commercial pollination services . . . [with] a cost/benefit ratio of over 68:1."

Unfortunately, this landmark study was riddled with methodological presumptions, which were soon challenged by Southwick and Southwick (1989 a and b). In addition to catching an arithmetic error leading to an overestimation of the 1985 U.S. honey bee pollination value by $945.9 million, the Southwicks pointed out the following faulty assumptions made by the Robinson et al. (1989) research, a project co-sponsored by the USDA Economic Research Service and National Honey Board:

1. It incorrectly assumed that with any yield reduction due to loss of pollination services there would be no change in the price of the crop (as does Prescott-Allen and Prescott-Allen 1986).
2. It incorrectly estimates yield losses due to loss of pollination both for seeds required to produce hay or vegetable crops, as well as the crops themselves. Only seed (and fruit) crops are affected by loss of pollination services, not yields of crops such as alfalfa or celery.
3. It presumes that there are no other introduced or wild pollinators that may substitute for honey bees should the managed bee colonies be eliminated or otherwise decline.

Southwick and Southwick (1989b) conclude by offering a case study of the U.S. almond crop, which draws upon introduced (*Osmia*) bees and wild pollinators in addition to honey bees:

> As the most egregious example of this flaw, consider the case of almonds (with their estimated value of $360.6 million per year . . .) The authors implicitly claim that there would be no almonds

without the service of honey bees. An important implication of this is that, in the absence of honey bees, almonds would become extinct since they could not produce. One must take great care in estimating values/losses of economic commodities.

Over the following four years, Southwick and Southwick (1992) refined a model that corrected these faulty assumptions and focused on the gains to consumers through lower prices for crops that are benefited by honey bees. Their model estimated that U.S. consumers realize $1.6–5.7 billion in annual social gains or surplus income that would be lost if honey bee services were reduced at different levels for 62 U.S. crops, including 20 fruit species, 17 vegetable species, and 5 oilseed species. Specifically, the Southwicks assessed for each crop the extent to which other (predominantly wild) pollinators could take up the slack in pollination services, should honey bees decline to the degree that their model predicted.

Perhaps inadvertently, the Southwicks created a model by which one could calculate the real and potential contributions of non-*Apis* pollinators not merely to crop yields, but to annual social gains of food consumers. If no native pollinators replaced managed honey bees in providing pollination services to the 62 U.S. crops in the model, the annual losses to consumers would likely be as high as $5.7 billion per year, and perhaps as much as $8.3 billion. However, if native and introduced wild pollinators substituted for honey bees in providing pollination services as fully as possible, these losses would be reduced to $1.6 billion per year. Therefore the potential value of non-*Apis* pollinators in the U.S. agricultural economy is on the order of $4.1 to 6.7 billion a year!

The Southwicks based their model on the assumption that managed honey bee colonies within reach of the croplands of the 62 commodities would eventually decline by 50 percent in the northern United States and by 100 percent in the southern states. By using yield loss and price increase figures like those for blueberries following pesticide poisoning of bees by organophosphates in the early 1970s (Kevan 1977), the Southwicks were able to assess the relative value of managed (mobile) honey bee colonies versus wild pollinators in native habitats adjacent to croplands. In addition, they carefully assessed trends in alfalfa seed crop production. If alfalfa leafcutter bees and other native pollinators had not been recruited to replace declining honey bees, there would be a 70 percent crop reduction in U.S. alfalfa yields, costing consumers $315 million a year. If, however, other pollinators are fully managed and their fieldside habitats are protected from pesticides and land clearing, they could reduce this loss by $275 million a year.

It should be pointed out that we lack sufficient reliable data to extend the Southwicks' model to cover all 103–108 crop species that feed the world, since the relative contributions of honey bees versus other pollinators remain unknown for most of these crops. Furthermore, while the crop-by-crop assessments made by the Southwicks used the best available information on pollination ecology, there are tremendous differences between growing regions and even between cultivars with regard to the effectiveness of specific pollinators.

With the recent untimely death of Ed Southwick, it is unclear who will take up the refinement of the Southwick brothers' model to supply us with the next approximation of the economic value of services provided by wild pollinators, either nationally or globally. Nevertheless, it is all too clear that their 1992 predictions of honey bee declines in the United States are coming true. Recently, however, we have begun collaborative research with Lawrence Southwick in an initial attempt with economic theory rigor to assess the contribution of non-*Apis* bee species to crop pollination in the United States.

The Impending Pollination Crisis

A feature article in *Science* magazine recently heralded, "Pollination worries rise as honey bees decline" (Watanabe 1994, p. 1170). In fact, the number of honey bee colonies in the United States has been declining since 1947, when it peaked at 5.9 million (Hoff and Willett 1993); by 1992, the number had slipped below 3 million, and by 1994 to less than half its peak (2.8 million). If this downward trend has gone on for nearly a half century, why does it currently merit headlines in *Science* and the *Washington Post*?

Not only is the current downward trend more precipitous than ever before—there was a 20 percent decline in the number of managed honey bee colonies in the United States between 1990 and 1994—but there are more impending threats challenging the remaining colonies of honey bees in North America (Watanabe 1994). As early warning signs of this pollination crisis emerged, the National Association of State Departments of Agriculture (NASDA 1991, p. 1) reminded policymakers that "no comprehensive national strategy exists to deal with the threat[s]" that face pollinators and their stewards.

These threats include diseases that now affect over 62 percent of all honey bee colonies within the fifty U.S. states: American and European bacterial foulbrood; fungal chalkbrood; and nosema, caused by a protozoan. Collectively, these diseases account for more than $32 million of losses to beekeepers, based on average colony replacement costs of $60, annual pollination service income of $15, and honey production income of $25 per colony (NASDA 1991). Ironically, NASDA has never calculated the losses of yields to crop producers and consumers as a result of this annual loss of 320,000 honey bee colonies due to diseases.

Nevertheless, these diseases have been historically present in the United States for some time; the threat of parasitic mites is, by contrast, relatively new. Tracheal mites (*Acarapis woodi*) were first found in the United States in 1984 and are now present in most states, resulting in an $80 million annual loss in managed colonies by 1991 (NASDA 1991). Varroa mites (*Varroa jacobsoni*) are external parasites first found in Florida and Wisconsin in 1987; they now occur in more than thirty states, resulting in another $80 million of income loss and replacement costs to beekeepers (NASDA 1991). Again, this loss of 160,000 colonies to parasites has never been used to calculate additional reductions in societal gains due to lower crop yields resulting from pollinator limitations.

Africanized honey bees, first released in Brazil in 1956, were discovered in Texas in 1990, and had reached more than eighty-five counties in Texas, New Mexico, Arizona, and California by June of 1995. American beekeepers are already abandoning some of their colonies because of the dangers of handling colonies in which Africanized bees are hybridizing with European honey bees, and because of liability issues raised by their neighbors. In Venezuela, the spread of Africanized honey bees resulted in nearly all part-time and hobbyist beekeepers dissolving their operations (NASDA 1991); if the Southeast and Western states within the potential range of Africanized honey bees lost all their part-time and hobbyist beekeepers, it would result in a loss of roughly 6,000 colonies valued at $600,000 (Hoff and Willett 1993). If all full-time operations were abandoned in these two regions as well, the 133,000 colonies annually maintained by part-time, full-time, and hobbyist beekeepers in the 1980s would be permanently lost, reducing the total number of colonies in the United States by 30 percent.

Despite the increasing mobility of hives and mandated monitoring of agrichemicals, beekeepers claim that there continue to be many pesticide and herbicide poisonings of honey bees each year. However, official U.S. government statistics report that only between 200 and 250 beekeepers claim poisoning of their colonies each year; of these claims, only 2 percent of them receive any reimbursement for their losses. Of those reporting pesticide poisoning, 30 percent report that their colonies have suffered reductions of half or more of the bees they have maintained (Hoff and Willett 1993). No less than 15,000 colonies are annually dramatically affected by pesticide poisoning. If we take into account the fact that pesticide use in the United States has doubled since *Silent Spring* was published (Carson 1962; Curtis and Profeta 1993), then pesticide effects on feral honey bees and other wild pollinators—undocumented in government statistics—are likely to be pervasive as well.

If those threats did not destabilize agricultural pollination services enough, in 1994 the USDA ended its price-support programs for honey, some of which began as early as 1951. As a result, subsidies were completely eliminated that had aided 92 percent of the full-time beekeepers, 84 percent of the part-time beekeepers, 65 percent of hobbyists in the late 1980s (Hoff and Willett 1993).

Cumulatively, all of these threats and economic changes have devastated the vital and undervalued pollination services provided by beekeepers, especially in states where Africanized honey bees have already arrived. For example, Arizona lost 44 percent of its honey bee colonies between 1986 and 1994, and its honey production is the lowest it has been since the 1970s, in the era of rampant bee poisoning by DDT, other organochlorines, and newly released organophosphates. In addition, the average yield from the remaining 47,000 colonies in Arizona was fifty-nine pounds, a 23 percent decrease from the year before, suggesting that mites and diseases were reducing yields in extant hives.

Watanabe (1994, p. 1170) suggests that these honey bee declines are already translating into reduced yields due to pollinator scarcity:

> The population declines are raising concerns that farmers won't have enough of the helpful insects to pollinate their crops. Take what [apiculturist] Morse sees in his own research pumpkin patch near Ithaca, New York. Flowers remain pollen-laden 5 hours after they open, even though by then they should be stripped off by feral bees. This lack of pollinating activity bodes poorly for New York's $13-million pumpkin crop, Morse says. California almond growers, whose crop is worth upwards of $800 million this year, are also experiencing serious bee shortages. Indeed, this year, for the first time, they had to bring in bees from Florida, Texas, South Dakota, and other states to pollinate their crops, instead of relying on local bees.

This anecdotal evidence suggests that both native, wild pollinators and introduced, semi-managed pollinators—not just managed honey bees—have been scarce in both the Northeast and the Southwest United States, particularly since the unusually severe weather of winter 1995–1996. In New York, squash and gourd bees in the genera of *Xenoglossa* and *Poponapis* should have been present even where managed *Apis* colonies have declined. In California, semi-managed *Osmia* (e.g., *Osmia lignoria*, the "Blue Orchard Bee") and other introduced bees should be providing sufficient services to almond orchards, unless their numbers have dwindled as well. Unfortunately, censuses of wild pollinators have seldom been done to provide benchmarks by which to measure this rapidly changing scene. There is an urgent need to census and monitor all potential pollinators in wild habitats adjacent to agricultural crops that are not being well serviced by declining honey bees, in case there develops a need to set aside pesticide-free habitat for them. In summary, it is clear that these "alternative" native bee pollinators must, by default, play a greater role in providing pollination services to agriculture as honey bees decline, assuming that their own wild and managed populations can be sustained.

Noneconomic Values

There are other, noneconomically appraised services that pollinators provide to the biotic communities in which they (and we) reside. It may not ever be possible to put a price tag on these services, except in the sense of surveying members of a human community for how much they are willing to pay to guarantee that the services are maintained through habitat set-asides, pesticide-spraying setbacks, or pesticide poisoning loss reimbursements to stewards of managed pollinator populations (Pearce and Moran 1994). Among these noneconomically appraised values are:

1. The value that pollinators have provided over evolutionary time in fostering the adaptive radiation of angiosperms to its present levels of diversity (Buchmann and Nabhan 1996; Wilson 1992).
2. The value that a diversity trapline and migratory pollinators have provided in structuring the spatial and temporal patterns of phenological events in biotic communities (Howell 1974; Heithaus 1974).
3. The value that pollinators have provided in capturing and redistributing floral components of global primary productivity (Roubik 1993).

Because of an explosion of new information regarding the latter value, we will briefly highlight one (nonanthropocentric) means by which this service can be appreciated.

Biologists—including many pollination ecologists—as well as the public in general, are largely unaware of the essential ecological services performed by pollinators apart from pollination in the narrow sense: the act of placing pollen on receptive stigmas. In their daily wanderings, bats, bees, beetles, birds, and butterflies, along with thousands of other species of highly vagile pollen vectors, collect and redistribute foodstuffs, then scatter their nitrogen-rich waste products. Until recently, pollinators have been largely unrecognized as consumers and distributors of energy-rich floral biomass or as dispersers of nitrogenous wastes.

Smithsonian scientist David Roubik can be largely credited with recognizing these additional roles for pollinators in biotic communities and their physical environments. He has demonstrated that a number of social bees in addition to *Apis*—including *Melipona* and *Trigona*—function as foraging superorganisms that effectively redistribute floral biomass for great distances around each of their perennial nesting cavities (Roubik 1993). This redistributed biomass may include not only nectar and pollen, but water, fragrant volatile oils, and resins (including saps and gums) as well.

Just how important on a regional or global scale is this redistribution of materials by social bees? Roubik (1993) has assessed the movement of materials by stingless social bees in Panama, where they occur in densities of up to twenty thousand individuals per hectare, and where total bee populations occur as high as forty-four thousand individuals per hectare. As herbivorous visitors to tropical flowers, these stingless bees remove slightly more than 3 percent of the primary productivity of Panamanian tropical forests, thereby consuming or redistributing more energy than do any of the following sets of herbivores: leafcutter ants, vertebrate frugivores, vertebrate folivores (including browsing game animals), and flower-visiting bats and birds. Only underground invertebrates and soil microbes collect, consume, or convert more tropical rainforest resources than do pollinating bees.

Until recently, materials such as nectar, pollen, resins, and even whole flowers have seldom been accounted for in calculations of primary forest productivity. Roubik has argued that the direct accounting of the costs of reproduction by rainforest trees must include the energetic value of the precursors to fruit and seeds, including floral attractants and rewards. In animal-pollinated neotropical forests such as those on Barro Colorado Island, Panama, aboveground net primary productivity is estimated at about 2.3 × 108 kilojoules per hectare. In such forests, social bees recycle 106 or 107 kilojoules per hectare per year, making them an indispensible component of the complex trophic webs within these communities. Roubik (1993) concludes that if bees were the only consumers of tropical forest plant products, the annual net primary productivity of these forests would be at least 3.2 percent greater than presently estimated for places such as Barro Colorado Island.

To place this in perspective, Roubik (1993) has compared the rainforest productivity passing through floral visitors with those passing through a farmer's hands in a tropical rice paddy or cornfield (Norman et al. 1984):

> The total harvest of food and material per hectare of a neotropical forest by all organisms using nectar, pollen and resin is thus between five and ten million kilojoules per hectare each year, an amount nearly half of the energy available for human consumption from a tropical hectare's annual rice or corn crop . . .

Kevan (1975) has pointed out that pollination as a process is often the weakest link in our understanding of how ecological communities function. A recent survey of visitors to the National Zoo's Pollinarium exhibit revealed that three-quarters of the respondents simply thought of pollen as a kind of allergenic dust or nuisance, or simply didn't recognize its role in plant reproduction. On the other hand, recently developed ecotourism efforts are sensitizing laypeople to the aesthetic and cultural values associated with monarch butterflies, nectar-feeding bats, hummingbirds, and giant Asian bees. The Arizona–Sonora Desert Museum where we work has launched a Forgotten Pollinators Public Awareness Campaign to overcome the negative attributes often associated with invertebrates and bats, so that pollinators can be more positively valued culturally and economically. This is necesssary, as table 8.2 suggests, because many pollinators are threatened, and with their loss, more obligately dependent plants will be endangered as well (Washitani et al. 1994; Buchmann and Nabhan 1996). While we expect that it will take some time before humankind can place the value of such rapidly disappearing ecological interactions in perspective, we hope that society's resource management agencies will find a more immediate way to heed the simple recommendation of pollination ecologist Peter Lesica (1993, p. 193): "Management activities that threaten pollinator populations should be avoided in order to protect populations of . . . endangered plant[s]." Without such changes in management, we will lose both economically valuable and ecologically valuable interactions between plants and animals, some of which have taken millennia to develop.

Acknowledgments

We dedicate this contribution to the memory of Dr. Edward E. Southwick, 1943–1995, an apiarist/ecologist and economist of bees, and columnist for the *American Bee Journal.* We thank G. Daily, M. Ingram, C. Fox, R. Naylor, D. Hancocks, D. Roubik, M. Burd, M. Robinson, N. Pratt, P. Kevan, R. and C. Prescott-Allen, P. Feinsinger, O. T. Kizer, K. Manning, and especially Mrill Ingram, for their assistance in preparing this manuscript. This work was funded by Agnese Haury, the Wallace Genetics Foundation, Wallace

Global Fund, W. Alton Jones Foundation, the Geraldine R. Dodge Foundation, and the Arizona–Sonora Desert Museum.

References

Aizen, M., and P. Feinsinger. 1993. "Forest fragmentation, pollination, and plant reproduction in a Chaco dry forest, Argentina." *Ecology* 75: 330–339.

Buchmann, S., and G. P. Nabhan. 1996. *The Forgotten Pollinators.* Washington, D.C.: Island Press.

Burd, M. 1994. "Bateman's principle and plant reproduction: The role of pollen limitation in fruit and seed set." *Botanical Review* 60: 81–109.

Carson, R. 1962. *Silent Spring.* Boston: Houghton-Mifflin.

Crane, E., and P. Walker 1984. *Pollination Directory for World Crops.* Bucks, England: International Bee Research Association.

Curtis, J., and T. Profeta, eds. 1993. *After Silent Spring.* New York: Natural Resources Defense Council.

Ehrenfeld, D. 1976. "The conservation of non-resources." *American Scientist* 64:648–656.

Ehrenfeld, D. 1981. *The Arrogance of Humanism.* New York: Oxford University Press.

Felger, R. S., and G. P. Nabhan. 1978. "Agroecosystem diversity: A model from the Sonoran Desert." In N. Gonzalez, ed., *Social and Technological Management in Dry Lands*, pp. 129–150. Boulder: American Association for the Advancement of Science Symposium 10 and Westview Press.

Harlan, J. 1975. *Crops and Man.* Madison, Wisc: American Agronomy Society and Crop Science Society of America.

Heithaus, E. R. 1974. "The role of plant-pollinator interactions in determining community structure." *Annals of the Missouri Botanical Garden* 61: 675–691.

Hendrix, S. D. 1994. "Effects of population size on fertilization, seed production and seed predation in two prairie legumes." *North American Prairie Conference* 13: 115–119.

Hoff, F. L., and L. S. Willett. 1993. *The U.S. Beekeeping Industry.* Washington D.C.: USDA Agricultural Research Service, pp. 1–72.

Howell, D. J. 1974. "Pollinating bats and plant commmunities." *National Geographic Society Research Reports, 1974 Projects* 173: 311–334.

Jennersten, O. 1988. "Pollination of *Dianthus deltoides* (Caryophyllaceae): Effects of habitat fragmentation on visitation and seed set." *Conservation Biology* 2:359–366.

Kevan, P. G. 1975. "Pollination and environmental conservation." *Environmental Conservation* 2, no. 4: 293–298.

Kevan, P. G. 1977. "Blueberry crops in Nova Scotia and New Brunswick: Pesticides and crop reductions." *Canadian Journal of Agricultural Economics* 25, no. 1: 61–64.

Lamont, B. R., P. G. L. Klinkhamer, and E. T. F. Witkowski. 1993. "Population fragmentation may reduce fertility to zero in *Banksia goodii*: A demonstration of the Allee effect." *Oecologia* 94, no. 3:446–450.

Lesica, P. 1993. "Loss of fitness from pollinator exclusion in *Silene spaldingii* (Caryophyllaceae)." *Madrono* 40, no. 4: 193–201.

MacGregor, S. E. 1976. *Insect Pollination of Cultivated Crop Plants*. USDA Agricultural Handbook 496. Washington, D.C.: U.S. Department of Agriculture.

Menke, H. F. 1952. "A six million dollar native bee in Washington state." *American Bee Journal* 92: 334–335.

Nabhan, G. P. 1995. "The dangers of reductionism in biodiversity conservation." *Conservation Biology* 9, no. 3:479–481.

Nabhan, G. P. 1996. *Pollinator Redbook, Volume One: Global List of Threatened Vertebrate Wildlife Species Serving as Pollinators for Crops and Wild Plants*. Tucson: Arizona-Sonora Desert Museum and Forgotten Pollinators Campaign Monographs.

National Academy of Sciences. 1975. *Underexploited Tropical Plants with Promising Economic Value*. Washington D.C.: National Academy Press.

NASDA (National Association of State Departments of Agriculture). 1991. *Honey Bee Pest: A Threat to the Vitality of U.S. Agriculture*. Washington, D.C., pp. 1–14.

Pearce, D., and D. Moran. 1994. *The Economic Value of Biodiversity*. London: Earthscan.

Prescott-Allen, C., and R. Prescott-Allen. 1986. *The First Resource: Wild Species in the North American Economy*. New Haven: Yale University Press.

Prescott-Allen, R., and C. Prescott-Allen. 1990. "How many plants feed the world?" *Conservation Biology* 4, no. 4: 365–374.

Rathcke, B. J., and E. Jules. 1994. "Habitat fragmentation and plant/pollinator interactions." *Current Science* 65: 273–278.

Renner, S. 1995. "Floral rewardlessness in the angiosperms." *Association for Tropical Biology Program and Abstracts*: 14.

Robinson, W. S., R. Nowogrodski, and R. A. Morse. 1989a. "The value of honey bees as pollinators of U.S. crops." *American Bee Journal* 129, no. 6: 411–423; 129, no.7: 477–487.

Robinson, W. S., R. Nowogrodski, and R. A. Morse. 1989b. "Pollination parameters." *Gleanings in Bee Culture*: 148–152.

Roubik, D. W. 1978. "Competitive interactions between neo-tropical pollinators and Africanized honeybees." *Science* 201: 1030–1032.

Roubik, D. W. 1993. "Direct costs of forest reproduction, bee-cycling and the efficiency of pollination needs." *BioScience* 18, no. 4: 537–552.

Roubik, D. W. 1995. *Pollination of Cultivated Plants in the Tropics*. Rome: Food and Agriculture Organization.

Southwick, E. E., and L. Southwick, Jr. 1989a. "Pollination puzzle." *Gleanings in Bee Culture* 498.

Southwick, L., Jr., and E. E. Southwick. 1989b. "A comment on 'Value of honey bees as pollinators of U.S. crops'." *American Bee Journal* 129: 805–807.

Southwick, E. E., and L. Southwick, Jr. 1992. "Estimating the economic value of honey bees (Hymenoptera: Apidae) as agricultural pollinators in the United States." *Economic Entomology* 85, no. 3: 621–633.

Suzan, H., G. P. Nabhan, and D. T. Patten. 1994. "Nurse plant and floral biology of a rare night-blooming cereus, *Peniocereus striatus* (Brandegee) Buxbaum." *Conservation Biology* 8, no. 2: 461–470.

Tepedino, V. J. 1979. "The importance of bees and other insect pollinators in maintaining floral species composition." *Great Basin Naturalist* 3:139–151.

Washitani, I., R. Osawa, H. Nimai, and M. Niwa. 1994. "Patterns of female fertility in heterostylous *Primula seiboldii* under severe pollinator limitation." *Journal of Ecology* 82: 571–575.

Watanabe, M. E. 1994. "Pollination worries rise as honey bees decline." *Science* 265: 1170.

Wilson, E.O. 1992. *The Diversity of Life.* Cambridge: Harvard and Belnap Press.

Chapter 9

Natural Pest Control Services and Agriculture

Rosamond L. Naylor and Paul R. Ehrlich

Natural pest control services maintain the stability of agricultural systems worldwide and are crucial for food security, rural household incomes, and national incomes in many countries. These services include the control of pests by their natural enemies—predators, parasites, and pathogens—and climatic-related controls affecting crop rotations, fallows, and flooding that interrupt herbivorous pest reproduction cycles and help to constrain competition by noncrop plants. Natural pest controls represent an important ecosystem service whose value has been recognized only recently in the gradual move from chemical pest control to integrated pest management in many agricultural regions. Yet application of synthetic pesticides remains the dominant form of pest control by human beings in agriculture, and their increasing use on a global scale is further reducing the viability of natural pest controls.

The ongoing elimination of this ecosystem service through habitat destruction and the intensification of agricultural systems is creating large costs to humanity in the form of foregone agricultural output, increased production instability, and higher input costs. Investments in plant breeding and biotechnology research designed to enhance host plant resistance in agricultural crops are needed now more than ever before and are seriously straining resources in national and international agricultural research systems. Moreover, there is a rising cost to society in the form of direct damages to health and ecosystems from increasing pesticide applications in agriculture.

In this chapter, we discuss both the function and the value of natural pest controls in agriculture. We describe several regional examples of agricultural systems in which the ecosystem service has broken down (at least temporarily). The loss of the ecosystem service has threatened food security in some of the cases and has resulted in high costs to farmers and to society in virtually all of them. We then estimate the aggregate value of the ecosystem service on a global scale, in terms both of the benefits it provides and the costs of replacing it when natural pest controls fail. Although the estimate is very rough, it provides a basis for identifying further data needs and for promoting a policy dialogue on pest management and the preservation of natural pest controls in agriculture.

Background on Pest Control

The agricultural revolution, some ten thousand years ago, set humanity on the road to cities, science, the industrial revolution, and total domination of the planet. In the process, many natural ecosystems have been transformed into agricultural ecosystems—those managed by *Homo sapiens* to maximize the production of desired plant or animal products. Today, about 10 percent of earth's productive land surface is used as croplands, and another 30 percent is permanent pastures (FAO 1994). Both cropland and pastures are highly modified from their original state as natural (nonmanaged) ecosystems. Croplands generally have had their plant diversity greatly reduced, and large herbivores (which would compete for the crops) and top predators that depend on them have been widely exterminated. Indeed, many modern agroecosystems can be viewed as simplified ones in which succession is kept in early seral stages, thus preventing takeover by forest, brushland, or perennial grassland that would reduce the production: respiration ratio and with it desired plant yields.

The transformation of agriculture from low-energy, extensive systems to high-energy, intensive systems has progressively worsened the battle between humanity and crop pests. Farmers must now inject increasing amounts of energy into agroecosystems to limit production losses to weeds that compete with crops, to pathogens that spread crop diseases, and to small herbivores that eat crops, such as rodents, insects, fungi, snails, and nematodes. These pests, especially herbivorous insects, are humanity's most important competitors for food and fiber, destroying an estimated 25–50 percent of crops before and after harvest (Pimentel 1979, Pimental et al. 1989). In this century the basic strategy used by *Homo sapiens* to exterminate these agricultural pests has been the large-scale use of poisons. Now roughly 2,500,000 tons of synthetic pesticides are applied annually to crops worldwide, 600,000 tons of which are applied in the United States alone (Pimentel et al. 1993a, 1993b).

It has been a poor strategy. Plants have been evolving chemical defenses against herbivores for millions of years, and herbivores have responded by evolving ways of avoiding or detoxifying the plant poisons. One result of this co-evolutionary "race" (Ehrlich and Raven 1965; Ehrlich 1970) has been to help create a herbivore fauna not easily conquered by the broadcast use of chemicals. Misuse of our chemical weapons has caused pesticide resistance to become a ubiquitous problem. At present more than 500 insect and mite species have evolved resistance to one or more pesticides, more than 100 weeds have become herbicide resistant, and about 150 plant pathogens are resistant to chemical weapons used against them (World Resources Institute 1994). But most important from the perspective of this chapter, the natural enemies of the pests have been decimated. Predatory insects, for example, do not have the same level of evolutionary experience with poisons as do the herbivorous pests. They also, on average, have much smaller populations (a characteristic of food chains, dictated by the second law of thermodynamics, is that a decreasing amount of energy is available to support populations at each successive trophic level), and many depend on substantial prey populations for their persistence. Thus spraying of poisons is much more likely to wipe out the organisms that control the pests than the pests themselves, or to so reduce their predator populations that resurgence of pests can cause considerable damage before control is reestablished.

As a result, use of broad-spectrum pesticides tends to severely damage the natural pest control service of ecosystems. Destruction of predator populations leads to explosions of their prey—the "promotion" of non-pest species to pest status. For instance, in California in the late 1970s, twenty-four of the twenty-five most important agricultural pests had been created by the pesticide industry (National Research Council 1989). Promotion of previously innocuous insects to pest status through misuse of pesticides has long been recognized as powerful evidence of the efficacy of this particular ecosystem service. The Cañete Valley cotton disaster in Peru (discussed below) is a classic case. The scale of the pest control service is difficult to imagine. It has been estimated that some 99 percent of potential crop pests are under natural ecosystem control (De Bach 1974).

Another line of evidence that shows the power of the natural pest control service is the efficacy of natural controls when they are reestablished on an exotic pest. A classic case was the introduction of *Opuntia* cactus into Australia by early settlers (Ehrlich 1986). The *Opuntia* took over some twenty-five million hectares of New South Wales and Queensland by 1925, infesting half of the land so thoroughly that it was unproductive for any form of human use. In its South American homeland, the *Opuntia* was not a pest, and so entomologists looked there for natural enemies. One such enemy, a small moth called *Cactoblastis cactorum,* had larvae that fed on *Opuntia.* The moth was introduced into Australia in 1926, and within five years had removed *Opuntia* from pest status and restrained it to a distribution of scattered clumps ever since. The story of the control of exotic rabbits in Australia by the Myxoma virus is similar (Fenner and Ratcliffe 1965), as is that of the control of the cottony-cushion scale in California by the Vedalia beetle and other parasites (van den Bosch and Messenger 1973).

The intensification of agriculture, now being carried out at an extremely high rate in many developing countries, tends both to encourage pests and to be destructive of natural pest control services. Polycultures have been replaced in many regions by cropping systems in which one crop dominates or by complete monocultures—often with low cultivar diversity. These homogeneous systems present pests with large, genetically uniform targets for attack in which refugia for predators (forest patches, hedgerows) are scarce (Altieri 1994; Gliessman 1990). Simple pressure on the land, and the introduction of irrigation, has led to greater cropping intensities (more crops per year and higher yields per crop) and the skipping of fallow periods that previously had helped to control pest populations (Samways 1994; Heinrichs 1994b). Fertilizer applications, particularly nitrogen, have risen dramatically, providing nutrients for insects and increasing biomass production and canopy density—thus enlarging both the habitat and food supplies of crop pests. Now pests are often able to breed year-round, while important predators such as insectivorous birds may retain restricted breeding seasons (Schulze and Mooney 1994; Altieri 1994).

Moreover, human transport adds important new pests to agroecosystems that do not include their natural enemies. More than two centuries ago the Hessian fly was accidentally brought to North America from Europe to become a plague of wheat farmers. A little more than a decade ago the golden apple snail from the Amazon basin was deliberately introduced to Asia and promptly began competing with farmers for their rice (Halwart 1994). The snail has become a major pest in Asian rice systems and, as discussed below, will continue to cause significant economic damage unless some native or new association natural predator(s) can be introduced.

The value of natural pest control services has increasingly been recognized by agriculturalists as the failures of pesticide use have become more obvious. Insects, weeds, snails, and plant diseases claim about the same share of global crop production today as in the pre-chemical era (World Resources Institute 1994). Even with the expansion of research on host plant resistance, pesticide use has been increasing at a higher rate than agricultural production. Yet without pesticides and host plant resistance in intensive agricultural systems—where natural pest control services have diminished—pest losses might be 30 percent higher than they already are (World Resources Institute 1988).

Successes in "biological control," as in the cases of *Opuntia*, rabbits, and cottony cushion scale, have encouraged scientists to seek ways of using this ecosystem service to their advantage and have led to the development of integrated pest management (IPM). In IPM, a mix of control strategies is employed (van den Bosch and Messenger 1973, Huffaker and Messenger 1976). Efforts are made to maintain pools of natural enemies, pest-resistant crop strains are grown when feasible, mixed-crop rotations, fallowing, and sanitation measures are taken where possible to interrupt pest life cycles and reduce pest habitat, and pesticides are used only when pest populations reach critical sizes despite other measures. In effect, IPM is a strategy mainly designed to reestablish the natural pest-predator balance that has tended to stabilize natural ecosystems and traditional crop production.

Evidence for the Efficacy of Natural Controls

Numerous examples can be found that demonstrate the value of natural pest control services for agricultural output and stability. Some of the most valuable examples relate to intensive production of rice, which contributes to 23 percent of the calories and 16 percent of the protein consumed globally (FAO 1992). Rice production has more than doubled since the Green Revolution that began in the mid-1960s (FAO, various years), yet the vulnerability of intensive rice systems to pest infestations, and hence yield instability, has remained high. Estimates by Cramer (1967), Pathak and Dhaliwal (1981), and Heinrichs (1994a) indicate that insects are responsible for more than a 30 percent yield loss in managed rice systems of Asia, where more than 90 percent of the world's rice is produced and consumed. Modern rice ecosystems, planted with monocultures of uniform varieties throughout the year and provided with high inputs of fertilizers, seem to provide ideal conditions for pest development (Way and Heong 1994).

The Brown Planthopper

The brown planthopper (*Nilaparvata lugens*) has been the most notorious insect pest of the modern Green Revolution in rice (Kenmore et al. 1984; Rombach and Gallagher 1994; Denno and Perfect 1994). It was a major threat to rice cultivation in the 1960s and 1970s, when modern varieties were first disseminated, and is still considered the single most important insect problem in rice today. Cumulative losses from the brown planthopper have been estimated in the hundreds of millions of dollars (Denno and Perfect 1994) and accounted in 1990 for some 6 percent of the total value foregone from all biotic and abiotic factors that reduce rice production (Herdt 1991). By the late 1970s, the value of these losses was already estimated at $300 million (Dyck and Thomas 1979).

Insecticide-induced resurgence is a leading factor contributing to the increase and severity of brown planthopper outbreaks in Asian rice. Nearly every recorded outbreak of brown planthopper in tropical rice systems has been associated with prior use of insecticides and the consequent disruption of population-regulating factors such as natural enemies (Kenmore 1980; Kenmore et al. 1984; Dobel and Denno 1994). Rola and Pingali (1993) show that the worst outbreaks came during the most widespread and intensive use of insecticides in these agricultural systems (1973–76). During the early 1970s, government programs throughout Asia encouraged multiple prophylactic insecticide applications as part of the Green Revolution technology package. Neither the planting of new varieties nor increasing fertilizer levels has been shown to induce brown planthopper outbreaks.

The widespread use of broad-spectrum insecticides has contributed in several ways to brown planthopper resurgence. Foremost, these insecticides kill not only the brown planthopper but also its natural parasites and predators like spiders, crickets, and small hemipterans (Kenmore et al. 1984; Rombach and Gallagher 1994; Dobel and Denno 1994; Ooi and Shephard 1994; Cronin and Strong 1994). The brown planthopper has a faster rate of regeneration and natural increase than most of its natural enemies. Early applications of broad-spectrum insecticides, such as the organophosphates and pyrethroids, have been shown to disrupt the rice arthropod community structure and to favor certain specialized herbivores like the brown planthopper (Cohen et al. 1994). Schoenly et al. (1994) estimated, for example, that broadcast sprays of deltamethrin caused an increase of about four million herbivores per hectare per sampling date. The favoring of herbivorous over predacious arthropods through insecticide use can thus lead to secondary pest outbreaks, where relatively rare insects prior to spraying become primary pest species. During the period of the worst brown planthopper outbreaks, Reissig et al. (1982) found that sixteen of thirty-nine insecticides tested caused brown planthopper resurgence.

The prevalence of the brown planthopper in rice is also highly dependent on the temporal availability of host plants. Intensification of rice production through continuous irrigation and the abandonment of crop rotations in many areas has exacerbated pest outbreaks. In addition, asynchronous planting (the staggering of planting dates in a given rice-growing region) has extended the time available for insects like the brown planthopper to increase and has reduced the length of fallow that they have had to endure. Since the brown planthopper is monophagous, it is more abundant where rice is highly concentrated. It is also constantly coevolving with rice to overcome its chemical and physical defenses (Heinrichs 1994b). Rice varieties that have been bred to have host plant resistance to the brown planthopper have suffered less damage than nonresistant varieties; however, the newly incorporated genes have not been able to remain effective for long periods in most cases (Heong et al. 1995; Bosque-Perez and Buddenhagen 1992). To date, more than fifty thousand accessions have been screened for resistance to the brown planthopper at the International Rice Research Institute (IRRI), and more than four hundred resistant accessions have been identified (Heinrichs 1994a).

Host plant resistance is just one part of a much larger IPM strategy that is now required to control the brown planthopper in rice. Lists of potential biological control agents exist (Rombach and Gallagher 1994), but few studies until recently (see Denno and Perfect 1994) report on their effectiveness or ecological role in controlling the pest. Biological control—the deliberate introduction of native and introduced enemies—has been shown to be effective in isolation but can be destroyed by mismanagement of pesticides and has thus had variable success throughout Asia (Benrey and Lamp 1994).

One of the most notable successes in IPM has been in Indonesia, where in response to brown planthopper resurgence, a presidential decree in 1986 banned fifty-seven of the sixty-six insecticides used on rice (Kenmore 1991; Denno and Perfect 1994; Rombach and Gallagher 1994). Pesticide subsidies as high as 80 percent were phased out over a two-year period, and some of these resources were used instead to support an IPM program. Since that time, Indonesia's rice harvest has risen by over 15 percent, national insecticide use has declined by 60 percent, and more than a quarter of a million farmers have been trained in IPM techniques. Between 1986 and 1990, the economic benefits of IPM were calculated at well over $1 billion in savings to both farmers and the Indonesian treasury.

Despite these efforts, the brown planthopper remains a pest in Indonesian rice production (McBeth 1995). IPM training is expanding but has still reached a minority of rice farmers, and asynchronous planting prevails in many rice-growing regions. Although pesticides are not subsidized by the government, farmers in various locations still spray prophylactically or before pest outbreaks show signs of becoming serious. Brown planthopper outbreaks are much better contained now than in the past, but their continuous appearance suggests that further investments in IPM training and extension services are warranted.

The Golden Apple Snail

The value of natural pest control services in rice agriculture is also reflected in the damage that has resulted from the importation of the golden apple snail (*Pomacea canaliculata*) into Asia from South America (Naylor 1996). The golden apple snail was introduced intentionally in 1980 with the expectation that it could be cultivated as a high-protein food source for local consumption and as an export commodity for high-income countries in Europe and North America. It has since invaded Asian rice systems, where it is dispersed through irrigation networks and feeds voraciously on young rice seedlings in the absence of its natural predators.

Infestations of the golden apple snail during the past fifteen years have been reported in Taiwan, Japan, the Philippines, China, South Korea, Malaysia, Thailand, Indonesia, Vietnam, Laos, and Papua New Guinea (Mochida 1991; FAO 1993; Halwart 1994). In virtually all of these cases, the snail was introduced intentionally into cement tanks, maintained ponds, and backyard soil pits; however, a combination of low market value and negligence has resulted in the release and escape of the snail into irrigation ditches and public waterways throughout the regions into which it was introduced. Since its release, the snail has fed on a wide range of aquatic plants of economic value to Asian farmers, including young rice seedlings, taro, swamp cabbage, lotus, mat rush, Chinese mat grass, wild rice, Japanese parsley, water chestnuts, and azolla (Mochida 1988, 1991; Adalla and Morallo-Rejesus, 1989). In addition, it has damaged maize (Cruz 1987) and citrus (Buendia 1988). By far the greatest damage has occurred in irrigated rice ecosystems, which provide an ideal environment for the dispersal and growth of the snail (Mochida 1988, 1991).

Unfortunately, there are little reliable data on the extent of infestation, crop damage, and yield loss in most rice-growing regions where the snail has been introduced. Naylor (1996) provides estimates of the economic costs of the snail in Philippine rice systems, where a wide range of field-survey data have been collected. The total direct damage in 1990 based on the market costs of yield loss, replanting, and typical control measures (such as hand-picking and the use of molluscicides and insecticides) was estimated to be $25–45 million. The cumulative economic damage, calculated on a present value basis assuming discount rates of 5 percent, 10 percent, and 15 percent, ranged from $425 million to almost $1.2 billion. The latter estimate is conservative, because it assumes that snail infestations are held at their 1990 level into the future.

Like most introductions of exotic species, the golden apple snail invasion occurred in human-altered and simplified ecosystems, presumably with a smaller number of species that might be predators or competitors of the snail than in the natural, undisturbed ecosystems they replaced. Local predators of the golden apple snail in Asian rice are limited mainly to those that are relatively omnivorous, including assorted rats, snakes, birds, frogs, dragonflies, spiders, and ants (Madambaand Camaya 1987; Acosta and Pullin 1989). Its main competitor, the native "kuhol" snail, is not closely related, and therefore the natural predators of snails in this system have not been effective at controlling the exotic golden apple snail (Acosta and Pullin 1989). Moreover, the golden apple snail reproduces about ten times faster than the native species and eats a varied diet of aquatic plants much more voraciously than the native snail (Adalla and Morallo-Rejesus 1988; Mochida 1989, 1991). In many areas where the golden apple snail has been introduced, native snail populations and their natural predators are declining.

Efforts are now under way to identify natural enemies from South America or new association natural enemies that might be used to control golden snail infestations in Asia (Halwart 1995), just as *Cactoblastis cactorum* was used to control *Opuntia* in Australia. In the meantime, an attempt is being made to stabilize golden snail populations in some agroecosystems by handpicking, applying molluscicides and insecticides, pasturing ducks, and raising fishes in paddies (ducks and certain fishes feed directly on the snail) (Halwart 1993; Naylor 1996; Warburton and Pingali 1993a, b). These methods are effective at reducing snail populations only at certain periods of their life cycle, however, and thus have limited success in controlling overall population growth. It is clear that large-scale economic damage from the snails will persist until some semblance of the natural pest control ecosystem service is established.

Cañete Valley Cotton Disaster

Natural pest control has a high value in commercial nonfood crops as well as food crops, as is clearly demonstrated in the case of cotton. Cotton is one of the world's most important commodities in terms of output and value, particularly in developing countries where roughly two-thirds of the world's cotton is grown (FAO 1994). It is also one of the most chemical-intensive crops, consuming about 10 percent of pesticides worldwide (Conway and

Pretty 1991). The experience with cotton and pesticides in the Cañete Valley of Peru represents a classic example of pest resurgence and the value of natural pest management. This episode, which began in the late 1940s shortly after the introduction of DDT launched an era of synthetic pesticides throughout the world, should have provided ample warnings to the global community of overuse and misuse of pesticides in agriculture.

Problems in the Cañete Valley began when sugar and other crops were replaced by a large-scale cotton production monoculture that was made possible by the expansion of irrigation in the 1930s and 1940s (De Loach 1970; Barducci 1972). Cotton was grown on up to 90 percent of the cultivated area of the valley in any single year, severely stressing the ability of natural enemies to control the pest load. Widespread insecticide use began in the late 1940s on virtually all of the area, which led to the typical pesticide treadmill. Chemical treatment for the tobacco budworm led to a rapid build-up of the cotton aphid. Resistance of the cotton aphid to insecticides developed, causing chemical applications to increase and a number of new, serious insect pests (such as leafworms, leafrollers, mealybugs, and bollworms) to emerge (Repetto 1985). By the mid-1950s, eight or nine pesticide applications per season were recommended to farmers, and up to twenty-one applications per season were used on large plantations (Barducci 1972).

Between 1950 and 1954, cotton yields increased from 494 kg/ha to 728 kg/ha. Cotton farmers concluded that if more pesticides were applied, yields would increase even further (Ehrlich et al. 1977). Resistance developed at that time, and yields fell by over 50 percent to 332 kg/ha in 1956 (Barducci 1972). Farmers realized that they simply could no longer control pests with synthetic organic pesticides. As a result, these chemicals were replaced with much greater use of biological control agents. Pesticide applications declined to 2.35 per crop season on average in 1957–58, resulting in large cost savings to farmers as well as yield gains. By 1963, cotton production had increased to 800 kg/ha. Subsequent research has identified the many predator insect species that had controlled potential pest outbreaks in the Cañete Valley prior to widespread applications of broad-spectrum pesticides (Braden 1979; Dover and Croft 1984; Barducci 1972).

The challenge in introducing and maintaining natural pest management in the Cañete Valley has been twofold: to educate individual farmers in IPM, and to ensure that all farmers in the valley are jointly pursuing an IPM strategy. As in the case with the brown planthopper in monoculture rice systems, monophagous pests produced from insecticide use by one cotton grower can become a potential threat to other farmers with crops at the susceptible growth stages.

The story of the Cañete Valley has been replayed in several other cotton-growing regions (World Resources Institute 1994). Between the 1950s and the 1970s, cotton growers in Central America increased the average number of insecticide applications per crop season from ten to forty; rising costs and declining yields due to pest resurgence, resistance, and secondary pest outbreaks forced many farmers into bankruptcy. Similarly, in northeast Mexico, resistance of tobacco budworms, a secondary pest that emerged in an effort to eradicate the boll weevil, devastated the cotton industry. Cotton crop acreage was reduced from 300,000 hectares in the 1960s to 500 hectares in 1970, forcing many cotton workers to leave the region. IPM is now practiced in many of the more productive cotton regions, including the Imperial Valley and the San Joaquin Valley of California, both of which have suffered large losses from primary and secondary pest infestations during the past thirty years (Harper and Zilberman 1989; Burrows 1983).

Bavarian Hedgerows

Previous examples suggest that crop monocultures are difficult environments in which to have effective natural pest control, either because the predator-pest balance is disrupted by excessive pesticide use, or because the system lacks adequate resources to maintain natural enemies. Annual crop monocultures often do not provide adequate sources of food (e.g., pollen, nectar, prey), shelter, breeding opportunities, and nesting sites for the effective performance of predators, which tend to be polyphagous and thus have broad habitat requirements (Altieri 1994; Rabb et al. 1976; Root 1975). A number of studies document the importance of adjoining wild vegetation in providing alternative food and habitat to natural enemies that move into nearby crops (van Emden 1965; Wainhouse and Coaker 1981; Altieri and Schmidt 1986). Hedgerows, for example, provide polyphagous predators with alternative sources of food, such as plant flowers, aphids, and other herbivores, and thus increase their chances of survival and effective pest management in intensive agricultural regions or in areas where fallow periods reduce access to pest herbivores.[1]

The role of Bavarian hedgerows in agriculture serves as an excellent illustration of the natural pest control service. Hedgerows are the most diverse woody vegetation in Germany, containing some thirty woody species, several of which are insect-pollinated (Schulze and Gerstberger 1994). They thus serve as a major habitat for herbivorous insects and aphids; the degree of herbivory in these systems is, in fact, larger than in any other community of woody species in Central Europe (Zwolfer and Stechmann 1989). These insects support a large number of predators and parasites in their natural setting. At the time when the grain crop is developing in the surrounding fields, the natural enemies are present and control the potential damage of aphids on the cereals. As a result, northeast Bavaria is one of the few regions in Germany where spraying pesticides against wheat aphids is not necessary.

The service that Bavarian hedgerows provide is increasingly vulnerable, however, to changes in agricultural practices. Modern machinery has required larger fields; wood from the hedgerows is no longer used for cooking or local woodcraft; and other food sources, such as berries, are no longer collected from the hedgerows. These changes have led to an escalating eradication of hedgerows in recent decades that is disrupting the natural pest-predator balance (Schulze and Gerstberger 1994). Moreover, the service that they have provided is largely irreplaceable on a time scale of decades. Zwolfer et al. (1984) have shown that, because populations of insects in hedgerows are very stable and local, hedgerows that were newly planted in open fields did not contain the full set of herbivores and predators even after forty years of growth. Without the natural pest control service, damage to cereals and the consequent use of pesticides almost surely will increase in the region.

Value of the pest control Service

The individual cases described in the previous section indicate that the loss in natural pest control services can result in damages amounting to millions of (current) dollars in a given crop cycle, and potentially billions of dollars on a cumulative basis when a pest incident persists. Moreover, the substitution of synthetic pesticides for natural pest controls can result in pest resurgence and secondary pest outbreaks that reduce the fundamental stability of agricultural systems. Nonetheless, the use of pesticides permits further intensification and growth in crop output as pest loads increase and predator habitats decrease. The costs that society is willing to pay to replace the diminishing natural pest control service with synthetic pesticides and other measures, such as host plant resistance, hand weeding, and flooding, provide a lower-bound estimate of the value of the ecosystem service.

Replacement Values

Global sales in pesticides have increased from $2.7 billion in 1970 to over $21 billion by the mid-1990s (World Resources Institute 1994). The distribution of expenditures on pesticides depends on the susceptibility of crops to pests and the cosmetic value of crops; for example, cotton and rice are intensive users of pesticides, whereas wheat, which is less vulnerable to pests, is not an intensive user in spite of its dominance in many cropping systems (USDA 1994). Woodburn (1993) estimates that the world market for pesticides used on rice is upwards of $3 billion—the highest among all food crops. Because of the added expense to farmers, pesticide use per crop area remains the largest on high-valued crops like fruits and vegetables, although these crops constitute a much lower share of global agricultural area than the staple grains.[2]

There has been an additional cost to society of pesticide applications in many countries where the chemicals have been subsidized. Repetto (1985) shows that total subsidies in developing countries range from 15–90 percent of the retail cost of pesticides, with the annual value running into the hundreds of millions of dollars in some cases.[3] These estimates are based on both direct and indirect subsidies, the latter including: below-market interest for loans on pesticides through state-controlled banks; low prices on imported chemicals due to over-valued exchange rates; and tax advantages to agrochemical companies on the import, sale, and distribution of pesticides. Based on conservative assumptions that poor countries consume one-third of the synthetic pesticides used globally (World Resources Institute 1994) and that they receive a median subsidy of 40 percent (Repetto 1985), the full (nonsubsidized) value of pesticide sales worldwide is estimated to be roughly $25 billion.

Another important component of the pesticide cost that society has had to bear is the indirect damages associated with their use. These damages include negative health effects (worker safety, exposure, and poison residues in food and drinking water), ecosystem effects, and the development of resistance to the chemicals in vectors of human disease such as malaria. A small share of pesticides applied on a given field—less than 0.1 percent for many insecticides—actually reaches the target pest (World Resources Institute 1994; Pimentel and Levitan 1988). The rest, by definition, becomes an environmental contaminant. Pesticide losses into the groundwater, surface water, atmosphere, and soils have a broad spectrum of damaging effects on wildlife, plant life, soil and water organisms, and humans. Some of the compounds, such as DDT and dieldrin, can persist in the environment for decades after their use; residues of DDT, for example, have yet to disappear and are found as far away from intensive agriculture regions as the North Pole (Graedel et al. 1993). Furthermore, many of the residues are hormone mimics or immunosuppressants, contributing to a potentially massive public health problem (Colborn et al. 1996).

The health costs of pesticide use include carcinogenic, reproductive, and immune-system damages, external damages to the skin and eyes, and acute toxicity (Repetto and Baliga 1996; Pingali and Roger 1995). The World Health Organization estimates that as many as 25 million people suffer health damages from pesticides each year (Jeyaratnam 1990). Within this category, more than 3 million people receive a very high level of pesticide exposure, and approximately 220,000 die from pesticide poisoning annually. The number of deaths in fact may be much higher, since it is believed that most pesticide poisonings are unreported. Over 90 percent of the pesticide poisonings and 99 percent of pesticide-poisoning deaths reported each year occur in the developing world. Many pesticides that have been banned in industrial countries are now used in developing countries, and there is a great deal of misuse in application, storage, handling, and disposal (World Resources Institute 1994, 1988; Rola and Pingali 1993). For example, recent farm surveys in the Philippines and Vietnam indicated that almost 40 percent of the insecticides used were in the "highly or extremely" hazardous chemical group belonging to WHO category I, and up to 80 percent of the insecticide sprays were being applied inappropriately to the targeted pests (Heong et al. 1995).

The ecosystem and health damages of global pesticide use cannot be calculated precisely—they are extremely wide ranging, both acute and chronic in their effects, and have no established market value. It seems clear, however, that the indirect costs are at least as great as the direct costs of pesticide consumption given that the vast majority of the chemicals do not actually reach their target pest. A conservative estimate of $25 billion for external damages can thus be used in the calculation of the replacement value of natural pest control services in addition to the $25 billion in direct pesticide costs.

The ongoing effort to replace natural pest control services with human controls also involves investment costs for plant breeding research on host plant resistance. In this case, the costs to society have had positive returns. Host plant resistance has played an important role both in raising yield stability and reducing the need for pesticides in many intensive agriculture systems (Panda and Khush 1995; Khush 1994; Heinrichs 1994a; Bosque-Perez and Buddenhagen 1992). Each year the international agricultural research centers[4] spend about $70 million on germplasm enhancement and breeding (CGIAR 1995). This research involves multiple-trait breeding, but most of it is associated with host plant resistance in one way or another. In total, therefore, the Consultative Group on International Agricultural Research (CGIAR) spends some $50 million directly or indirectly each year on host-plant resistance (McCalla 1995). Within the CGIAR, this is not a trivial number. An estimated 46 percent of rice breeding efforts are devoted to insect resistance (Heinrichs 1994a), and almost 30 percent of wheat breeding research is focused on pest resistance (Byerlee 1994).

Moreover, research expenditures by the CGIAR represent only a small fraction—2 percent or less—of global research funds spent by the private sector and national agricultural research institutions on plant breeding. With roughly $15 billion spent on global agricultural research each year (Anderson et al. forthcoming), an estimated $4 billion is spent annually on research related to host plant resistance.

Assuming that pesticide applications and host plant resistance are the two main tools for controlling pests in intensive, largely monoculture systems, then the total cost of efforts to replace natural pest control services in these systems is estimated at $54 billion annually.[5] This is a conservative calculation, because it excludes the costs of several other human controls, such as hand weeding, early season flooding in irrigated systems to suppress weed growth, and hand collection of pests (such as in the case of the golden apple snail above). Given the high ecosystem value of scarce water supplies (see chapter 11) and the rising real cost of labor in many regions of the world, these alternative forms of control will become increasingly expensive in the future.

Marginal Values

Annual replacement values of $54 billion or more underscore the importance of preserving this ecosystem service. At the margin, each additional unit of natural pest control lost in agriculture undoubtedly has a high cost. Calculating the marginal cost is virtually impossible, however, due to the difficulty in identifying a baseline and measuring a unit change in the natural pest control service. Given the inherent simultaneity, or two-way causality, between pesticide use and the loss of natural controls, it may be more insightful to turn the question of marginal values on its head and ask: what are the marginal benefits to society from an incremental reduction in pesticide use?

Breaking the positive feedback loop between pesticide use and the decline in natural predators, in particular, requires the introduction of a variety of cultural practices embedded in an IPM strategy, such as crop rotations, fallows, and the use of biological controls. The Indonesian example of implementing an IPM program to control the brown planthopper (discussed above) certainly demonstrates the marginal benefits of reducing pesticide use and reestablishing the natural pest control service. In that case, the benefits to a single country for a single crop exceeded $1 billion in a four-year period. Based on the magnitude of this result, one can only project that replacing pesticides with natural pest controls on a global scale would lead to marginal benefits in the tens of billions of dollars annually.

Valuing Nonmarginal Losses

At the limit—that is, in the absence of natural controls all together—the 60–70 percent of food production that is not currently destroyed by pests would be. This conclusion is suggested by the near total destruction of some crops in the absence of human controls (McEwen 1978). Examples include the complete loss of rice production to hopperburn in areas severely infested by the brown planthopper, such as Guadalcanal in the Solomon Islands and certain locations in Java and Sumatra of Indonesia (Rombach and Gallagher 1994), by the takeover of huge areas of Australia by *Opuntia*, and so on. In the absence of controls, natural or anthropogenic, herbivores will expand their populations until they consume their resources (e.g., Klein 1968).

It seems likely, therefore, that in the absence of natural controls, at least commercial crops, and probably all crop production, would become impossible. The number of herbivores attacking crops would be greatly multiplied, as numerous benign species would be "promoted" to pest status. In addition, attempts at using more pesticide applications to substitute for lost natural controls would quickly make most pests highly resistant. Few, if any, crops would lack at least one completely uncontrollable pest.

In this worst-case scenario—that is, if all crops were lost—the entire worth of today's crop production, estimated at roughly $1.4 trillion dollars, would be the market value of this ecosystem service in 1995. The total value of the service, including market and nonmarket costs, would of course be infinite, since humanity would be eliminated by such an outcome. A more conservative estimate involves removing such factors as winter or dry season fallows from the category of natural controls and assuming (again, perhaps, conservatively) that crops followed by long fallows could produce about half of their yield before pest populations would recover sufficiently to totally destroy them. Thus, if the global cropping index (number of crops per year on a given piece of land) were to fall from an estimated 1.3 to 1.0, yields were to fall to half that with other natural or anthropogenic controls, and prices were to remain constant, the value of the loss would be in the order of $540 billion.

But obviously prices would not remain constant if yields were to fall by one-half and the cropping index were to decline. Prices would rise substantially, increasing both the value of the ecosystem service and the return on investments to replace natural pest controls. For example, investments in biotechnology research on resistance breeding, which is already well under way (Panda and Khush 1995), would likely increase, as would investments in research on biological controls, allelopathy, and pesticides. The dynamics of the system on both the production and the consumption sides are impossible to predict. The only certainties in this case are that food security would be threatened; poor consumers, who have few substitution possibilities among staple foods, would be at a much greater risk of hunger; and farm management practices would have to be altered to take advantage of existing natural pest control services.

Conclusion

Valuations of the global pest control ecosystem service ranging between $54 billion and $1 trillion are uncomfortably large. It is clear that many of the estimates we have made or accepted are extremely crude—losses to pests and indirect costs of pesticide use, for example. Anyone wishing to make a more precise estimate, however, faces immense conceptual and empirical problems even for a one-year snapshot. Moreover, assumptions about the dynamics of these systems through time will likely cause the upper and lower bounds to diverge rather than converge. Whatever the "precise" estimate, there is no question that the value of natural pest control services is extremely large by any standard and deserving of much more attention than it is currently receiving.

We question, in fact, whether there would be any utility to making a more precise estimate. The dollar value of a service that permits humanity to avoid famines in which millions or perhaps billions of people would die is only worth calculating to impress those to whom all values can be made monetary. Furthermore, many other ecosystem services are involved in providing *Homo sapiens* with food security—soil generation and preservation, supply of fresh water, climate amelioration, and pollination, to name a few. Therefore, placing a monetary value on pest control services alone faces a nearly insuperable problem of double-counting.

Indeed, we believe that the best summary of the value of the ecosystem services supporting agriculture was a comment made by a distinguished economist after a seminar at the Federal Trade Commission some years ago. The speaker had denigrated the potential economic impact of global warming by pointing out that agriculture and forestry "accounted for only 3 percent of the U.S. gross national product." The comment was: "What does this genius think we're going to eat?"

Acknowledgments

The authors would like to thank Drs. Walter Falcon, Donald Kennedy, and Gretchen Daily for their helpful comments, and The Pew Charitable Trusts for supporting this research through the Pew Scholars Program in Conservation and the Environment.

Notes

1. It should be noted that hedgerows may also be a source of pests and are therefore not entirely risk free in terms of pest management.

2. Global arable and permanent cropland in 1991 was 1.4 billion hectares; roughly 50 percent of this total was composed of cereals (USDA 1993).

3. These estimates are based on a study of nine countries: Indonesia, China, Pakistan, Ghana, Senegal, Egypt, Honduras, Ecuador, and Colombia. This set of countries was chosen for its diversity; the countries have very different agricultural and economic systems, and they span a large geographic range. The median subsidy was 44 percent of retail costs.

4. These centers are under the Consultative Group for International Agricultural Research (CGIAR) and are composed of institutions such as International Rice Research Institute (IRRI) and Centro Internacional de Mejoramiento de Maiz y Trigo (CIMMYT).

5. In this estimate, the natural pest control services are assumed to include cultural practices such as crop diversification and the maintenance of fallows. The value of these practices will be discussed below.

References

Acosta, B., and R. Pullin. 1989. *Summary Report of a Workshop on the Environmental Impact of the Golden Snail (Pomacea Sp.) on Rice Farming Systems in the Philippines.* Manila, Philippines: International Center for Living Aquatic Resources Management.

Adalla, C. B., and B. Morallo-Rejesus. 1989. "The golden apple snail, Pomacea Sp.: a serious pest of lowland rice in the Philippines." In I. Henderson, ed., *Slugs and Snails in World Agriculture.* Thorton Heath: British Crop Protection Council, pp. 417–422.

Altieri, M. A. 1994. *Biodiversity and Pest Management in Agroecosystems.* New York: Food Products Press and Howorth Press.

Altieri, M. A., and L. L. Schmidt. 1986. "The dynamics of colonizing arthropod communities at the interface of abandoned organic and commercial apple orchards and adjacent woodland habitats." *Agriculture,Ecosystems, and Environment* 16: 29–43.

Anderson, J. R., P. G. Pardey, and J. Roseboom. Forthcoming. "Sustaining growth in agriculture: A quantitative review of agricultural research investments." *Agricultural Economics* 10(2): 107–123.

Barducci, T. B. 1972. "Ecological consequences of pesticides used for the control of cotton insects in Cañete Valley, Peru." In M. T. Farvar and J. P. Milton, eds., *The Careless Technology: Ecology and International Development.* Garden City, New York: Natural History Press, pp. 423–438.

Benrey, B., and W. Lamp. 1994. "Biological control in the management of planthopper populations." In R. E. Denno and T. J. Perfect, eds., *Planthoppers: Their Ecology and Management.* New York: Chapman & Hall, pp. 519–550.

Bosque-Perez, N. A., and I. W. Buddenhagen. 1992. "The development of host-plant resistance to insect pests: Outlook for the tropics." *Proceedings of the 8th International Symposium on Insect-Plant Relationships.* Boston: Kluwer Academic Publishers, pp. 235–249.

Braden, L. 1979. "Integrated pest management in the developing world." *Annual Review of Entomology* 24: 225–254.

Buendia, L. M. 1988. "The golden apple snail: A rice menace." *Monitor* 16: 6.

Burrows, T. M. 1983. "Pesticide demand and integrated pest management: a limited dependent variable analysis." *American Journal of Agricultural Economics* 65, no. 4 (November): 806–810.

Byerlee, D. 1994. "Modern varieties, productivity and sustainability: Recent experience and emerging challenges." Mexico: Centro Internacional de Mejoramiento de Maiz y Trigo (CIMMYT).

CGIAR (Consultative Group on International Agricultural Research). 1994. *CGIAR Annual Report, 1993–1994.* Washington, D.C.: World Bank.

Cohen, J. E., K. Schoenly, K. L. Heong, H. Justo, G. Arida, A. J. Barrion, and J. A. Litsinger. 1994. "A food-web approach to evaluating the impact of insecticide spraying on insect pest population dynamics in a Philippine irrigated rice ecosystem." *Journal of Applied Ecology* 31(4): 747–763.

Colburn, T., D. Dumanoski, and P. Myers. 1996. *Our Stolen Future.* New York: Dutton Press.

Conway, G., and J. Pretty. 1991. *Unwelcome Harvest: Agriculture and Pollution.* London: Earthscan Press.

Cramer, H. H. 1967. *Plant Protection and World Crop Production.* Leverkusen, Germany: Farbenfabriken Bayer AG.

Cronin, J. T., and D. R. Strong. 1994. "Parasitoid interactions and their contribution to the stabilization of auchenorrhyncha populations." In R. E. Denno and T. J. Perfect, eds., *Planthoppers: Their Ecology and Management.* Chapman & Hall, New York, pp. 400–428.

Cruz, R. 1987. "Golden snail infests rice and corn fields in six towns." *Manila Journal* (September 9): 4.

De Bach, P. 1974. *Biological Control by Natural Enemies.* London: Cambridge University Press.

De Loach, C. J. 1970. "The effect of habitat diversity on predation." *Proceedings from the Tall Timbers Conference on Ecological Animal Control by Habitat Management,* no. 2 (February 26–28): 223–241.

Denno, R. F., and T. J. Perfect, eds. 1994. *Planthoppers: Their Ecology and Management.* New York: Chapman & Hall.

Dobel, H., and R. Denno. 1994. "Predator-planthopper interactions." In R. E. Denno and T. J. Perfect, eds., *Planthoppers: Their Ecology and Management.* New York: Chapman & Hall, pp. 325–399.

Dover, M., and B. Croft. 1984. *Getting Tough: Public Policy and the Management of Pesticide Resistance.* Washington, D.C.: World Resources Institute.

Dyck, V. A., and B. Thomas. 1979. "The brown planthopper problem." In *Brown Planthopper: Threat to Rice Production in Asia.* Los Banos, Philippines: International Rice Research Institute, pp. 3–17.

Ehrlich, P. 1970. "Coevolution and the biology of communities." In K. L. Chambers, ed., *Biochemical Coevolution.* Corvallis, Oregon: Oregon University Press, pp. 1–11.

Ehrlich, P. 1986. *The Machinery of Nature.* New York: Simon and Schuster.

Ehrlich, P., and P. Raven. 1965. "Butterflies and plants: A study in coevolution." *Evolution* 18: 586–605.

Ehrlich, P., A. Ehrlich, and J. Holdren. 1977. *Ecoscience: Population, Resources, Environment.* San Francisco: W. H. Freeman.

FAO (Food and Agricultural Organization). Various years. *Production Yearbook.* Rome: FAO.

FAO (Food and Agricultural Organization). 1992. *Food Balance Sheets* (on tape). Rome: FAO.

FAO (Food and Agricultural Organization). 1993. "Golden apple snail present in Papua New Guinea." *Plant Protection Bulletin* 41, no. 1: 35–36.

FAO (Food and Agricultural Organization). 1994. AGROSTAT.PC., version 3.0.

Fenner, F., and F. N. Ratcliffe. 1965. *Myxomatosis.* Cambridge, England: Cambridge University Press.

Gliessman, S., ed. 1990. *Agroecology: Researching the Ecological Basis for Sustainable Agriculture.* New York: Springer-Verlag.

Graedel, T. E., T. S. Bates, A. F. Bouwman, D. Cunnold, J. Dignou, I. Fung, D. J. Jacob, B.K. Lamb, and J. A. Logan. 1993. "A compilation of inventories of emissions to the atmosphere." *Global Biogeochemical Cycles* 7(1): 1–26.

Halwart, M. 1993. "Fish in rice fields." In P. P. Milan and J. Margraf, *Philippine Freshwater Ecosystems*, pp. 54–69. Leytes, Philippines: VISCA-gtz Tropical ecology Program.

Halwart, M. 1994. "The golden apple snail *Pomacea canaliculata* in Asian rice farming systems: Present impact and future threat." *International Journal of Pest Management* 40(2): 199–206.

Halwart, M. 1995. Personal communication. Fishery Resources Officer, Inland Water Resources and Aquaculture Service, Food and Agricultural Organization, Rome.

Harper, C R., and D. Zilberman. 1989. "Pest externalities from agricultural inputs." *American Journal of Agricultural Economics* 71(3): 692–702.

Heinrichs, E. A. 1994a. "Host Plant Resistance." In E. A. Heinrichs, ed., *Biology and Management of Rice Insects*, pp. 517–548. London: Wiley, and Los Banos, Philippines: International Rice Research Institute.

Heinrichs, E. A., ed. 1994b. *Biology and Management of Rice Insects.* London: Wiley, and Los Banos, Philippines: International Rice Research Institute.

Heong, K. L., P. S. Teng, and K. Moody. 1995. "Managing rice pests with less chemicals." *GeoJournal* 35(3): 337–349.

Herdt, Robert. 1991. "Research priorities for rice biotechnology." In G. S. Khush and G. H. Toenniessen, eds., *Rice Biotechnology*, pp. 19–54. Wallingford, England: CAB International.

Huffaker, C. B., and P. S. Messenger, eds. 1976. *The Theory of Biological Control.* New York: Academic Press.

Jeyaratnam, J. 1990. "Acute pesticide poisoning: A major global health problem." *World Health Statistics Quarterly* 43: 139–143.

Kenmore, P. E. 1980. "Ecology and outbreaks of a tropical insect pest of the Green Revolution, the rice brown planthopper, *Nilaparvata lugens*." Ph.D. dissertation. University of California, Berkeley.

Kenmore, P. E., F. O. Carino, C. A. Perez, V. A. Dyck, and A. P. Gutierrez. 1984. "Population regulation of the rice brown planthopper, *Nilaparvata lugens,* within rice fields in the Philippines." *Journal of Plant Protection: Tropics* 1: 19–37.

Kenmore, P. E. 1991. *Indonesia's Integrated Pest Management—A Model for Asia.* FAO Rice IPC Programme, FAO, Manila, Philippines.

Khush, G. S. 1994. "Modern varieties—their real contribution to food supply and equity." *GeoJournal* 35(3): 275–284.

Klein, D. R. 1968. "The introduction, increase, and crash of reindeer on St. Matthew Island." *Journal of Wildlife Management* (32): 350–367.

Madamba, C., and E. Camaya. 1987. "The golden snail: Boon or bane." *Proceedings of the 18th Anniversary and Annual Convention of the Pest Control Council of the Philippines*, vol. 1, pp. 42–50.

McBeth, John. 1995. "Grain games; grain pains." *Far Eastern Economic Review.* June 29: 63–64.

McCalla, A. 1995. Personal communication. Director, Agricultural and Rural Development Department, The World Bank.

McEwen, F. L. 1978. "Food production: The challenge for pesticides." *BioScience* 28: 773–784.

Mochida, O. 1988. "Nonseedborne rice pests of quarantine importance." In *Rice Seed Health*, pp. 117–129. Los Banos, Philippines: International Rice Research Institute.

Mochida, O. 1991. "Spread of freshwater *Pomacea* snails (*Pilidaae, Mollusca*) from Argentina to Asia." *Micronesica Supplement* 3: 51–62.

National Research Council. 1989. *Alternative Agriculture.* Washington, D.C.: National Academy Press

Naylor, R. 1997. "Invasions in agriculture: The cost of the golden apple snail in Asia." *AMBIO* (in press).

Ooi, P. A. C., and B. M. Shephard. 1994. "Predators and Parasitoids of Rice Insect Pests." In E. A. Heinrichs, ed., *Biology and Management of Rice Insects*, pp. 585–612. London: Wiley, and Los Banos, Philippines: International Rice Research Institute.

Panda, N., and G. S. Khush. 1995. *Host Plant Resistance to Insects*. Wallingford, England: CAB International.

Pathak, M. D., and G. S. Dhaliwal. 1981. "Trends and strategies for rice insect problems in tropical Asia." *IRRI Rice Research Paper Series*, no. 64. International Rice Research Institute, Los Banos, Philippines.

Pimentel, D. 1979. "Benefits and costs of pesticide use in United States food production." *BioScience* 28:778–784.

Pimentel, D., and L. Levitan. 1988. "Pesticides: Where do they go?" *Journal of Pesticide Reform* 7(4): 2–5.

Pimentel, D., L. McLaughlin, A. Zepp, B. Lakitan, T. Kraus, P. Kleinman, F. Vancini, W. Roach, E. Graap, W. Keeton, and G. Selig. 1989. "Environmental and economic impacts of reducing U.S. agricultural pesticide use." *Handbook of Pest Management in Agriculture*: 223–278.

Pimentel, D., H. Acquay, M. Biltonen, P. Rice, M. Silva, J. Nelson, V. Lipner, S. Giordano, A. Horowitz, and M. D'Amore. 1993a. "Assessment of environmental and economic impacts of pesticide use." In D. Pimentel and H. Lehman, eds., *The Pesticide Question: Environment, Economics, Ethics*, pp. 47–84. London: Chapman and Hall.

Pimentel, D., C. Kirby, and A. Shroff. 1993b. "The relationship between cosmetic standards for foods and pesticide use." In D. Pimentel and H. Lehman, eds., *The Pesticide Question: Environment, Economics, Ethics*, pp. 85–105. London: Chapman and Hall.

Pingali, P., and P. Roger, eds. 1995. *Impact of Pesticides on Farmer Health and the Rice Environment*. Boston: Kluwer Academic Publishers.

Rabb, R. L., R. E. Stinner, and R. van den Bosch. 1976. "Conservation and Augmentation of Natural Enemies." In C. B. Huffaker and P. Messenger, eds., *Theory and Practice of Biological Control*, pp. 233–254. New York: Academic Press.

Reissig, W. H., E. A. Heinrichs, and S. L. Valencia. 1982. "Insecticide induced resurgence of the brown planthopper, *Nilaparvata lugens,* on rice varieties with different levels of resistance." *Environmental Entomology* 11, no. 1: 165–168.

Repetto, R., and S. Baliga. 1996. *Pesticides and the Immune System: The Public Health Risks*. Washington, D. C.: World Resources Institute (March).

Repetto, R. 1985. *Paying the Price: Pesticide Subsidies in Developing Countries*. Washington, D.C.: World Resources Institute (December).

Rola, A., and P. Pingali. 1993. *Pesticides, Rice Productivity, and Farmers' Health: An Economic Assessment*. Los Banos, Philippines: International Rice Research Institute, and Washington, D.C.: World Resources Institute.

Rombach, M. C., and K. D. Gallagher. 1994. "The Brown Planthopper: Promises, Problems, and Prospects." In E. A. Heinrichs, ed., *Biology and Management of Rice Insects,* pp. 693–712. London: Wiley, and Los Banos, Philippines: International Rice Research Institute.

Root, R. B. 1975. "Some consequences of ecosystem texture." In S. A. Levin, ed., *Ecosystem Analysis and Prediction*. Philadelphia: Index of Applied Mathematics.

Samways, M. J. 1994. *Insect Conservation Biology*. London: Chapman & Hall.

Schoenly, K., J. E. Cohen, K. L. Heong, G. Arida, A. T. Barrion, and J. A. Litsinger. 1994. "Quantifying the impact of insecticides on food web structure of rice arthropod populations in Philippine farmers' irrigated fields." In G. A. Polis and K. O. Winemiller, eds., *Food Webs: Integration of Patterns and Dynamics*, pp. 343–351. London: Chapman & Hall.

Schulze, E. D., and H. A. Mooney, eds., 1994. *Biodiversity and Ecosystem Function.* Berlin: Springer-Verlag.

Schulze, E. D., and P. Gerstberger. 1994. "Functional Aspects of Landscape Diversity: A Bavarian Example." In E. D. Schulze and H. A. Mooney, eds., *Biodiversity and Ecosystem Function*, pp. 453–466. Berlin: Springer-Verlag.

USDA (U.S. Department of Agriculture). 1993. *World Agriculture: Trends and Indicators, 1970–91.* USDA Statistical Bulletin Number 861. Washington, D.C.: U.S. Department of Agriculture (November).

USDA (U. S. Department of Agriculture). 1994. *Agricultural Resources and Environmental Indicators.* Washington, D.C.: U.S. Department of Agriculture (December).

van den Bosch, R., and P. S. Messenger. 1973. *Biological Control.* New York: Intext Press.

van Emden, H. F. 1965. "The role of uncultivated land in the biology of crop pests and beneficial insects." *Scientific Horticulture*17: 121–136.

Wainhouse, D., and T. H. Coaker. 1981. "The distribution of carrot fly (*Psila rosae*) in relation to the fauna of field boundaries." In J. H. Thresh, ed., *Pests, Pathogens and Vegetation: The Role of Weeds and Wild Plants in the Ecology of Crop Pests and Diseases*, pp. 263–272. Boston: Pitman.

Warburton, H., and P. Pingali. 1993a. "Problems of golden snail management in the Philippines: a discussion of the damage caused by this pest, the availability and feasibility of control measures, and its wider impact on the rice farming system." Los Banos, Philippines: International Rice Research Institute (mimeo).

Warburton, H., and P. Pingali. 1993b. "The farmer and the golden snail: How Philippine farmers cope with a newly introduced pest." Los Banos, Philippines: International Rice Research Institute (mimeo).

Way, M. J., and K. L. Heong. 1994. "The role of ecosystems biodiversity in the dynamics and management of insect pests of tropical irrigated rice: A review." *Bulletin of Entomological Research* 84:567–587.

Woodburn, A. 1993. *Rice—The Crop and Its Agrochemicals Market.* Midlothian, England: Allan Woodburn Associates, and Cambridge, England: Managing Resources.

World Resources Institute. 1988. *World Resources: A Guide to the Global Environment.* Oxford, England: Oxford University Press.

World Resources Institute. 1994. *World Resources: A Guide to the Global Environment.* Oxford, England: Oxford University Press.

Zwolfer, H., G. Bauer, G. Hensinger, and D. Stechmann. 1984. "Die tierokologische Bedeutung und Bewertung von Hecken." *Berichte der Bayerischen Akademie fur Naturschutz und Landschaftspflege, Beiheft* 3/2: 1–155. Cited in E. D. Schulze and

P. Gerstberger. 1994. "Functional Aspects of Landscape Diversity: A Bavarian Example." In E. D. Schulze and H. A. Mooney, eds., *Biodiversity and Ecosystem Function*, pp. 453–466. Berlin: Springer-Verlag.

Zwolfer, H., and D. Stechmann. 1989. "Struktur und Funktion von Hecken in tierokologischer Sicht." *Verh Ges Okol* 17: 643–656. Cited in E. D. Schulze and P. Gerstberger, 1994, "Functional Aspects of Landscape Diversity: A Bavarian Example." In E. D. Schulze and H. A. Mooney, eds., *Biodiversity and Ecosystem Function*. pp. 453–466. Berlin: Springer-Verlag.

Part III

Services Supplied by Major Biomes

Chapter 10

Marine Ecosystem Services

Charles H. Peterson and Jane Lubchenco

The sea and all it provides to help support human society is too often taken for granted. When human population size was low and industrialization of societies was limited, this lack of appreciation had global, if not always local, defensibility. Under the influences of the present large, industrialized, and technologically empowered human society, however, the need is urgent to recognize and acknowledge explicitly the many ways in which ocean ecosystems serve to provide present and future economic value. Without such detailed valuation, costs of various activities that degrade and threaten the continued provision of ocean ecosystem services to human societies may not be adequately considered in formulation of public policy (Ehrlich and Ehrlich 1992) and may be borne by society as a whole rather than being more fairly paid for and benefiting from the degradation. The goal of this chapter is to identify the specific ecosystem services that oceans provide so that future work can determine their economic value. Although our focus here is on economic valuation, we do not mean to imply that economic value represents the sole or even primary justification for conservation of ocean ecosystems. Ethical arguments also have considerable force and merit (Fairweather 1993).

Scope and Working Definitions

Although economists may not normally draw a distinction between goods and services, we partition the two by separating ocean ecosystem goods for discussion by Kaufman and Dayton (1996). These authors not only detail the wealth of goods produced by the oceans, including especially provision of economically valuable fishery products worldwide, but they also explain clearly how continued provision of these goods requires that the natural functioning of the ocean ecosystems that produce them be sustained indefinitely. In other words, one important ecosystem service is the biological food-web production process that results in making goods available for exploitation. Here we first review some general flaws in present management of both fisheries and environmental quality. We then identify and discuss specific services of ocean ecosystems, including: (1) global materials cycling, (2) transformation, detoxification and sequestration of pollutants and societal wastes, (3) support of the coastal ocean-based recreation, tourism, and retirement industries, (4) coastal land development and valuation, and (5) provision of cultural and future scientific values.

We adopt a broad definition of the oceans that includes estuaries. Such a broad definition seems necessary because of the extensive use of estuarine and coastal nurseries by marine organisms. In addition, evaluating the interconnections between the land and the sea is critical to achieving an understanding of important marine ecosystem services.

Fisheries and Environmental Management

While most industrialized nations have developed management schemes designed to protect water quality and the services provided by aquatic ecosystems, intrinsic flaws exist in the management process because of failure to deal properly with the uncertainty associated with scientific advice and the problem of comparing costs and benefits on differing time scales. The management process provides an arena in which inputs from natural sciences and socioeconomic sources are examined to reach some acceptable policy or specific plan. Unfortunately, the short-term costs of establishing a regulation to protect the environment are relatively easily quantified and immediate, whereas the costs of not protecting environmental quality and not preserving natural ecosystem services are less readily quantified and possess longer time horizons (Malone et al. 1993). This same inequity in the character of the costs has led to widespread overharvest of marine fish stocks and dramatic long-term loss of income to fishermen. For example, overfishing and the resultant moratorium on fishing for northern cod off eastern Canada led to twenty-seven or twenty-eight thousand unemployed in New Foundland and Labrador in 1992 or a rate of about 30 percent unemployment (Rose 1995) as fishermen lost the valuable fisheries production services of the ocean ecosystem. Fisheries management has repeatedly mortgaged the future for short-term gain, even while espousing a devotion to maximizing sustainable yield (Ludwig et al. 1993). This has occurred in part because the absence of private property rights over most fisheries removes a potential incentive for their conservation (the tragedy of the commons: Hardin 1968).

If future costs of diminished ecosystem services are discounted in any formal benefit-cost analysis by a factor greater than the inflation rate to account for the time value of money, then the questions of intergenerational equity also arise in development of policy. Some would argue that inclusion of this portion of the discount rate in comparing economic costs that accrue on different time scales has the effect of weighting future costs much less than present costs. In response, one could show that productive capital is also passed on to future generations, raising their standard of living. Whether the future costs and benefits are fairly balanced is not at all clear.

Even if economic analysis were able to construct fair and balanced estimates of present and future costs and benefits of alternatives, there are strong arguments for adopting a risk-averse environmental policy. For example, costs of environmental clean-up and remediation are very large in contrast to pollution prevention, so that if we learn in the future that we have underestimated the extent of some permitted degradation of ecosystem services, costs of clean-up will likely be greater than what prevention would have cost. More important, our scientific uncertainties about ecosystem processes also imply a need for a precautionary principle in environmental management (see Perrings 1991) because of the potential for an unexpected and possible irreversible collapse of ecosystem functions on which humans rely. Some functions of natural ecosystems are not fully replaceable by any mitigation actions, a further argument for caution in formulation of environmental policy (Gren et al. 1994).

The Ocean's Role in Global Materials Cycling

The earth's biosphere is affected by and dependent on the large-scale global geochemical processes that cycle the materials necessary for life itself. The terrestrial biosphere is connected to the land, the atmosphere, and the sea through fundamental processes that move and transform elements. The political and cultural subdivisions of the human populations and the short time scales of human lifetimes and political contemplation can lead to a failure of human societies to consider the relationships of life on earth to these fundamental processes occurring on global spatial scales and on time scales longer than a few years. Yet recent scientific study has revealed how radically human society is changing processes on global scales and how rapid rates of anthropogenic change are occurring, compared to time scales of natural change (NRC 1983, 1987).

A complete review of the global geochemistry of elements essential for life on earth lies outside the scope of this chapter, but some discussion of the most alarming current disruption of natural global geochemical processes seems appropriate. Ocean ecosystems play a major role in the global geochemical cycling of all the elements that represent the basic building blocks of living organisms, carbon, nitrogen, oxygen, phosphorus, and sulfur, as well as other less abundant but necessary elements. Of these, the anthropogenic impacts on carbon and the carbon cycle are of most pressing concern.

Ocean ecosystems are important participants in the global carbon cycle, such that in the absence of life in the sea, the equilibrium partitioning of carbon among rock (the lithosphere), the atmosphere, and ocean waters would be dramatically altered (Sarmiento et al. 1995). Berner et al. (1983) provide a nice account of the role of ocean ecosystems in the CO_2 cycle. Carbon is sequestered in continental rocks in two major forms, sedimentary organic matter (kerogen) and solid-phase carbonates. The kerogen is derived from the sedimentary remains of soft tissues of organisms, whereas carbonates in rocks come mostly from skeletons of marine plants and animals. Through interactions with the atmosphere, these carbon compounds in rocks are weathered (chemically degraded). Kerogen is oxidized to return carbon back to the atmosphere directly as CO_2. Carbonates are weathered through exposure to rainwater, which is weakly acidic (carbonic acid) as a consequence of dissolved carbon dioxide. This weathering yields dissolved bicarbonate ions, calcium, magnesium, and other cations.

After dissolved bicarbonate is returned to the sea in rainwater runoff, biological uptake produces particulate carbonate again. This incorporation of dissolved bicarbonates into skeletal tissues of marine plants and animals provides the vehicle for transfer of (bi)carbonate dissolved in ocean waters back into the sediments. Its burial there in sedimentary strata and ultimate transformation into rocks reduces the pool of atmospheric carbon dioxide and oceanic dissolved bicarbonate by storage in a solid phase in the earth's crust. Tectonic processes release CO_2 as a gas created from subjecting the sedimentary carbonates to high pressure and temperature. This completes the crude outline of the earth's natural carbon cycle. The marine ecosystem provides the service in this cycle of biologically transforming dissolved bicarbonate into particulate carbonates in the form of skeletons available for burial. If the sea were devoid of biota, the transfer of CO_2 from the atmosphere to the sea floor through biological production would cease and atmospheric CO_2 concentrations would rise (Berner et al. 1983, Sarmiento et al. 1995).

This service provided by ocean ecosystems represents but one example of how the ocean biota role in geochemical cycling is vital to life on land. We develop it in some detail because the consequences of release of greenhouse gases like CO_2 through fossil fuel burning are so immediate and so serious that the importance of the biological partitioning of CO_2 into the ocean also grows. Enhanced atmospheric greenhouse gases imply dramatic variations in global temperatures, changes in rainfall and land productivity patterns, and sea level rise and coastal flooding (see NRC 1983, Fischer 1984, NRC 1987).

Transformation, Detoxification, and Sequestration of Wastes

The oceans are used by human society as a repository for unwanted materials that we create and release onto land, into streams and rivers, and even into the atmosphere. Oceans are also directly used as dumping grounds for various societal wastes. The aquatic ecosystems of rivers, estuaries, and the ocean act upon these materials in a variety of ways to transform them, in some cases to detoxify them, and in other cases merely to sequester them.

Transformation—The Case of Nutrients

A universal example of how human society uses aquatic ecosystems to treat its wastes is provided by a review of the disposal of sewage wastewater (see NRC 1993). Modern secondary sewage treatment produces tremendous loadings of inorganic nutrients (nitrogen and phosphorous) in aquatic systems. Nitrogenous nutrients originating largely from fossil fuel burning are also injected into estuarine and coastal waters through acid rain (Paerl 1993). The nutrients are processed by the aquatice system, where they are removed from the water through uptake by plants, especially phytoplankton but also riparian vegetation of wetlands.

The marginal economic value of using aquatic ecosystems to scrub nutrients from sewage wastewater could be estimated by using the standard engineering formulae for calculating costs of various additional levels of treatment. For example, assuming a population of over ten thousand people, for a flow of five million gallons per day, the costs of construction alone for a treatment plant with some nutrient removal capabilities would be $4.2 million (1996 dollars) more than the $23.9 million for constructing the analogous Advanced Treatment I plant without nutrient removal capability (EPA 1995). Increased treatment would also imply greater operating costs not included in this sample calculation.

The allowable loading of these nutrients into the aquatic system is limited by the capacity of the aquatic ecosystem to degrade microbially the organic matter produced (Nixon 1995). This process of increasing the rate of supply of organic matter to a system is termed eutrophication (Nixon 1995). Excessive eutrophication causes reduction in ecosystem services through at lease two consequences, anoxia and nuisance algal blooms. An overload of organic production induces oxygen depletion, anaerobic microbial production of toxic hydrogen sulfide, and massive mortality of estuarine and marine animals. Thus, a conversion to hypereutrophic conditions transforms the entire aquatic ecosystem into one no longer supporting normal production of valuable fishes and invertebrates and no longer oxydizing the organic wastes discharged into it and produced in it by nutrient discharge (Elmgren 1989).

Eutrophication also stimulates growth of nuisance algae, such as blue-greens and dinoflagellates (Paerl 1998, Smayda 1990). These nuisance algae are often toxic to estuarine and marine animals, and in some instances threaten human health (Cosper et al. 1989). Red tide dinoflagellates can produce and release as aerosols vertebrate neurotoxins, causing long-lasting neurological injury to people who breathe the fumes. Stomach upset and disruptions of the gastrointestinal system are common symptoms of exposure to red tides. Paralytic shellfish poisoning is caused by human ingestion of shellfish exposed to toxic algae. Nuisance algal blooms often discolor the waters with reds, yellows, or browns, and release foul odors. The economic impacts of fish kills and losses to aquaculture businesses from red tides and other nuisance algal blooms are large and growing worldwide in frequency and severity (Paerl 1993).

Bivalve molluscs within the estuary act as a filter with a potential for removing excess algal production induced eutrophication. For example, Newell (1988) calculated crudely that at historic levels of natural abundance, the oysters of Chesapeake Bay filtered a volume of water equal to the complete volume of the bay in a three-day period. The effect of such biological filtration is clearly to improve water clarity by removal of suspended materials and to transfer production from the pelagic to the benthic realms in the system (Dame 1994). Filter feeding by benthic animals can thus be viewed as one important sort of top-down control of the estuarine system that may compensate for the bottom-up enhancement induced by excess nutrient addition. Unfortunately, mismanagement of the American oyster in Chesapeake Bay and other major estuaries of the northeast and mid-Atlantic coast has lead to a decline of almost two orders of magnitude in oyster abundance, diminishing one important estuarine ecosystem service. Growing eutrophication implies that restoration of oysters and other long-lived suspension-feeding bivalves may be an appropriate form of biomanipulation (see Carpenter et al. 1995) to enhance this particular ecosystem service of estuaries, now so much more in demand (Lenihan and Peterson 1996).

Detoxification

Some of society's wastes are detoxified by naturally functioning marine ecosystems, thereby representing yet another service provided free of charge to society. For example, petroleum hydrocarbons are spilled and released into the environment with great frequency. Many of the component compounds of petroleum carry important health risks to humans who are exposed to them. When in association with sediment particles, components of petroleum hydrocarbon are deposited on the floor of the estuary and ocean, where naturally occurring microbes detoxify these compounds and ultimately degrade them into carbon dioxide and water (Cerniglia and Heitcamp 1989). This is a service rendered by the microbial community of marine ecosystems. It results from aerobic processes because oxygen is the source of electrons for the degradation process catalyzed by the microbes. By inducing anoxia, eutrophication of our estuarine and marine waters endangers the valuable ecosystem service of microbial detoxification of petroleum hydrocarbons released into the environment.

Sequestration

Other important classes of toxic materials produced by industrialized human societies are not so readily degraded and transformed by ocean ecosystem processes. These include many artificial organic pollutants, such as DDT, PCB, and dioxins. Since these materials are not naturally produced, it is less likely that the microbial community has the capacity to utilize them as organic substrates, and indeed they are extremely persistent in the environment. Heavy metals, such as mercury, lead, copper, tin, zinc, and arsenic, represent another important class of pollutant released into the marine environment by industrialized societies. To some degree, the estuarine and marine ecosystems serve to transform heavy metals by binding them with sediments in a fashion that renders them biologically unavailable (Cross and Sunda 1978). However, often these pollutants are not transformed into harmless compounds by marine ecosystem processes but instead present biological hazards, placing wildlife and humans at risk (Long and Morgan 1990).

The oceans are intentionally used by many human societies as dumping grounds for various wastes, including especially toxic and harmful by-products of industrial society, such as nuclear wastes deposited in the Arctic Ocean from the former Soviet Union. In many instances, the motivation for ocean dumping is to allow these materials to be sequestered by the ocean environment in a place that will retain them far from any possible contact with humans. This sequestering function of ocean ecosystems could conceivably be performed by a sea devoid of life; however, organically mediated sedimentation onto the sea floor helps bury and isolate much of this waste and thereby perform the intended ecosystem service. Unfortunately, complete isolation and sequestration of these materials in bottom sediments in a form that is biologically unavailable may not always be achieved (Long and Morgan 1990).

Value of Ocean Ecosystems to Tourism, Recreation, and Retirement

Naturally functioning ocean ecosystems provide direct economic value to several coastal industries in developed nations worldwide. Rarely is the dependence of those industries upon the ecosystem services clearly defined. Some limited information is available from damage assessments following large-scale pollution events, such as the *Exxon Valdez* oil spill. Otherwise, the economic work linking ecosystem function and structure to human enterprises and their valuation remains to be done.

Ecotourism

Excluding commercial fishing, the coastal industry most obviously tied to a naturally functioning ocean ecosystem is probably the tourism industry. Tourism is said to be the world's largest business (Miller and Auyong 1991). Ocean ecosystems of several sorts make huge contributions to the tourism economies of coastal regions. The three most important sets of examples of ocean ecosystem services feeding economically vital ecotourism industries and economies involve coral reef systems, polar ocean systems, and coastal estuaries and wetlands.

Coral Reef Contributions

The economic welfare of many coastal nations in the tropics is dependent on the ability to offer tourists various ocean ecosystem amenities, prominent among them opportunity for sealife viewing. This exemplifies a nonconsumptive use value of natural ocean ecosystems. For example, the majority of Caribbean islands have economies based on ecotourism, in which viewing reef fishes as well as the corals and associated invertebrates by snorkeling, diving, and glass-bottomed boats plays a major role. Even before the explosive growth of the industry and tourism more generally in the past two decades, spending by visitors accounted for 55 percent of the total GNP of the Bahamas and averaged 17 percent of GNP for the eastern and southern Caribbean nations in 1977 (Beekhuis 1981). Such regions have a vital economic interest in preserving the functioning of the ocean ecosystems that produce these diverse, colorful marine animals.

While certain components of the coral reef ecosystem are valued directly because of their visual, aesthetic appeal, namely the reef fishes, corals, and colorful benthic invertebrates, they in turn are supported by a nexus of ecosystem interactions required to sustain them. Most ecologists agree that the complexity of interactions and degree of interrelatedness among component species is higher on coral reefs than in any other marine environment (Hughes et al. 1992). This implies that the ecosystem functioning that produces the most highly valued components is also complex and that many otherwise insignificant species have strong effects on sustaining the rest of the reef system.

Coral reef ecosystems and the tourism that they generate are now seriously endangered by degradation of the corals, which form the structural habitat of this system. Although the causes of loss of coral reef habitat are numerous and not always known, many anthropogenic factors contribute (d'Elia et al. 1991). Coral bleaching has been linked to global atmospheric changes, both to global warming and to enhanced UV exposure as the ozone layer has been depleted and as enhanced tropical douldrums have smoothed the sea surface and promoted deeper penetration of damaging UV (Glynn 1991, Gleason and Wellington 1993). Exploitation of reef fishes, many of which are important herbivores, and eutrophication through discharge of sewage, have led to a profusion of algae overgrowing and killing corals (Hughes 1994). Sediment erosion from improper development on coastal lands has lead to coral mortality from turbidity and burial (Roberts 1993). Outbreaks of corallivorous consumers that have denuded vast reef areas may be linked to human disruptions of the ocean ecosystem. Countries such as Australia and perhaps Belize that recognize the economic importance of sustaining the coastal reef ecosystem and act effectively to protect it will prosper, while others may suffer dramatic losses of income from assuming that this ecosystem service will be provided indefinitely without management to protect it. The science-based management plans created and enforced by the Great Barrier Reef Park Authority in Australia will return huge economic rewards for sustaining the ecosystem structure, composition, and function on which such a valuable tourism industry is based (Kelleher and Kenchington 1992).

Polar Ocean Ecosystem

The economic value of ecotourism in the coastal marine environment is not limited to tropical countries with coral reefs. Coastal marine environments in high latitudes are characterized by ecosystems in which the top carnivores, charismatic seabirds and marine mammals, are abundant. The economic value of tourism to regions like Alaska is immense. Some of this is a reflection of the geological vistas, including glaciers and other inanimate components of the coastal ecosystem. However, much is related to the provision of ready opportunity to view puffins, auklets, murres, seals, otters, sea lions, killer whales, and other beloved marine wildlife. The significance of abundant wildlife and a relatively pristine ecosystem to the ecotourism of such polar regions as Alaska is reflected in economic studies of the impacts of the *Exxon Valdez* oil spill: in 1989, the revenues from visitors coming to Alaska fell 8 percent in south-central and 35 percent in southwest Alaska below the previous summer (McDowell Group 1990), which represented a $19 million loss in visitor spending. (This is an underestimate because it ignores expected increases in tourism.) Ecotourism organized to exhibit polar wildlife and coastal ecosystems is also a rapidly growing economic enterprise in New Zealand, where tours progress southward to Antarctica.

Estuarine Ecosystems

A third type of coastal marine environment that deserves special mention for its value in supporting ecotourism is the estuary, with its tidal flats, wetlands, marshes, and mangroves. Estimates of the value of the local economy of coastal wetlands through aggregate provision of recreational opportunities, fish production, storm protection, and water treatment range from $800 to $9,000 per acre (Anderson and Rockel 1991, Kirby 1993). This habitat with its high primary productivity sustains large populations of attractive and readily viewed shorebirds and waterbirds. When they can be viewed in a natural setting of lush coastal vegetation, the probing shorebirds, ducks, flamingoes, herons, egrets, gulls, and terns of this coastal marine ecosystem represent an important natural asset underlying substantial coastal tourism industries worldwide. Some specific estuarine systems come immediately to mind as the most important examples, including Kakadu in the Northern Territory of Australia, the coast of Namibia, the Everglades in Florida, and the Copper River Delta in Alaska. While these may be the most spectacular illustrations of the value of coastal estuarine ecosystem services to the tourism industry, similar economic contributions also exist in other regions with more diversified economies. Because of competing demands within the coastal region, where human population is most concentrated, and because most of those competing uses are incompatible with sustaining populations of the birds and wildlife of the estuary, proper planning in this environment needs to evaluate and consider the value of services derived from the naturally functioning ecosystem. Economically valuable tourism industries could be lost if newly permitted uses disrupted the ecosystem services on which the tourism depends.

Local Tourism, Recreation, and Retirement

Our discussion of the value of marine ecosystem services to tourism has been focused until now on ecotourism. Ecotourism represents just one part of the total contribution of naturally functioning coastal ecosystems to coastal tourism industries. Local tourism is a mainstay of many coastal economies in developed countries worldwide, and one of the important amenities that helps value one tourist destination more highly than another is the availability of various, usually nonconsumptive, uses of natural coastal marine ecosytems. Many coastal tourists look for opportunities to go sportfishing, bird watching, or whale watching, to practice nature photography, or simply to enjoy immersion in an undegraded coastal setting. Each of these opportunities depends on sustaining function of the coastal marine ecosystem and provision of its services.

The ocean ecosystems offer other important recreational activities to tourists and residents beyond those associated with observing and enjoying local plants and animals. People extract satisfaction from water sports and activities, such as sailing, surfing, boating, and swimming in the sea. These recreational activities are also services provided by marine ecosystems from which people derive satisfaction, and they therefore have utilitarian value to tourists and residents that could be quantified. This class of ecosystem services depends largely on the abiotic components of the marine ecosystem, namely presence of a fluid surface. One could reasonably ask whether an ocean devoid of life might not continue to provide this class of services to human society. The answer is that the satisfaction derived by the majority of participants in these recreational activities is dependent on the total quality of the experience and would be diminished in the absence of graceful pelicans and dolphins or in the presence of waters characterized by foul odor, obvious discoloration, or noxious organisms like stinging jellyfish. Thus there is a biotic contribution even to this class of recreational services offered by marine ecosystems, such that policy that affects the structure and function of ocean ecosystems has potential impact on the value of those services that take advantage of the physics of the ocean service.

Ocean Ecosystem Services

The nonconsumptive (or passive use) value (see the discussion of these distinctions in the chapter by Goulder and Kennedy) of a naturally functioning ecosystem can be translated in monetary value by contingent valuation analysis. When applied to valuation of natural resource damages, this approach utilizes a random survey approach to sample people's willingness to pay to prevent ecological harm of a certain sort or alternatively willingness to accept compensation for that injury to the natural ecosystem. This approach was used to construct a conservative estimate of the passive non-use value of the damages done to the marine ecosystem by the *Exxon Valdez* oil spill (Carson et al. 1994). The estimate is extremely conservative because each time a choice had to be made, the more conservative option was selected. For example, the degree of ecological damage was intentionally understated in the survey document. Survey respondents were told that ecological damages were restricted to a loss of 75–150,000 out of 1.5 million seabirds and 580 sea otters and 100 harbor seals. They were also told that these populations would return to normal within three to five years at most and that no other long-term damage would occur to the ecosystem. Furthermore, willingness to pay was used as the measure, which is typically lower than willingness to accept estimates (Hanemann 1991). The survey included only people who resided outside of Alaska, thereby involving negligible contribution from those who included any consumptive or nonconsumptive uses in their responses. Under these conditions, the median household willingness to pay to avoid another similar injury to the marine ecosystem of the Prince William Sound region of central Alaska was $31, which expands to a value of $2.8 billion summing over all households in the United States (Carson et al. 1992). This analysis does not represent valuation of the entirety of the passive non-use value of this one ecosystem, but rather just the loss of marginal value associated with the oil spill, yet the number is large. In addition, this reflects only one type of economic importance placed by the public on naturally functioning marine ecosystems, namely the existence value, ignoring all the provision of goods and other use values of the ecosystem.

The specific economic enterprises relying on tourism in the coastal regions are extensive. The tour operators represent only the tip of the iceberg of the financial value of tourism. Indeed, one method to placing a value on the economic contribution of tourism from ecosystem services is to sum the multitude of travel costs incurred by people to participate in these tourism opportunities. Tourism contributes to the transportation industries, including airlines, rails, buses, boat transport, and automobile support services. The lodging industry benefits directly and massively from tourism on the coast. This includes not only hotels and motels, but also condominiums and private rental housing. Tourists spend money on meals, supplies, and recreational equipment while visiting. The infusion of new money into a local coastal economy from tourists has cascading indirect economic benefits as those funds support jobs, investment, and other services in the local region.

Coastal Real Estate Development and Land Valuation

The quantity and quality of amenities provided by the coastal marine ecosystem also has impacts on coastal property values. Localities in demand as tourist destinations and residential areas providing attractive recreational opportunities experience heightened demand for and thus valuation of real estate. For example, comparisons of land values before and after implementation of Maryland's Chesapeake Bay Critical Area and New Jersey's Pinelands regulations revealed increases of 5–17 percent for developed and 5–25 percent for undeveloped land within the protected area (Beaton 1988). In developed and some developing countries, appeal to a highly mobile and discriminating population of retirees has great potential for dictating coastal property values. Such demographic movements have transformed South Florida and contributed immensely to its economy. For example, total economic activity in the marine recreational boating industry in Florida increased by 80 percent between 1980 and 1985 (Milton and Adams 1987). To the degree that demand for such coastal real estate is based upon amenities made available by the local marine ecosystem, this process contains intrinsic contradictions. Too many people in a coastal region can degrade the local environment and prevent the local marine ecosystem from continuing to provide the services that helped attract people initially. Consequently, recognition of the dependence of the existing local economy of a region on provision of ecosystem services is vitally important so that growth management can be used to prevent loss of the supporting ecosystem services in the future.

Cultural Value and Future Scientific Values

While difficult to quantify except perhaps by some form of contingent valuation procedures, marine ecosystems have cultural value in the present and potential for realization of scientific value to society in the future. For many groups of native peoples in industrialized countries, there is explicit legal acknowledgment of their rights to a healthy and productive natural ecosystem. For example, the various Native American cultures in Alaska, including Aleutic, Eskimo, and other native peoples, possess a traditional culture that is intrinsically dependent upon the natural ecosystem, including the marine realm. A long tradition of subsistence is based on the use of goods derived from the marine ecosystem that are extracted by the taking of plants and animals for foods, clothing, shelter, fuel, medicines, and other purposes. But in addition to the provision of material goods, the natural ecosystem provides the basis of culture in these societies. The transmission of cultural information about the habits of marine animals and about the ecosystem processes that organize nature forms a centerpiece of traditional society and culture for these and many other native peoples. The natural world and the integrity of natural ecosystems also form an explicit or implicit part of the religious beliefs and cultural heritage of essentially all human religions and cultures. Such values need recognition.

One component of the wealth of society is the body of scientific knowledge that society has accumulated and that supports numerous advances in the human condition. These achievements include, of course, medical discoveries, but also improved basic understanding of the functioning of the natural ecosystems that enable technological progress to occur. Such scientific advances are achievable through exploitation of opportunity that resides in the undiscovered information contained in natural ecosystems. In a real sense, the natural ecosystem is a repository of information, a capital resource that when tapped in the future will create economic wealth and improve the welfare of human society. Although the scope and application of future scientific discoveries are impossible to predict, it is clear that failure to preserve this information bank that is the natural ecosystem represents irretrievable loss of natural capital that would generate tangible future economic value.

Lest we become transfixed by the task of placing economic value on natural ecosystems, we must recognize that the most compelling basis for the preservation of our natural heritage is still probably ethical. Preservation of species, maintenance of biodiversity, and sustaining of natural processes feels morally right. Passing on the legacy of nature to future generations should motivate most conservative actions. In this chapter, however, we have illustrated ways in which natural ocean ecosystems also contribute directly and indirectly to aspects of human enterprise that have economic value. We hope that by explicitly identifying some of the most economically important of these ocean ecosystem services, we can stimulate inclusion of their contributions to human enterprise in future benefit-cost analyses. Such economic analyses represent but one of several inputs to development of environmental policy.

Conclusions

Consideration of how ocean ecosystems provide economic benefit to human society always includes the value of fisheries production but rarely reflects complete analysis of other important services provided by ocean ecosystems. The oceans play a critical role in the global materials cycling that sustains climate and life support systems on land. In the absence of ocean biota, for example, the biological pump that injects carbonates into sediments would cease and atmospheric carbon would increase in the form of greenhouse gas CO_2, with dramatic disruptions to human society from resultant climate change and sea level rise. In combination with rivers and estuaries, ocean ecosystems serve to process, transform, detoxify, and sequester many of society's waste products. For example, if the nutrient removal from sewage wastewater now conducted by aquatic ecosystems and wetland processes were to be achieved through engineering, costs of treatment would increase tremendously. Abuse of the nutrient scrubbing service of aquatic ecosystems has a cost, however, in that nutrient overloading creates eutrophication and disrupts provision of natural ecosystem services by removing oxygen, causing fishkills, and inducing toxic algal blooms.

In addition to these geochemical functions of oceans, ocean ecosystems act to sustain valuable human business enterprises. Tourism has been identified as the world's largest business, and in coastal regions much of that tourism depends on amenities and values provided by ocean ecosystems. The beauty of the diverse and colorful coral reef animals in the tropics; the majesty of the killer whales, the penguins and puffins, and other abundant marine mammals and seabirds in polar oceans; and the profusion of wonderful waterbirds nested within a backdrop of greenery in coastal wetlands worldwide support exceptionally valuable ecotourism industries. Provision of opportunities to use and enjoy the coastal marine ecosystems contributes substantially to the value of local tourism, recreation, and retirement industries. Coastal land valuation is enhanced by preservation of natural functions of ocean ecosystems.

Traditional economic analysis does not readily quantify many of the more important human societal values vested in naturally functioning marine ecosystems. The natural ocean ecosystem forms the cultural core of several indigenous human societies and is important to religious beliefs of many. The ocean ecosystem can also be viewed as a capital resource containing opportunity for future scientific discovery that will enhance the wealth and welfare of human society. Successful resolution of policy questions involving intergenerational equity and avoiding the trap of sacrificing long-term sustainability to avoid short-term costs is needed to ensure perpetual transmission of the legacy of value in nature. Despite a focus on economic valuation, the ethical arguments for conservation of ocean ecosystems should not be overlooked: for most people, conservation represents the right thing to do, and satisfying that moral imperative has value too.

Acknowledgments

We thank J. Grabowski, L. Kaufman, A Maxson, and K. Smith for guidance to relevant literature. R. T. Barber, G. C. Daily, M. E. Hay, H. S. Lenihan, F.

Micheli, K. Turner, and anonymous referees made helpful suggestions on the concepts and presentation. Financial support was provided by the Pew Charitable Trusts and the National Science Foundation.

References

Anderson. R., and M. Rockel. 1991. "Economic valuation of wetlands." Discussion paper #065. American Petroleum Institute, Washington, D.C.

Beaton, W. 1988. *The Cost of Government Regulations, vol. 2. A Baseline Study of the Chesapeake Bay Critical Area.* Chesapeake Bay Critical Area Commission, Annapolis, Md.

Beekhuis, J. V. 1981. "Tourism in the Caribbean: Impacts on the economic, social, and natural environments." *Ambio* 10:325–331.

Berner, R. A., A. C. Lasaga, and R. M. Garrels. 1983. "The carbonate-silicate geochemical cycle and its effect on atmospheric carbon dioxide over the past 100 million years."*American Journal of Science* 283: 641–683.

Carpenter, S. R., S. W. Chisholm, C. J. Krebs, D. W. Schindler, and R. F. Wright. 1995. "Ecosystem experiments." *Science* 269: 324–327.

Carson, R. T., R. C. Mitchell, W. M. Hanemann, R. J. Kopp, S. Presser, and P. A. Raud. 1994. "Contingent valuation study of lost passive use: Damages from the *Exxon Valdez* oil spill. "Discussion paper 94–18. Resources for the Future, Washington, D.C.

Cerniglia, C. E., and M. A. Heitcamp. 1989. "Microbial degradation of PAH in the aquatic environment." In U. Varanasi, ed., *Metabolism of Polycyclic Aromatic Hydrocarbons in the Aquatic Environment.* CRC Press, Boca Raton, Fla.

Colborn, T. 1995, "Environmental estrogens: Health implications for humans and wildlife." *Environmental Health Perspectives*103 (Supplement 7): 135–136.

Cosper, E. M., V. M. Bricelj, and E. J. Carpenter, eds. 1989. *Novel Phytoplankton Blooms: Coastal Marine Studies.* Springer-Verlag, Berlin.

Dame, R. F. 1994. "The role of bivalve filter feeder material fluxes in estuarine ecosystems." In *Bivalve Filter Feeders in Estuarine Processes*, pp. 245–269. NATO ASI Series V.G33. Springer-Verlag, Heidelberg.

Ehrlich, P. R., and Ehrlich, A. H. 1992. "The value of biodiversity." *Ambio* 21: 219–226.

Elmgren, R. 1989. "Man's impact on the ecosystem of the Baltic Sea: Energy flows today and at the turn of the century." *Environmental Science and Technology* 9: 635–638.

EPA (Environmental Protection Agency). 1995. *1996 Clean Water Needs Survey Manual.* U.S. Environmental Protection Agency, Washington, D.C.

Fairweather, P. 1993. "Links between ecology and ecophysiology, ethics and the requirements of environmental management." *Australian Journal of Ecology* 18:3–20.

Fischer, A. G. 1984. "The two phanerozoic supercycles." In W. A. Berggren and J. A. Van Couvering, eds., *Catastrophes and Earth History*. Princeton University Press, Princeton, N.J.

Gleason, D. F., and G. M. Wellington. 1993. "Ultraviolet radiation and coral bleaching." *Nature* 365:836–838.

Glynn, P. W. 1991. "Coral reef bleaching in the 1980s and possible connections with global warming." *TREE* 6:175–178.

Gren, I. M., C. Folke, K. Turner, and I. Bateman. 1994. "Primary and secondary values of wetland ecosystems." *Environmental and Resource Economics* 4:55–74.

Hanemann, W. M. 1991. "Willingness to pay and willingness to accept: How much can they differ?" *American Economic Review* 81: 635–647.

Hardin, G. 1968. "The tragedy of the commons." *Science* 162:1243–1248.

Hughes, T. P. 1994. "Catastrophies, phase shifts, and large-scale degradation of a Caribbean coral reef." *Science* 1547–1551.

Hughes, T. P., D. Ayer, and J. H. Connell. 1992. "The evolutionary ecology of corals." *TREE* 7:292–295.

Kelleher, G., and R. Kenchington. 1992. *Guidelines for Establishing Marine Protected Areas*. IUCN, The World Conservation Union, Gland, Switzerland.

Kirby, K. 1993. "Wetlands not wastelands." *Scenic America Technical Information Series* 1(5):1–8.

Lenihan, H.S., and C.H. Peterson. 1996. "How the interaction between reef habitat degradation and water quality decline induces oyster loss." *Ecology*. In review.

Long, E. R., and L. G. Morgan. 1990. *The Potential for Biological Effects of Sediment-Sorbed Contaminants Tested in the National Status and Trends Program*. National Oceanographic and Atmospheric Administration Technical Memo NOS OMA 52, NOAA, Seattle, Washington.

Ludwig, D., R. Hilborn, and C. Walters. 1993. "Uncertainty, resource exploitation, and conservation: Lessons from history." *Science* 260: 17, 36.

Malone, T.C., W. Boynton, T. Horton, and C. Stevenson. 1993. "Nutrient loadings to surface waters: Chesapeake Bay case study." In *Keeping Pace with Science and Engineering. Case Studies in Environmental Regulation*, M. R. Uman, ed., pp. 8–38. National Academy Press, Washington, D.C.

McDowell Group. 1990. *An Assessment of the Impact of the Exxon Valdez Oil Spill on the Alaskan Tourism Industry. Phase I: Initial Assessment*. Exxon Valdez Oil Spill Trustees, Anchorage, Alaska.

Miller, M. L., and J. Auyong. 1991. "Coastal zone tourism. A potent force affecting environment and society." *Marine Policy* 15:75–99.

Milton, J. W., and C. M. Adams. 1987. "The economic impact of Florida's recreational boating industry in 1985." Technical paper no. 50. Florida Seagrant Program, Gainesville, Fla.

Newell, R. I. E. 1988. Ecological changes in Chesapeake Bay: Are they the result of overharvesting the American oyster, *Crassostrea virginica*? In *Understanding the Estuary: Advances in Chesapeake Bay Research*, pp. 536–566. Chesapeake Bay Research Consortium, Baltimore, Md.

Nixon, S. W. 1995. "Coastal marine eutrophication: A definition, social causes, and future concerns." *Ophelia* 41:199–219.

NRC (National Research Council). 1983. *Changing Climate, Report of the Carbon Dioxide Assessment Committee.* National Academy Press, Washington, D.C.

NRC. 1987. *Responding to Changes in Sealevel: Engineering Implications National* Academy Press, Washington, D.C.

NRC. 1993. *Managing Wastewater in Coastal Urban Areas.* National Academy Press, Washington, D.C.

Paerl, H.W. 1988. "Nuisance phytoplankton blooms in coastal, estuarine and inland waters." *Limnology and Oceanography* 33: 823–847.

Paerl, H.W. 1993. "Emerging role of atmospheric nitrogen deposition in coastal eutrophication: Biogeochemical and trophic perspectives. *Canadian Journal of Fisheries and Aquatic Science* 50: 2254–2269.

Perrings, C. 1991. "Reserved rationality and the precautionary principle: Technical change, time, and uncertainty in environmental decision making." In *Ecological Economics: The Science and Management of Sustainability*, R. Costanza, ed. Columbia University Press, New York.

Roberts, C. M. 1993. "Coral reefs: Health, hazards, and history." *TREE* 8:425–427.

Rothschild, B. J., J. S. Ault, P. Golletquer, and M. Heral. 1994. "Decline in Chesapeake Bay oyster populations: A century of habitat destruction and overfishing." *Marine Ecology Progress Series* 111: 29–39.

Sarmiento, J. L., R. Murnane, and C. LeQuere. 1995. "Air-sea CO_2 transfer and the carbon budget of the North Atlantic." *Phil. Trans. R. Soc. Lond. B.* 348:211–219.

Smayda, T. J. 1990. "Novel and nuisance phytoplankton blooms in the sea: Evidence for a global epidemic," In E. Graneli, B. Sundstrum, L. Edler, and D. M. Anderson, eds., *Toxic Marine Phytoplankton*, pp. 29–40. Elsevier, Amsterdam.

Chapter 11

FRESHWATER ECOSYSTEM SERVICES

Sandra Postel and Stephen Carpenter

It is no coincidence that early human civilizations sprang from river valleys and floodplains. Sufficient quantities of freshwater have underpinned the advancement of human societies since their beginning. Today, we rely on the solar-powered hydrological cycle not only for water supplies, but also for a wide range of goods and life-support services, many of which are hidden and easy to take for granted.

Only a small portion of earth's water wealth consists of liquid water that is fresh enough to drink, grow crops, and satisfy other human needs. Of the total volume of water on the planet (an estimated 1,386,000,000 cubic kilometers, or km^3), only 2.5 percent is fresh—and two-thirds of that is locked in glaciers and ice caps. Merely 0.77 percent of all water is held in lakes, rivers, wetlands, underground aquifers, soil pores, plant life, and the atmosphere (Shiklomanov 1993).

Of particular importance to the sustenance of earth's biological richness is precipitation on land, an estimated ~110,000 km^3 per year (L'Vovich et al. 1991). This water is made available year after year by the hydrological cycle and constitutes the total terrestrial renewable freshwater supply. Natural systems, such as forests, grasslands, and rivers, as well as many human-dominated landscapes, such as croplands and pasture, depend upon this rainfall and are finely tuned to natural precipitation patterns.

In some sense, this water is infinitely valuable, since without it land-based life as we know it would disappear. In this chapter, however, we focus not on the entire hydrological cycle, but on the benefits to the human enterprise

Table 11.1. Services provided by rivers, lakes, aquifers, and wetlands

Water Supply
Drinking, cooking, washing, and other household uses
Manufacturing, thermoelectric power generation, and other industrial uses
Irrigation of crops, parks, golf courses, etc.
Aquaculture
Supply of Goods Other Than Water
Fish
Waterfowl
Clams and mussels
Pelts
Nonextractive or Instream Benefits Flood control
Flood control
Transportation
Recreational swimming, boating, etc.
Pollution dilution and water quality protection
Hydroelectric generation
Bird and wildlife habitat
Soil fertilization
Enhanced property values
Non-user values

provided by freshwater systems—primarily, rivers, lakes, aquifers, and wetlands. We attempt to estimate the total value of selected goods and services provided by these systems and, where data exist, offer some estimates of marginal values as well (see Goulder and Kennedy, chapter 3, this volume).

The benefits provided by freshwater systems fall into three broad categories: (1) the supply of water for drinking, irrigation, and other purposes; (2) the supply of goods other than water, such as fish and waterfowl; and (3) the supply of nonextractive or "instream" benefits, such as recreation, transportation, and flood control. Table 11.1 provides a more complete listing of the services that rivers, lakes, wetlands, and underground aquifers provide to the human economy.

Water Supply Services

Once precipitation falls on land, it divides into two parts—evapotranspiration (representing the water supply for all nonirrigated vegetation) and runoff (overland flow back toward the sea via rivers, streams, and underground aquifers). Through their role in the hydrological cycle, rivers, lakes, and underground aquifers provide a renewable source of freshwater for the human economy to tap. They are the principal source of freshwater for irrigation, households, industries, and other uses that require the removal of water from its natural channels.

Human demands for this water have increased rapidly in recent decades as a result of population growth, changes in diet, and higher levels of material consumption: withdrawals or extractions of water from the aquatic environment have more than tripled since 1950 (Shiklomanov 1993). Today, the volume of water removed from rivers, lakes, and aquifers for human activities worldwide totals some 4,430 km^3 per year (Postel et al. 1996). Because accessing this water typically requires the construction of dams, reservoirs, canals, groundwater wells, and other infrastructure, there is a direct and tangible economic cost associated with it; this water supply service is not totally free. However, the full value of the service comes to light by considering the cost of replacing natural sources of freshwater with the next best alternative.

Unlike oil, coal, or tin, for which substitutes exist, freshwater is largely nonsubstitutable. The next best alternative is water processed by technological desalination—the removal of salt from seawater, the function performed naturally by the hydrological cycle. Worldwide, desalination accounts for less than 0.1 percent of total water use (Wangnick Consulting 1990). It is a highly energy-intensive process and therefore an expensive supply option. The cost of desalination is in the neighborhood of $1–2 per cubic meter ($m^3$) (OTA 1988)—four to eight times more than the average cost of urban water supplies today (World Water/World Health Organization 1987), and at least 10–20 times what most farmers currently pay (Postel 1992). Not surprisingly, some 60 percent of the world's desalting capacity is in the Persian Gulf, where fossil energy sources are abundant and freshwater is scarce. Through desalination, countries in this region have essentially been turning oil into water to satisfy drinking and other household needs.

Clearly, if the world's total demand for water had to be met through desalination, water use would be substantially lower than it is today because of the higher supply price. We make no adjustments to the demand picture other than to assume that water not consumed during use is reused and recycled, so that only the volume of water currently consumed (in contrast to used) would need to be desalted. This amounts to an estimated 2,010 km^3/year after subtracting for reservoir losses (Postel et al. 1996), which would be greatly reduced if water was no longer stored for long periods of time. Assuming an average cost of $1.50/$m^3$, desalinating this volume of water would cost on the order of $3,000 billion/year—roughly 12 percent of current gross world product.

Since we are focused only on the water supplied by freshwater systems, we capture only a portion of the total value of the natural desalting service provided by the hydrological cycle. Were we to include in our replacement cost calculation the water evapotranspired in situ by the trees harvested for lumber and fuel, by the grasslands used for grazing livestock, by the croplands watered only by rainfall, and by all other vegetation that supports human activity, we would produce a cost figure about nine times larger (Postel et al. 1996). As such, our figure represents a lower-bound estimate of the value of earth's renewable water supply overall, but an upper-bound estimate of the value of freshwater systems for irrigation, industrial, and municipal water supply. As freshwater resources are depleted or degraded in quality, as is happening in many parts of the world, desalination will be used incrementally as a costly replacement source.

Supply of Goods Other Than Water

In addition to supplying water, aquatic ecosystems provide many other goods of value to the human economy. Among the most important are fish, waterfowl, shellfish, and pelts.

The global freshwater fishery harvest offers a lower-bound estimate of the commercial value of freshwater fish. The annual harvest in 1989–91 was about fourteen million tons, and was valued at some $8.2 billion (FAO 1994). This figure does not include the values of the distribution economy or other components of the total economic impact of fishing.

Perhaps surprisingly, the value of sport fisheries often exceeds that of commerical fisheries—in some areas by one hundredfold or more (Talhem 1988). Sport fishing is a substantial recreational pursuit in the United States. In 1991, thirty-one million anglers fished an average of fourteen days each in the United States (U.S. Department of Interior 1991). Expenditures—including equipment, travel costs, etc.—totaled about $16 billion. The full economic impacts of freshwater angling, however, are far larger than direct expenditures (Felder and Nickum 1992). These impacts include changes in income or employment resulting from angling, spending on intermediate goods and services by firms that benefit directly from angling, and the economies supported by those firms. In the United States alone, the total economic output of freshwater fishing in 1991 was approximately $46 billion.

Waterfowl hunting in the United States in 1991 involved ~3 million hunters who, on average, spent about seven days each hunting migratory ducks and geese (U.S. Department of Interior 1991). Expenditures for these activities totaled $670 million. This figure underestimates the total economic value of waterfowl hunting, however, because it does not include secondary economic impacts.

Although the total global value of fish, waterfowl, and other goods extracted from freshwater systems cannot be estimated from available data, it certainly exceeds $100 billion per year and may be several times that amount. Moreover, the marginal value of these benefits is increasing in many places, as more people desire to spend time and money on these outdoor activities.

A wide variety of human activities threaten to diminish the benefits derived from living resources extracted from aquatic ecosystems. Overexploitation threatens to permanently diminish fish stocks. Toxic pollutants can render fish and other aquatic organisms unsafe to eat or reduce their productivity (Levin et al. 1989). Eutrophication, which can be caused by erosion, sewage inputs, or loss of riparian ecosystems, is correlated with undesirable shifts in fish communities (Carpenter et al. 1996). And to the extent that exotic species are introduced to develop sport fisheries, unexpected costs may result—such as collapse of native fish stocks and the spread of disease—that offset the benefits of the new fishery (Magnuson 1976, Moyle et al. 1987).

Nonextractive or Instream Benefits

Freshwater provides a host of services to humanity without ever leaving its natural channel or the aquatic system of which it is a part. These are the services most easily taken for granted, because they are provided with minimal or no investment or action on our part. They are also the services most rapidly being lost, since water and land management decisions frequently do not adequately value them or take them into account.

Most instream benefits have strong "public good" characteristics that make it difficult to capture their full value in the marketplace. For example, rivers, lakes, and reservoirs can provide environmental and recreational benefits to many people simultaneously (known in the economics lexicon as "nonrivalry in consumption"). It is also frequently difficult or impossible to exclude anyone from enjoying the benefits of public good resources, whether they pay for that enjoyment or not (known as "nonexcludability") (Colby 1989a; see also chapter 3, this volume).

The value of at least some instream services provided by aquatic systems depends on cultural and societal factors, which makes it impossible to derive an estimate of their total global value. Recreational uses, for example, may be valued highly in wealthy countries but very little in poor countries, where people do not have as much free time or money to enjoy leisure activities. By contrast, flood-recession farmers, fishers, and pastoralists may value certain instream services more than the rich, because they depend directly on them for their livelihoods. The value placed on protection of habitat for fish, birds, and other wildlife also may vary with the cultural and economic setting in which the aquatic habitat resides. What follows is a discussion of a few of the nonextractive or instream benefits provided by freshwater systems, along with some estimates of their value—either by way of rough global figures, or by regional or local examples.

Pollution Dilution

In late 1994 and early 1995, an estimated forty thousand migratory birds died at a reservoir in central Mexico. Scientists identified the cause to be an extremely high concentration of untreated human sewage in the water body, which allowed botulism bacteria to spread and poison the food eaten by ducks and other migratory waterfowl. During the months when most of the birds died, the reservoir reportedly consisted almost entirely of raw sewage (Dillon 1995). Given the vast quantities of sewage produced by the world's 5.7 billion people (Population Reference Bureau 1995), such incidents might be commonplace were it not for a key environmental service performed by freshwater systems: the dilution of pollutants.

Freshwater remaining in its natural channels helps keep water quality parameters at levels safe for fish, other aquatic organisms, and people. Today, some 1.2 billion people—about one out of every three in the developing world—lack access to safe supplies of drinking water, and 1.7 billion lack adequate sanitation services (Christmas and de Rooy 1991). As a result, water-borne diseases are primary killers of the world's poorest. The number of deaths due to unsafe water and inadequate sanitation—which include at least 2 million children each year—would be far higher were it not for the dilution of pollution by freshwater systems.

The old adage "Dilution is the solution to pollution" described the basic approach to pollution control up until about 1970, when, in response to pollution episodes like the Cuyahoga River catching fire in the United States, laws began to be passed requiring that cities and industries treat their waste before releasing it into the environment. Large sums were spent to restore and protect water quality. Virtually all countries, however, still depend heavily upon the diluting capacity of natural waters. Even in the OECD countries, domestic wastewater treatment is estimated to cover only about 60 percent of the population (Biswas 1992). Information for developing countries is sparse, but treatment coverage is certainly far lower. Moreover, few regions control for farm runoff and other dispersed pollution sources that add substantial quantities of sediment, pesticides, and fertilizers to water bodies. Dilution alone is certainly not sufficient to protect water quality or human health where pollution is highly concentrated or toxic, or where people lack access to safe drinking water supplies or adequate sanitation. But without the dilution function, things would be much worse.

One way of gauging the value of dilution as an instream service is to estimate what it would cost to remove all nutrients and contaminants from wastewater technologically. The combined cost of primary and secondary treatment is on the order of 8¢/m^3 (Bouwer 1992).

Costs of the advanced treatment needed to meet strict standards for the reuse of wastewater are considerably higher— in the range of 15–42¢/m^3, depending on the size and type of operation (Richard et al. 1991). Currently, municipal use worldwide totals ~300 km^3/year, while industrial use totals ~975 km^3/year; consumption in each sector amounts to an estimated 50 km^3 and 90 km^3/year, respectively (Shiklomanov 1993). If there was no diluting service whatsoever, and all of the municipal wastewater (which we assume equals 80 percent of the unconsumed municipal use, or 200 km^3/year) required advanced treatment at an average cost of 25¢/m^3, the treatment would cost ~$50 billion. Much industrial water is used for cooling, and therefore does not get severely contaminated. If we assume that one-third of the unconsumed industrial water (or 295 km^3) required advanced treatment at an average cost of 35¢/m^3, the total annual cost of this treatment would be just over $100 billion. The combined cost of $150 billion/year likely underestimates the total value of the dilution function, because a portion of agricultural drainage water would also require treatment to remove nitrates, pesticides, and other contaminants, a cost we do not attempt to estimate here.

Society already pays some of this price because pollution loads often exceed what nature can absorb, process, or dilute. But were the natural dilution service to be completely absent, the economic costs of keeping water pollution at harmless or tolerable levels would rise greatly. The risk today is that as increasing quantities of water are diverted from rivers and other water bodies to satisfy rising water demands, less water remains instream to provide this important ecosystem service. Decisions to divert water from its natural channels need to take into account the increased treatment costs that may be incurred as a result, as well as the potential costs to downstream water users of lower-quality water.

Transportation

In many parts of the world, inland waterways offer convenient and relatively inexpensive pathways for the transport of goods from one place to another. One way of valuing this instream service would be to estimate the cost of the next best alternative means of freight transportation in each area where navigation is used, and then to calculate the total cost-savings from navigation—an extremely difficult task since the next best alternative and its cost would vary from place to place. An easier approach is to examine the revenue derived from transportation by freshwater, averaged over all types of goods transported, exclusive of taxes. (Ideally, we should subtract from such figures the cost of maintaining navigation channels in order to arrive at a more accurate value of the ecosystem service, but we do not do that here.) In the United States, such revenues total $360 billion per year (U.S. Department of Transportation 1993, 1994), and in Western Europe they total $169 billion per year (U.N. Environment Programme 1992, United Nations 1994).

Unfortunately, consistent or reliable figures for transportation revenues are not available for Asia, Africa, or South America. However, the major rivers of these continents are important arteries for commerce. In China, for example, waterways accounted for 9 percent of the cargo shipped in 1988 (Burki and Yusuf 1992).

Thus, the combined revenue derived from transportation by water in the United States and Western Europe—$529 billion per year—provides a lower-bound estimate of the value of this instream service. The additional value from water transport in other geographic areas, along with the benefit of waterways for human travel (which is not included in these revenue figures), would raise the total value of this important instream service considerably. These transportation benefits are placed at risk by river diversions that reduce flows to levels too low to support navigation, by land-use practices that result in siltation of waterways, and by other activities that impair the use of freshwater systems for shipping.

Recreation

Freshwater systems provide numerous and varied opportunities for recreation—including swimming, sports fishing, kayaking, canoeing, and rafting. Like most other instream benefits, these recreational services have "public good" characteristics that make it difficult to capture their full value in the marketplace. In countries such as the United States, where enjoyment of the outdoors is on the rise, a large group of people benefit from these recreational services, but the total value of their enjoyment is difficult to measure. There is no charge levied or donation made that fully captures their collective willingness to pay.

Fortunately, economists have attempted to estimate the value of freshwater systems for recreation in some specific locales. Colby (1989a, b) summarized some of these findings for the western United States and finds that as of the mid-to-late 1980s the estimated economic value of recreational water uses there ranged between $4/acre-foot (AF) and $80/AF (or $3-65/1,000 m^3). (See table 11.2.) Studies of Colorado's Cache la Poudre River, for example, suggest that the value of an additional AF of flow during low-flow periods is $21 for fishing and $15 for shoreline recreation (Colby 1989a). The value of an additional flow unit in this river drops to zero at higher flow levels, since at these times flows are adequate for recreational uses. Likewise, another study cited by Colby (1989a) of a river in northern Utah found that the value of an additional unit of instream flow is zero until river flows drop to half of peak levels but reach $80/AF when flows are down to 20–25 percent of peak levels. These findings confirm what is intuitively obvious: that what recreationists value is the maintenance of a minimum flow in the river that safeguards recreational uses.

Instream recreational uses of water also generate substantial additional benefits to local economies in the form of recreation-related expenditures, such as boating, fishing, and camping equipment. One study (cited in Colby 1989a), for example, found that boaters on a twenty-mile stretch of the Wisconsin River spurred more than $800,000 in sales by local businesses during the summer season. Such sales are a key source of livelihood for small towns and Native American reservations in the western United States.

Table 11.2. Estimated nonmarket recreational water values, selected examples

Use	Description	Estimated Value
Fishing	Additional AF during low flows; Colorado	$21/AF[a]
Shoreline recreation	Additional AF during low flows; value drops to 0 during high flows; Colorado	$15/AF
Reservoir recreation	Leaving water in high mountain reservoirs for an additional two weeks in August; Colorado	$48/AF
River recreation	Additional AF when flows were 20–25% of peak levels; northern Utah	$80/AF
Fishing	Additional AF above the 35% flow level; Colorado mountain streams	$21/AF
Kayaking	Same as above	$5/AF
Rafting	Same as above	$4/AF

[a]Acre-foot; 1 AF = 1,234 m^3.

Sources: Colby 1989a, Moore and Willey 1991.

How do these instream recreational water values compare with the lowest-value offstream uses of water, which are typically in agriculture? This is difficult to answer because irrigation water frequently is heavily subsidized. A common way of estimating agricultural water values is through the farm budget method. After subtracting from total farm revenues all of the non-water production costs, a residual amount remains that represents the maximum amount the farmer could pay for water without suffering a net loss. Saliba and Bush (1987) applied such an approach to determine irrigation water values on the west side of California's San Joaquin Valley and came up with values ranging from $20/AF for safflower production to over $53/AF for melons. Howe and Ahrens (1988) used a version of the method for the upper Colorado River basin and concluded that the value of water in wheat production was no more than $25/AF; in barley, oats, and potato production, no more than $15/AF; in oats production, no more than $10/AF; and in production of corn for silage, no more than $4/AF. Finally, one study in the early 1980s cited in Colby (1989b) suggested that 80 percent of the irrigation water values in the western United States were below $55/AF.

An important conclusion thus emerges: at least during low-flow periods, the marginal value of water for instream recreational uses appears to be equal to or greater than the marginal value of water used in a substantial portion of irrigated agriculture in the western United States. The key policy message is similar to that for pollution dilution: Were these instream recreational values properly taken into account, fewer diversions for offstream uses would be economically justified. And a corollary: If water markets were able to operate more freely and purchases of water for instream recreational uses were more feasible, water would likely shift out of agriculture to the protection of instream recreational services.

Provision of Habitat

The supply of vital habitat by aquatic ecosystems depends greatly upon the dynamic connection between water and land, physical processes such as water and sediment flows, and a host of biophysical conditions such as water quality, temperature, and food web relationships. Freshwater ecosystems contain abundant life, including 41 percent of the world's known fish species and most of the world's endangered fish species (Moyle and Cech 1996). Decades of large-scale water engineering have disrupted many critical ecosystem functions and processes, with consequences that are just beginning to be recognized.

The provision of habitat in many large river systems, for example, depends critically on the annual flood. Floodplains are not only highly productive biologically, they offer a variety of aquatic habitats, including backwaters, marshes, and lakes. During a flood, many aquatic organisms leave the river channel to make use of these floodplain habitats as spawning, breeding, and nursery grounds. As floodwaters recede, young fish, waterfowl, and other organisms get funneled back into the main channel, along with nutrients and organic matter from the floodplain. In turn, the floodwaters deposit a new supply of sediment that enhances the floodplain's fertility. In this way, so called "flood pulses" provide critical habitat and increase the productivity of both the floodplain and the main river channel (Johnson et al. 1995). Examples of large river-floodplain ecosystems that are world renowned for their wildlife and other habitat benefits include the Gran Pantanal of the Paraguay River in South America, which alone harbors 600 species of fish, 650 species of birds, and 80 species of mammals (Covich 1993); the Sudd swamps on the White Nile in Sudan; and the Okavango River wetlands in Botswana (Sparks 1995).

In addition, the timing, volume, and quality of water flowing in its natural channel greatly affect the supply of habitat for fish and other aquatic organisms. Migrating fish species, for instance, may require certain minimum flow volumes at particular points in their life cycle. And many species have specific temperature, water quality, and other needs that must be met if they are to survive in a given river system.

The value of natural river, lake, and wetland systems as habitat for fish, waterfowl, and wildlife is even harder to estimate than recreational values, since the beneficiaries and benefits are much less clear and direct. In some cases, these values become visible only when they are lost or destroyed. In the Aral Sea basin in Central Asia, for instance, what was once the world's fourth largest inland lake has lost two-thirds of its volume because of excessive river diversions for irrigated agriculture. Some 20 of the 24 native fish species have disappeared, and the fish catch, which totaled ~40,000 tons a year in the 1950s and supported 60,000 jobs, has dropped to zero (Glazovskiy 1991, Micklin 1992).

Wetlands have shrunk by 85 percent, which, combined with high levels of agricultural chemical pollution, has greatly reduced waterfowl populations. In the delta of the Syr Dar'ya River—one of the Aral Sea's two major sources of inflow—the number of nesting bird species has dropped from an estimated 173 to 38 (Micklin 1992). This region illustrates vividly how economic and social decline may follow close on the heels of ecological destruction.

In the western United States, the emergence of active water markets combined with growing public interest in preserving fish species, bird populations, and wildlife generally has begun to attach some market values to the critical habitat supplied by aquatic ecosystems. During 1994, there were nineteen reported water transactions in the western United States that had the purpose of securing more water for aquatic habitats, especially rivers and wetlands (Smith and Vaughan 1995). A sampling of such transactions during recent years gives at least a partial sense of water's current market value for habitat preservation or restoration in this part of the world:

- In 1994, the federal Bureau of Reclamation decided to lease just over 183,000 AF of water from contractors supplied by a large federal project in California in order to augment streamflows for migrating fish, supply more water to wildlife refuges, and increase freshwater outflows through the Sacramento–San Joaquin Delta. Most of this water will cost $50/AF (Smith and Vaughan 1995).
- A multi-agency program initiated and continuing at present is transferring water rights from farms within the Bureau of Reclamation's Newlands Project to Lahonton Valley wetlands, which include the Stillwater Wildlife Refuge. Two private conservation organizations—The Nature Conservancy and Nevada Waterfowl Association—have been involved in purchasing water rights for this transfer (Wigington, personal communication, 1996), with prices for permanent water rights estimated in the early stages of the program to be in the range of $200-300 per acre-foot (Shupe 1989; Smith and Vaughan 1991).
- In 1992, the San Luis–Kesterson Wildlife Refuge received 250 AF of groundwater from a consortium of users for a price of $20/AF for the purpose of maintaining wetlands at Kesterson (Smith and Vaughan 1992b).
- In 1994, the Bonneville Power Administration (a federal agency that is a major supplier of hydroelectric power in the Pacific Northwest) decided to lease 16,000 AF/year of Upper Snake River water from an Oregon farm primarily to increase streamflows for migrating salmon (apparently there are hydropower benefits as well). The annual lease is renewable for up to three years, and the water will cost BPA $50–80/AF (Smith and Vaughan 1995).

As these examples illustrate, the value of water for habitat protection in the western United States, as with the value of instream water for recreation, appears to equal or exceed that for some offstream uses, particularly in agriculture.

Option, Bequest, and Existence Values

Because of freshwater's central role in maintaining uniquely beautiful natural areas, critical habitat, or highly valued recreational sites, "non-user" values of water can be substantial. Estimating people's willingness to pay to preserve the option of enjoying a site in the future (option value), to ensure that descendants will be able to enjoy a site (bequest values), or simply to know that a site will continue to exist (existence values) is not easy. These values are important, however, particularly when irreversible decisions are to be made, such as constructing a dam that will flood a beautiful mountain canyon, or channeling through a wetland that will permanently destroy wildlife habitat. According to Colby (1989a), "existence, bequest and option values ranging from $40–$80 per year per non-user household have been documented for stream systems in Wyoming, Colorado, and Alaska." It is estimated that the total (user and non-user) benefits of preserving Mono Lake levels amount to about $40 per California household, 80 percent of which is attributed to option, bequest, and existence values (Colby 1989a).

Threats to Aquatic Ecosystem Services

For most of human history, water management has largely been an attempt to manipulate the hydrological cycle for human benefit. The pace and scale of water engineering schemes have increased greatly during this century, especially during its latter half. Worldwide, the number of large dams (those more than fifteen meters high) has climbed from just over five thousand in 1950 to approximately thirty-eight thousand today. More than 85 percent of large dams have been built during the last thirty-five years. Engineers have built thousands of kilometers of diversion canals, channels, and levees to divert water for human uses, to drain wetlands for farms and shopping malls, and to control floods. The human enterprise has massively changed the aquatic environment in a very short period of time, and the consequences are just beginning to come to light.

A myriad of human activities—from the construction of dams, dikes, and levees to uncontrolled pollution and climatic change—now threaten the aquatic ecosystem services that humanity depends on and benefits from in so many ways (see table 11.3.) Signs that the aquatic environment is in jeopardy abound. A substantial fraction of the rare and threatened species of North America are aquatic, and primarily freshwater. In North America, the American Fisheries Society estimates that 364 species or subspecies of fish are now threatened, endangered, or of special concern—the vast majority of them at risk because of habitat destruction (Williams et al. 1989). Throughout Canada, the United States, and Mexico, an estimated 20 percent of amphibians and fishes, 36 percent of crayfishes, and 55 percent of Unionid mussels are imperiled to some degree or are already extinct (Allan and Flecker 1993). As Covich (1993) has noted, "We have often ignored the high species richness associated with inland waters and have allowed many freshwater habitats to be dammed, channelized, drained, eroded, and pol-

Table 11.3. Threats to aquatic ecosystem services from human activities

Human Activity	Impact on Aquatic Ecosystems	Values/Services at Risk
Dam construction	Alters timing and quantity of river flows, water temperature, nutrient and sediment transport, delta replenishment; blocks fish migrations	Habitat, sports, and commercial fisheries; maintenance of deltas and their economies
Dike and levee construction	Destroys hydrologic connection between river and floodplain habitat	Habitat, sports, and commercial fisheries; natural floodplain fertility; natural flood control
Excessive river diversions	Depletes streamflows to ecologically damaging levels	Habitat, sports, and commercial fisheries; recreation; pollution dilution; hydropower; transportation
Draining of wetlands	Eliminates key component of aquatic environment	Natural flood control, habitat for fisheries and waterfowl, recreation, natural water filtration
Deforestation/ poor land use	Alters runoff patterns, inhibits natural recharge, fills water bodies with silt	Water supply quantity and quality, fish and wildlife habitat, transportation, flood control
Uncontrolled pollution	Diminishes water quality	Water supply, habitat, commercial fisheries, recreation
Overharvesting	Depletes living resources	Sport and commercial fisheries, waterfowl, other living resources
Introduction of exotic species	Eliminates native species, alters production and nutrient cycling	Sport and commercial fisheries, waterfowl, water quality, fish and wildlife habitat, transportation
Release of metals and acid-forming pollutants to air and water	Alters chemistry of rivers and lakes	Habitat, fisheries, recreation
Emission of climate-altering air pollutants	Has potential to make dramatic changes in runoff patterns from increases in temperature and changes in rainfall	Water supply, hydropower, transportation, fish and wildlife habitat, pollution dilution, recreation, fisheries, flood control
Population and consumption growth	Increases pressures to dam and divert more water, drain more wetlands, etc.; increases water pollution, acid rain, and potential for climate change	Virtually all aquatic ecosystem services

luted with nutrients, salts, silt, and chemicals. Biodiversity and ecosystem integrity are declining in a wide range of locations throughout the world. . . ."

Establishing direct links between human activities and losses of aquatic ecosystem services in specific locations is often difficult. In the Mississippi River valley, the draining of wetlands and alteration of river channels has destroyed a large portion of the river system's natural flood protection services. The loss of these services was partially responsible for the massive flooding that occurred during 1993, which caused property damages estimated at $12 billion (Myers and White 1993).

Gore and Shields (1995) link an 80 percent decline in the commercial fish harvest in the Missouri reach of the Missouri River with the loss of natural habitat from the channel and meander belts, along with a shortening of the river. They also connect an 87 percent drop in the average fall-run chinook salmon population in California's Sacramento River with a 43 percent reduction in the area of freshwater wetlands in the river valley between 1939 and the mid-1980s. In the Vistula River in Eastern Europe, where the commercial fish harvest has declined sharply, they note habitat changes that include the elimination of islands and braided reaches, as well as a 50 percent reduction in channel width.

In 1992, a committee of the Water Science and Technology Board of the U.S. National Research Council released a study broadly examining the state of aquatic ecosystems in the United States and the need and potential for their restoration. Among the study's findings (National Research Council 1992):

- The nation has lost ~117 million acres of wetlands over the past two centuries—a 30 percent loss of presettlement wetland area. Excluding Alaska, more than half of wetland area has been lost.
- More than 85 percent of the inland water surface is artificially controlled.
- More than half of the nation's perennial rivers and streams have fish populations that are adversely affected by turbidity, high temperatures, toxins, or low levels of dissolved oxygen. Almost 40 percent are affected by low flows, and 41 percent by siltation, bank erosion, and channelization.
- Approximately 2.6 million acres of lakes are impaired relative to their intended use, with non-point pollution from farming activities the leading cause.

No doubt, similar syntheses of the state of aquatic ecosystems and resources in other parts of the world would suggest severe degradation and impairment of ecological services as well. Moreover, with the possible exception of dam construction (Postel et al. 1996), there is little sign of any reduction in the human activities causing this degradation. Indeed, with population and consumption growing by record amounts annually, pressures on the aquatic environment are bound to increase. And the prospect of global climatic change from the build-up of greenhouse gases in the atmosphere adds a troubling wild card to the overall picture (Waggoner 1990). With our present network of dams, reservoirs, and other water infrastructure geared to present patterns of rainfall and runoff, climatic change could greatly impair virtually all of the ecosystem services that aid and underpin the human economy.

Conclusion

Rivers, lakes, aquifers, and wetlands provide a myriad of benefits to the human economy—including water for drinking, irrigation, and manufacturing; goods such as fish and waterfowl; and a host of non-extractive benefits, including recreation, transportation, flood control, bird and wildlife habitat, and the dilution of pollutants. These latter "instream" benefits are particularly difficult to measure, since many are public goods that are not quantitatively valued by the market economy, and they are values that would vary with culture and place. The total global value of all services and benefits provided by freshwater systems is thus impossible to measure accurately but would almost certainly measure in the several trillions of dollars.

In combination, the value of freshwater ecosystems and the numerous threats to them strongly suggest the need for a major international effort to prevent further degradation to these environmental support systems, as well as to restore a portion of the services that have been lost. The full economic impacts of dams and river diversions, the draining of wetlands, and other activities have often been underestimated because the resulting loss of ecosystem services has been overlooked. Better accounting of the nonmarket values of rivers, lakes, and wetlands would help ensure that land-use and water management decisions are both economically rational and environmentally sound. In the western United States, for example, the marginal value of water for recreation and habitat protection appears to equal or exceed that for irrigated agriculture, at least during low-flow periods. Public policies, including heavy irrigation subsidies and antiquated water rights systems, often are not in accord with this finding.

Much additional research is needed to establish the intricate connections between human activities and the loss of freshwater ecosystem services. However, given the rapid pace of ecosystem destruction and decline, the irreversible nature of many of these losses, and the high value of freshwater ecosystem services to the human economy, it would seem wise to err on the side of overprotection of freshwater systems from this point forward.

Acknowledgments

We thank Matthew Wilson for research assistance, the Pew Fellows Program in Conservation and Environment for financial support, and A. Covich, J. F. Kitchell, and D. E. Schindler for helpful comments on a draft of the manuscript.

References

Allan, J. D., and A. S. Flecker. 1993. "Biodiversity conservation in running waters." *BioScience* 43:32–43.

Biswas, A.K. 1992. "Sustainable water development: A global perspective." *Water International* 17: 68–79.

Bouwer, H. 1992. "Agricultural and municipal use of wastewater." Prepared for a meeting of the International Association of Water Pollution Research and Control, Washington, D.C.

Burki, S. J., and S. Yusuf, eds. 1992. "The sectoral foundations of China's development." World Bank Discussion Paper No. 148, China and Mongolia Department Series.

Carpenter, S. R., T. Frost, L. Persson, M. Power, and D. Soto. 1996. "Freshwater ecosystems: Linkage of complexity and processes." In H. Mooney, J. H. Cushman, E. Medina, O. Sala, and E.D. Shulze, eds., *Functional Roles of Biodiversity: A Global Perspective.* John Wiley and Sons, New York.

Christmas, J., and Carel de Rooy. 1991. "The decade and beyond: At a glance." *Water International* 16: 126–134.

Colby, B.G. 1989a. "The economic value of instream flows—Can instream values compete in the market for water rights?" In L. J. MacDonnell, T. A. Rice, and S. J. Shupe, eds., *Instream Flow Protection in the West,* Natural Resources Law Center, University of Colorado School of Law, Boulder, Colorado.

Colby, B.G. 1989b. "Estimating the value of water in alternative uses." *Natural Resources Journal* 29: 511–527.

Covich, A.P. 1993. "Water and ecosystems." In P.H. Gleick, ed., *Water in Crisis: A Guide to the World's Freshwater Resources,* pp.40–55. Oxford University Press, New York.

Dillon, S. 1995. "Scientists say raw sewage killed 40,000 birds." *The New York Times,* September 29.

FAO (United Nations Food and Agriculture Organization). 1994. *FAO Yearbook of Fishery Statistics* vol. 17.

Felder, A. J., and D. M. Nickum. 1992. *The 1991 economic impact of sport fishing in the United States.* American Sportfishing Association, Alexandria, Virginia.

Glazovskiy, N. F. 1991. "Ideas on an escape from the 'Aral crisis.'" *Post-Soviet Geography* 32: 73–89.

Gore, J. A., and F. D. Shields, Jr. 1995. "Can large rivers be restored?" *BioScience* 45: 142–152.

Howe, C. W., and W. A. Ahrens. 1988. "Water resources of the upper Colorado River basin: Problems and policy alternatives." In M. T. El-Ashry and D. C. Gibbons, eds., *Water and Arid Lands of the Western United States*, pp. 169–232. Cambridge University Press, New York.

Johnson, B. L., W. B. Richardson, and T. J. Naimo. 1995. "Past, present, and future concepts in large river ecology." *BioScience* 45: 134–141.

Levin, S. A., M. A. Harwell, J. R. Kelly and K. D. Kimball, eds. 1989. *Ecotoxicology: Problems and Approaches*. Springer-Verlag, New York.

L'Vovich, M. I., G. F. White, A. V. Belyaev, J. Kindler, N. I. Koronkevic, T.R. Lee, and G.V. Voropaev. 1990. "Use and transformation of terrestrial water systems." In B. L. Turner, ed., *The Earth as Transformed By Human Action*. Cambridge University Press, Cambridge.

Magnuson, J. J. 1976. "Managing with exotics—A game of chance." *Transactions of the American Fisheries Society* 105: 1–9.

Micklin, P. P. 1992. "The Aral crisis: Introduction to the special issue." *Post-Soviet Geography* 33: 269–282.

Moore, D., and Z. Willey. 1991. "Water in the American West: Institutional evolution and environmental restoration in the 21st century." *University of Colorado Law Review* 62: 775–825.

Moyle, P. B., and J. J. Cech. 1996. *Fishes*, 3rd edition. Prentice-Hall, Upper Saddle River, New Jersey.

Moyle, P. B., H. W. Li, and B. Barton. 1987. "The Frankenstein effect: Impact of introduced fishes on native fishes of North America." In R. H. Stroud, ed., *The Role of Fish Culture in Fish Management*, pp. 415–426. American Fisheries Society, Bethesda, Maryland.

Myers, F. M., and G. F. White. 1993. "The challenge of the Mississippi flood." *Environment* 35, no. 10.

National Research Council. 1992. *Restoration of Aquatic Ecosystems*. National Academy Press, Washington, D.C.

OTA (Office of Technology Assessment). 1988. *Using Desalination Technologies for Water Treatment—Background Paper*. U.S. Government Printing Office, Washington, D.C.

Population Reference Bureau. 1995. *1994 World Population Data Sheet*. Population Reference Bureau, Washington, D.C.

Postel, S. 1992. *Last Oasis: Facing Water Scarcity*. W.W. Norton & Co., New York.

Postel, S. L., G. C. Daily, P. R. Ehrlich. 1996. "Human appropriation of renewable freshwater." *Science* 271: 785–788.

Richard, D., et al. 1991. "Wastewater reclamation costs and water reuse revenue." Prepared for American Water Resources Association 1991 Summer Symposium *Water Supply and Water Reuse: 1991 and Beyond*, San Diego, California.

Saliba, B.C., and Bush, D.B. 1987. *Water Markets in Theory and Practice: Market Transfers, Water Values, and Public Policy.* Westview Press, Boulder, Colorado.

Schwarz, H. E., J. Emel, W. J. Dickens, P. Rogers, and J. Thompson. 1990. "Water quality and flows." In Turner, B.L., W. C. Clark, R. W. Kates, J. F. Richards, J. T. Mathews, and W. B. Meyer, eds., *The Earth as Transformed By Human Action*, pp. 253–264. Cambridge University Press, Cambridge, England.

Shiklomanov, I. A. 1993. "World freshwater resources." In P. H. Gleick, ed., *Water in Crisis: A Guide to the World's Freshwater Resources*, pp. 13–24. Oxford University Press, New York.

Shupe, S. J., ed. 1989. *Water Market Update* (March). Shupe & Associates, Santa Fe, New Mexico.

Smith, R. T., and R. Vaughan. 1991. "1990 annual transaction review: Growing diversity of water agreements." *Water Strategist* (January). Stratecon, Claremont, California.

Smith, R. T., and R. Vaughan. 1992a. "1991 annual transaction review: Water comes to town." *Water Strategist* (January). Stratecon, Claremont, California.

Smith, R. T., and R. Vaughan. 1992b. "San Luis/Kesterson Wildlife Refuge acquires water from Panoche Water District." *Water Intelligence Monthly* (December). Claremont, California.

Smith, R. T., and R. Vaughan. 1995. "1994 annual transaction review: Markets expanding to new areas." Water Strategist (January). Stratecon, Claremont, California.

Sparks, R. E. 1995. "Need for ecosystem management of large rivers and their floodplains." *BioScience* 45: 168–182.

Talhem, D. R. 1988. *Economics of Great Lakes Fisheries: A 1985 Assessment.* Technical Report No. 54. Great Lakes Fisheries Commission, Ann Arbor, Michigan.

United Nations. 1994. *Annual Bulletin of Transportation Statistics for Europe.* U.N. Statistical Bulletin.

United Nations Environment Programme. 1992. *Environmental Data Report 1991–1992.*

United States Department of Interior Fish and Wildlife Service, U.S. Department of Commerce, and the Bureau of the Census Economic and Statistics Administration. 1991. *The 1991 National Survey of Fishing, Hunting, and Wildlife-Associated Recreation.*

United States Department of Transportation. 1993. *National Transportation Statistics Annual Report: Historical Compendium, 1960–1992.* Bureau of Transportation Statistics Report BTS-93-1.

United Nations Department of Transportation. 1994. *Transportation Statistics Annual Report 1994.* Bureau of Transportation Statistics Report BTS-94-1.

Waggoner, P. E., ed. 1990. *Climate Change and U.S. Water Resources.* John Wiley, New York.

Wangnick Consulting. 1990. *1990 IDA Worldwide Desalting Plants Inventory.* International Desalination Association, Englewood, New Jersey.

Williams, J. E., J. E. Johnson, D. A. Hendrickson, and S. Contreras-Balderas. 1989. "Fishes of North America endangered, threatened, or of special concern: 1989." *Fisheries* 14 (November-December): 2–20.

World Water/World Health Organization. 1987. *The International Drinking Water Supply and Sanitation Decade Directory*. Thomas Telford, London.

Chapter 12

The World's Forests and Their Ecosystem Services

Norman Myers

The world's forests cover some thirty-four million square kilometers or roughly 27 percent of the ice-free land surface of the earth (FAO 1995). Their present expanse is a full one-third less than it was in historical times, and in both the tropical and boreal zones we are witnessing an accelerating decline of forests. The rigors of global warming are likely to bring on still further deforestation. If we carry on with business as usual and with altogether inadequate conservation measures, today's young people may eventually look out on a largely deforested world (Myers 1996). While this will mean a sizeable drop in supplies of timber and fuelwood, it will be much more significant in terms of ecosystem services lost. This chapter takes a look at what is at stake.

Forests supply ecosystem services of numerous sorts (Adamowicz et al. 1993). They stabilize landscapes (Woodwell 1993). They protect soils and help them to retain their moisture and to store and cycle nutrients (Ehrlich and Ehrlich 1992). They serve as buffers against the spread of pests and diseases (Woodwell 1995). By preserving watershed functions, they regulate water flows in terms of both quantity and quality (Bruijnzeel 1990), thereby helping to prevent flood-and-drought regimes in downstream territories (Sfeir-Younis 1986). They are critical to the energy balance of the earth (Woodwell 1993). They modulate climate at local and regional levels through regulation of rainfall regimes (Meher-Homji 1992) and the albedo effect (Gash and Shuttleworth 1992); and at planet-wide level, they help to contain global warming by virtue of the carbon stocks in their plants (especially trees) and soils (Woodwell and Mackenzie 1995).

Certain of these ecosystem services tend to be more prominent in tropical forests, and it is this biome that hence serves as the main focus for much of this chapter. Although these forests cover only 6 percent of earth's land surface, they receive almost half of earth's rainfall on land, making them exceptionally important for watershed functions. Sometimes this rainfall can be unduly heavy; at Cherrapunji in northeastern India, 22.5 meters of rain fell during just five months of the monsoon season in 1974, an average of 15 centimeters per day—an amount way beyond what we would normally encounter in other forest zones. In those tropical forests that receive over 3 meters of rainfall a year, a half-hour thunderstorm can produce 25 millimeters of rain, or forty times more water than a typical shower in the northeastern United States. Forests' linkages with the supply of water for public health are much more important in developing countries than elsewhere (since these countries often have poorly developed water supplies for household use), making tropical forests particularly pertinent. Tropical forests generally feature more vegetation than any other biome on earth, causing them to play an especially significant role as concerns the albedo effect and global warming. The reader will quickly recognize the ways in which tropical forests provide more ecosystem services than do other forests—which is not to say of course that other forests do not play a front-rank role in supply of ecosystem services.

Watershed Services

Deforestation of upland catchments often leads to disruption of hydrological systems, causing year-round water flows in downstream areas to give way to flood-and-drought regimes. This is especially the case in the humid tropics, where the forests exert a "sponge effect" and soak up moisture before releasing it at regular rates. While forest cover remains intact, rivers not only run clear and clean, they also flow throughout the year. When the forest is cleared, rivers start to turn muddy and become swollen or shrunken. In several major river basins of the humid tropics, notably those of the Ganges, Brahmaputra, Chao Phraya, and Mekong, rainy-season supplies of water tend to be released in floods, followed by months-long droughts.

How does this sponge effect actually operate? The multi-storyed structure of the forest, together with its abundant foliage, helps break the impact of a tropical downpour. Much of the water trickles down branches and tree trunks, or drips off leaves in a fine spray, so that when the rainfall reaches the ground, it percolates steadily into the soil or runs off into streams and rivers gradually. An undisturbed dipterocarp forest in Southeast Asia intercepts an average of at least 35 percent of rainfall, whereas a logged forest intercepts less than 20 percent, and a plantation of rubber or oilpalm trees, only 12 percent (Ba 1977). In the Tai forest of southwestern Ivory Coast in West Africa, rivers flowing from a primary forest release twice as much water halfway through the dry season, and between three and five times as much at the end of the dry season, as do rivers from coffee plantations (Dosso et al. 1981).

Control of Soil Erosion

The impact of tropical downpours causes more soil erosion in deforested areas than anywhere else on earth. The most prominent instance is probably Nepal, where the Department of Soil and Water Conservation estimates that between 30 and 75 tonnes of soil are washed away from each hectare of deforested land each year. This means that the country altogether loses as much as 240 million cubic meters of soil annually—a highly precious export that Nepal unwittingly dispatches to India (Cool 1980). The economic costs are substantial, especially to agriculture.

Almost as extreme is the deforestation-caused erosion in much of Ethiopia, where only forty years ago forest covered about 10 percent of the country, an amount that has now fallen to 3 percent. As a result, huge quantities of topsoil flow from the Ethiopian highlands each year, some of it being carried down the Blue Nile until it silts up the Roseires Dam many hundreds of kilometers westward over the border in Sudan.

To consider some economic repercussions of erosion, let us look at the case of Indonesia. Eroded territories, formerly forested, now exceed 400,000 square kilometers, or about one-fifth of national territory. The agricultural costs of this erosion are sizeable, especially on the island of Java, which, with an area of 132,470 square kilometers (roughly the same as New York State or Greece), and with 110 million people, possesses only 15 percent of its original forest cover. In the late 1980s Java's croplands were losing at least 770 million tonnes of topsoil per year, worth a rice output of more than 1.5 million tonnes and equivalent, at an average consumption rate of 100–130 kilograms per person per year, to the needs of 11.5–15 million people per year (Magrath and Arens 1989).

As noted, washed-away topsoil causes rivers to become burdened with suspended sediment, turning some of them into turgid streams of mud. The Ganges, for instance, carries an average annual load of sediment amounting to 1,544 tonnes per square kilometer of drainage basin, or fourteen times higher than that of the Mississippi. The Brahmaputra carries almost as much, 1,429 tonnes; while the Irrawadi carries well over 900 tonnes, the Indus over 500 tonnes, and the Mekong almost 500 tonnes (Reiger 1977). Siltation in the Ganges system is so pronounced that a number of riverbeds are rising at a rate of almost half a meter a year, grossly aggravating floods. Newly formed shoals have rendered several sectors of the main river unnavigable, while certain industrial installations in the downstream plain must suspend activities for several months of the year due to lack of water. The ports of both Calcutta and Dacca are silting up.

Deforestation-derived siltation proves a problem for water impoundments in several parts of the humid tropics. In the Philippines, the Ambuklao Dam is silting up so fast that its useful life is being reduced from its planned fifty-six years to thirty-two years, due to deforestation in the Agno River watershed (Wiens 1989; see also Myers 1988). The Mangla Dam in Pakistan, completed in 1967, receives so much silt from the Jhelum River watershed that its operational life is being reduced from more than one hundred years to less than fifty years. Also in Pakistan, the world's largest dam, the Tarbela Dam, is losing its storage capacity of 12 billion cubic meters at a rate that will leave the dam useless within just forty years. In Ecuador, the 100-million-cubic-meter Poza Honda Reservoir, constructed in 1971, is losing its capacity at a rate that will leave the installation useless within only another twenty-five years; a conservation program to reforest that half of the 175-square-kilometer watershed where forest cover has been eliminated would cost only $1.8 million, extending the reservoir's life to its planned fifty-year life span and producing benefits of at least $30 million (Fleming 1979). Siltation of hydropower and irrigation-system reservoirs worldwide, i.e., not just in the humid tropics but derived in major measure from deforestation in watersheds, is estimated to levy a cost of $6 billion a year (Mahmood 1987).

As a further form of "hidden" environmental costs of sedimentation, consider the impact on offshore fisheries. Encircling the Philippines' seven thousand islands with their eighteen thousand kilometers of coastline (only slightly less than the United States'), there are 44,000 square kilometers of coral reefs, which supply about one-tenth of the country's fisheries catch in both commercial and subsistence terms. Several parts of these fisheries are declining, due to deforestation far inland. Along other sectors of the Philippines' coastal zones, mangrove ecosystems are being suffocated by silt carried down from watershed catchments. In the late 1980s, the country was exporting $100 million of oysters, mussels, clams, and cockels each year—and again, these valuable fisheries were being depleted through the ecological "backlash effects" of deforestation. Also in the late 1980s, and in the particular locality of Bacuit Bay on Palawan island, logging on steep slopes had increased soil erosion 235 times above that for undisturbed forest, with a "silt smother" effect for the Bay's coral reef and its fisheries that reduced commercial revenues by almost half in the mid-1980s (Hodgson and Dixon 1988).

Linkages with Public Health

Everybody needs water every day for drinking, cooking, washing, and sanitation. A minimum amount is twenty liters a day—roughly the amount a rich-world person uses with every flush of the toilet. As the standard of living increases in developing countries, so does the demand for water, until a better-off person in an urban community may consume two hundred liters a day. There are now one hundred million more developing-world people without access to clean water or adequate sanitation than in 1970 (World Health Organization 1994). Were these two services to become more widely available, they would help to cut down on the toll of water-related diseases. Typhoid, cholera, amoebic infections, bacillary dysentery, and diahrrea, among related diseases, cause an estimated mortality of fifteen million people each year, plus morbidity relating to as much as 80 percent of all sicknesses.

Large numbers of these afflicted people live in the humid tropics. This means that tropical forests, by assuring dependable supplies of good-quality water for household use, make a substantial contribution to campaigns for better health. To meet basic needs for water and sanitation for the entire developing world would cost $300 billion over a program lifetime of ten years. Were a mere 1 percent of this sum, $3 billion a year, to be allocated to safeguarding watershed systems, it would represent as sound an investment as water piping, stand taps, sewers, and the other conventional equipment that development agencies envisaged for the Water and Sanitation Decade (1981–1990).

Yet in a tropical forest territory with some of the highest rainfall on earth, Peninsular Malaysia, water is now rationed for part of the year in Kuala Lumpur and several other urban areas. Water demand in the Peninsula is projected to double during 1990–2010; and as water supplies decline in face of rising demand, so costs increase for the Malaysian consumer. The price of water from a catchment with undisturbed forest increases twofold when the forest becomes subject to controlled logging and fourfold when the forest becomes subject to uncontrolled logging (Myers 1992a).

In similar fashion, public-health programs in Bangkok, Manila, Lagos, Abidjan, and several other conurbations of the humid tropics are being set back through deforestation-caused declines in quantity and quality of water supplies. Bangkok, for instance: With a populace of seven million people, water demand has now topped one billion liters a day. The city's population is projected to top twelve million in 2010, and its thirst to increase at least three times. At present, about one-third of the water supply comes from the giant Chao Phraya River, which rises twelve hundred kilometers away in the northern mountains of the country. The rest of the city's water comes from local wells (Phisphumvidhi 1981).

Both these sources of Bangkok's water are declining. The river's basin covers 177,550 square kilometers, or more than one-third of the entire country, and supports almost half the population. As a result of the disrupted river flow, viz. a regime of too much water followed by too little, the river is less and less capable of supplying the needs of Bangkok. So the citizens turn increasingly to their eleven thousand groundwater wells—with the result that, at present rates of extraction, many if not most of the underground stocks are expected to give out by 2000. Worse still, the massive pumping from subsurface reservoirs is causing the city to sink. Bangkok is literally declining, at a rate fourteen times faster than Venice. Because most of the city now lies a mere one meter above sea level, the present situation, if allowed to persist, could soon leave the city below sea level.

Regulation of Rainfall Regimes

Deforestation in the tropics can sometimes result in reduced rainfall (Salati and Nobre 1992; Myers 1992b). This is unusually significant for agriculture. In northwestern Peninsular Malaysia, the Penang and Kedah states have experienced disruption of rainfall regimes to the extent that twenty thousand hectares of paddy ricefields have been abandoned and another seventy-two thousand hectares have registered a marked production drop-off in this "rice bowl" of the Peninsula (Chan 1986). Similar deforestation-associated changes in rainfall have been documented in much of the Philippines, southwestern India, montane Tanzania, southwestern Ivory Coast, northwestern Costa Rica, and the Panama Canal Zone (Meher-Homji 1992).

Deforestation can also affect rainfall regimes at much wider levels. According to research with water isotopes (Salati and Nobre 1992), much of the moisture in Amazonia—between half and four-fifths in central and western sectors—remains within the eco-zones involved. That is, it is constantly transpired by plants into the atmosphere, where it gathers in storm clouds before being precipitated back onto the forest (mean recycling time, 5.5 days). Thus much of the Amazonian forest respresents a significant source of its own moisture: it does not have to depend on moisture advected from adjacent stretches of ocean.

Were Amazonia to be widely deforested, there could be a pronounced decrease in the amount of moisture that is being evapotranspired from it into the atmosphere, leading to a significant decline in rainfall. A decline of this scope alone would entrain profound and irreversible ecological changes in many parts of the basin. More important still, it could trigger a self-reinforcing process of growing desiccation for remaining forest cover, with declining moisture stocks followed by yet more desiccation, and so forth. Eventually the repercussions could extend outside Amazonia, even to southern Brazil with its major agricultural lands (Salati and Nobre 1992).

Whether a similar drying-out phenomenon could occur in other large tracts of tropical forest, notably that of the Zaire basin, is scarcely considered to date, let alone scientifically investigated.

The Albedo Connection

Much of the energy that converts surface moisture into water vapor comes from the sun's radiational heating of the land surface. The energy thus depends on surface albedo, or relevant degree of reflectant "shininess" of the land surface (Gash and Shuttleworth 1992). In turn, the albedo depends on vegetation, which absorbs more heat than does bare soil. Over thick vegetation, vigorous thermal currents take moisture (provided by the same plant cover) up into the atmosphere, where it condenses as rain. Because of its influence on convection patterns and wind currents, and hence on rainfall regimes, the albedo effect constitutes a basic factor in controlling climate. In tropical forest zones, it appears to be, together with surface roughness, the most important factor by which the land surface can affect climate.

When vegetation is removed from the earth's surface in large quantities, the result is often a self-promoting cycle of albedo enhancement, leading to a new stable state of less warm soil, lower rainfall, and sparser vegetation. There ensues a significant decrease in rainfall and evapotranspiration, as also in cloud cover. Were albedo-derived processes to become disrupted through tropical deforestation, the repercussions could well extend, via altered patterns of air circulation, throughout an entire region. If, say, the whole of Amazonia's almost three million square kilometers were to be deforested during a period of thirty-five to fifty years, giving way to grasslands and crops, the surface albedo is estimated to increase from 0.11 to 0.19, and rainfall to decrease by 0.5–0.7 mm per day, with both evapotranspiration and cloud cover being reduced (Henderson-Sellers and Gornitz 1984).

Climate Regulation: Global Warming

Still more important is the forests-climate linkage at global level, through forests' role as carbon sinks and hence their capacity to mitigate global warming (Apps and Price 1996; Woodwell and Mackenzie 1995). Forests currently hold some 1,200 gigatonnes (billion tonnes) of carbon in their plants and soils (out of 2,000 gigatonnes in all terrestrial plants and soils), by contrast with 750 gigatonnes in the atmosphere (Houghton et al. 1990; Woodwell 1993; see also Dixon et al. 1994). Around half of the forest carbon is located in boreal forests, more than one-third in tropical forests, and roughly one-seventh in temperate forests; over two-thirds of the total is contained in soils and peat deposits (Dixon et al. 1994). Boreal forests, being earth's largest terrestrial biome, probably contain more carbon than all the earth's proven fossil fuel reserves. They thus possess the scope for both the greatest change in the global carbon cycle and the greatest potential feedbacks on climate systems (cp. Nilsson 1995).

Another way to view the flywheel effect of the world's forests on carbon stocks is to note that they account for 65 percent of net plant growth and carbon fixation on land (Zak 1995). Just Siberia's forests absorb 10 percent of human emissions of carbon dioxide annually (Alexeyev 1991). All in all, the annual respiration and photosynthesis of forests transfer carbon dioxide equivalent to 12–14 percent of the atmospheric content (Woodwell 1995).

When forests are burned—as is the case with cattle ranching and small-scale agriculture in the humid tropics and with fires both wild and human-made in the boreal zone—they release their carbon. Of the roughly 7.6 gigatonnes of carbon emitted per year into the global atmosphere, and contributing almost half of greenhouse-effect processes, 1.6 gigatonnes (plus or minus 0.4 of a gigatonne) come from forest burning in the tropics (Houghton 1993, Myers 1989), almost all the rest stemming from combustion of fossil fuels. At the same time, forest expansion and growth in the temperate and boreal zones sequester 0.7 of a gigatonne (plus or minus 0.2 of a gigatonne) per year. All in all, then, there is a net flux to the atmosphere of 0.9 of a gigatonne (plus or minus 0.4 of a gigatonne) of carbon per year (Dixon et al. 1994).

Note that the tropical forest–burning component is increasing more rapidly than the fossil fuel component. It grew by 75–100 percent during the 1980s (Myers 1989; see also Houghton 1993), whereas fossil fuel combustion is rising little at present. Note also that forest burning releases another greenhouse gas, methane, much more potent as a global-warming agent than carbon dioxide, though regrettably its amount remains largely unknown (Vitousek and Matson 1992).

In addition, global warming itself will cause increased die-off and decomposition of forest biomass, in turn triggering a further release of carbon dioxide and methane (Apps and Price 1996). As much as one-third of the world's forests could be threatened in this manner (Houghton et al. 1995). This will likely apply especially to boreal forests, which feature an expanse of 9.2 million square kilometers, and (to repeat a key point) contain around half of all forest carbon worldwide (Dixon et al. 1994). The largest single tract is in Siberia, amounting to 5.5 million square kilometers or nearly twice as much as in Brazilian Amazonia. Its woody biomass is estimated to contain 40–60 gigatonnes of carbon, though only around half as much as in Amazonia; and its forest soils, detritus, and litter contain another two or three times as much carbon (Nilsson 1993). Since boreal forests are located in northern high latitudes, where temperatures will rise most in a greenhouse-affected world, they could soon start to undergo marked desiccation and die-off (except in those areas where there will be an offsetting increase in precipitation).

Global warming will also cause boreal forests to become more vulnerable to fires, whether humanmade or wild fires (Stocks 1991). In Canada, the warming trend of the past two decades has coincided with a sixfold increase in forest areas burned as compared with the century trend (Auclair and Carter 1993). Future global warming could increase the length of Canada's fire season by more than 20 percent and the severity of fires by 46 percent. All this would lead to the release of large amounts of carbon and methane (Woodwell 1995).

Were there to be progressive depletion of boreal forests along these lines, their expanse could decline by at least 40 percent and conceivably 60 percent (some estimates suggest 90 percent) within the next three to five decades. This would release between 1.5 and 3.0 gigatonnes of carbon per year over the period, probably more than is being emitted annually from tropical deforestation today and equivalent to 20–40 percent of all current anthropogenic emissions of carbon dioxide (Jardine 1994; Smith and Shugart 1993; see also Durning 1993; Woodwell 1995).

Worse, the incipient decline of boreal forests could well lead to an increased rate of die-off in remaining forests in other parts of the world, plus a decline of biomass in other biomes. The process could reinforce itself through multiple positive feedbacks, e.g., enhanced emissions of methane from permafrost areas and tundra zones (Houghton et al. 1995). This could result in a decline of 10 percent of all carbon held on land in plants and soils, resulting in turn in a release of 4.0 gigatonnes of carbon per year on average over a period of 50 years (Kolchugina and Vinson 1995). At the same time, there could be a 10-percent increase in plant respiration due to just a one degree C. rise in temperature, leading to a further release of 2.5 gigatonnes of carbon (Woodwell et al. 1995).

So the process would be unlikely to be linear. There could be all manner of environmental discontinuities that ecologists can only surmise about at present. It is plausible, i.e., there is a nontrivial risk, that we could eventually (or even soon?) face a "runaway" greenhouse effect as boreal forests decline, taking with them their crucial function in the global carbon budget. If we then wanted to halt global warming, we would find that much irreversible damage had already been done through the momentum of climate-change dynamics (Woodwell et al. 1995; see also Jardine 1994).

How fast are forests actually disappearing? The situation with tropical forests is well known. They are declining by rather more than 150,000 square kilometers per year (FAO 1992; Myers 1992b). The annual rate of deforestation doubled during the 1980s, and by 1996 it may well have reached 2.5 percent or even near 3.0 percent. Temperate forests, by contrast, are more or less in equilibrium.

As for boreal forests, they have recently started to decline in several sectors. In Siberia, logging, often clear-cut logging, and fires already destroy forty thousand square kilometers of forests per year, though by contrast with the situation in the tropics, these boreal forests generally regenerate themselves—albeit with a reduced capacity to store carbon for a while. Another sixty-five thousand square kilometers are depleted through the factors listed plus industrial pollution (Alexeyev 1991; Kolchugina and Vinson 1995). This amount is twice as much as recent annual deforestation in Brazilian Amazonia and four times as much as the area logged each year in boreal forests of Canada—where, however, virtually all forests have been assigned for eventual logging (Kurz and Apps 1995).

As a measure of the logging surge that could overtake Siberia, note the latent timber demand in neighboring China. A nationwide construction boom has brought on a severe timber shortage for a country with 21 percent of the world's population, 8 percent of the global economy but only 3 percent of the earth's forests. Consumption of three hundred million cubic meters of wood in 1991 exceeded the sustainable yield of the country's forests by 30 percent. As a result of over-logging, there will probably be no mature commercial trees left in two-thirds of China's forests by the year 2000. Timber imports doubled during the decade 1984–93, reaching more than ten million cubic meters of roundwood equivalent—a total that is predicted to soar to sixty million cubic meters by shortly after the year 2000 (FAO 1993a). Per capita consumption of two main timber products, sawnwood and panels, amounts to less than two-thirds of Asia's average and less than two-fifths of Indonesia's. Were China to increase its consumption to match Indonesia's, its share would amount to almost 60 percent of Asia's total—and if it ever matched Japan's, then 280 percent of Asia's total (Bochuan 1991; Ryan and Flavin 1995). In short, China seems poised to become the world's leading importer of wood, with all that will mean for forests in neighboring Siberia.

In addition to over-logging, some 250,000 square kilometers of Siberia's forests have been burned in recent years, due for the most part to fires running out of control following the decline of forestry management capacities after the Soviet Union's demise (Alexeyev 1991). (Compare burning in Canada: as much as seventy thousand square kilometers a year in the 1990s [Riley 1995].) So extensive are logging and burning in Siberia's forests that between 1988 and 1993 they certainly switched from being a net carbon sink to a net source (Nilsson 1995). Some observers (Isaev et al. 1993) consider that Siberia's carbon accumulation from the atmosphere may already have been cut by one-third, and in the eastern forests it may even have been eliminated. On top of this, certain boreal forests are experiencing acid precipitation, notably in northeastern North America and northern and central Europe. So extensive is the problem in Europe that there is a risk of commercial losses totaling $30 billion a year (Nilsson 1994).

Finally under this heading, let us note a further linkage between forests and global warming, this one taking advantage of forests' capacity to absorb carbon dioxide from the atmosphere. Reforestation of ten million square kilometers of deforested lands in the humid tropics could sequester between 100 and 150 billion tonnes of carbon over the next fifty to one hundred years (Houghton and Woodwell 1989; Myers and Goreau 1992). At least eight million square kilometers of deforested and degraded land are already available. Moreover, reforestation would offer many supplementary benefits in the form of watershed functions, soil protection, and the like. So promising is this proposal that it has gained support in principle from the Intergovernmental Panel on Climate Change, which envisages the planting of 120,000 square kilometers of trees a year for twenty years, making a total of 2.4 million square kilometers (Houghton et al. 1990). Already the Netherlands government is engaged in a twenty-five-year program to finance reforestation projects covering 1,250 square kilometers in South America, in order to offset carbon emissions from a new six-hundred-megawatt, coal-fired power station in the Netherlands. For other examples of such "forest-carbon sink" bargains, see Trexler 1991; and on the scope for carbon offsets, see Brown and Adger 1994; Grubb 1993.

Biodiversity Habitats

As indicated in chapter 14 of this book, "Biodiversity's Genetic Library," forests supply habitats for large numbers of species, populations, and other forms of biodiversity. Just tropical forests are estimated to harbor at least 50 percent and probably a much larger proportion of all species on earth. This biodiversity supplies abundant ecosystem services by virtue of its "genetic library" function. Since this has been dealt with in detail in chapter 14, it is not touched upon further here, except to note that the economic values in question can be exceptionally large and hence are included in the summary economic evaluation presented in "Overall Environmental Values" below.

Not covered in chapter 14 are biodiversity products in the form of non-wood products such as wild fruits and fibers, also subsistence-hunting meat (for an extended review, see Pimentel et al. 1996). In Nigeria's relict forests, local people still derive a renewable harvest of almost 100,000 tonnes of good-quality meat per year from animals such as grass-cutters (giant rats), small antelopes, and sundry monkeys. On average, this wild meat constitutes one-fifth of all animal protein in local people's diets—while in parts of Zaire the proportion rises to almost 27 percent and in Cameroon, Ivory Coast, and Liberia, to a massive 70 percent (Sale 1983). In Peruvian Amazonia, certain rural communities depend on wild meat for 80–85 percent of their animal protein. In Ecuador's sector of Amazonia, the renewable harvest of some forty species of mammals, plus birds, turtles, and fish, also caiman hides and primates for medicine, could generate as much as $200 per hectare, by contrast with commercial logging, which generates only $150 per hectare (1980 values) (Paucar and Gardner 1981; see also Grimes et al. 1993).

Some Economic Values

It is often easy to calculate the costs of a specific action to assist the forests' cause, e.g., the budget for a plantation forest. It is less easy, though no less pertinent, to calculate the concealed costs of inaction. Herewith a few examples to illustrate "both sides of the fence."

Watershed Functions

In Rwanda, the montane forest of the Volcanoes Park (home to one of the last populations of the mountain gorilla) covers only 1 percent of the country but acts as the sponge that absorbs and metes out about 10 percent of agricultural water for that severely overpopulated nation (McNeely and Miller 1984). In Java, deforestation-derived siltation of reservoirs, irrigation systems, and harbors levied damage costs worth $58 million in 1987, plus additional damages to coastal fisheries and water supplies for urban communities (Magrath and Arens 1989). The on-site soil conservation benefits of tree cover within India's forests are worth between $5 billion and $12 billion per year, or $100–240 per hectare (Chopra 1993), while the nationwide value of forest services in regulating river flows and containing floods is roughly assessed at $72 billion a year (Panayotou and Ashton 1992; see also Chopra 1993). Perhaps most pertinent of all: What price for watershed services in the year 2025 when a full three billion people in developing countries may well be suffering water shortages (Postel 1992)?

Wild Foods and Other Nonwood Products

As forests disappear, there is a decline in nonwood products that are often of unusual value to local people, especially in tropical forests. Collecting of wild cacao, acai, and rubber in the Amazon estuary is worth $79 per hectare per year (Anderson and Ioris 1992), while in another part of Brazilian Amazonia, just the collecting of Brazil nuts is worth $97 per hectare (Mori 1992). Harvesting of wild fruit and latex around Iquitos in Peruvian Amazonia is worth a whopping $6,330 per hectare per year, to be contrasted with sustainable timber harvesting worth $490 per hectare (Peters et al. 1989). For many further examples with respect to tropical forests, see Godoy et al. 1993.

These nonwood products can also be significant in forests outside the tropics. In parts of the Mediterranean basin—Greece, Italy, Spain, France, Algeria, Morocco, and Tunisia, the trade in cork, resin, mastic gum, honey, mushrooms, wild fruit, and wild game, added to the value of trees used in livestock production, had an estimated value of more than $1 billion in 1992 and a potential value (were all products to be developed to their full marketplace scope) of $5 billion (FAO 1993b).

Link-Up with Fisheries

At the Korup Park in Cameroon, forest-derived protection of fisheries has a net present value (discounted at 8 percent) of $3.8 million ($58 per hectare). This is to be compared with flood control benefits, $1.6 million, and soil fertility maintenance, $0.5 million; plus the opportunity cost of lost forest use by local people, $2.6 million (Ruitenbeek 1989). In the Philippines, watershed protection of fisheries is worth $6.2–$8.1 million per year (Hodgson and Dixon 1988). In Indonesia, mangrove protection of fisheries, plus some agriculture, is estimated at $536 million (Ruitenbeek 1992). Old-growth forests in the Pacific Northwest of North America protect habitats for 112 fish stocks—and the salmon industry alone is worth $1 billion per year (U.S. Forest Service 1993).

Climate Connections

As for the stabilizing effect of forests in the global climate system, we can use a "central" value of $20 of global-warming damage for every tonne of carbon released (Brown and Pearce 1994; Fankhauser 1995). If we then apply this figure to tropical forests, we find that converting open forests to agriculture or pasture would result in damage of roughly $600–$1,000 per hectare; conversion of closed secondary forest, $2,000–$3,000 per hectare; and conversion of primary forest to agriculture, roughly $4,000–$4,400 per hectare (these estimates allow for carbon fixation in the subsequent land use) (Brown and Pearce 1994). This attribute offers a far higher rate of return than any alternative form of current land use in tropical forests. Alternatively reckoned, to replace the carbon storage function of tropical forests (never mind temperate and boreal forests) could cost $3.7 trillion (Panayotou and Ashton 1992; see also Brown and Adger 1994).

Overall Environmental Values

Remarkably enough in light of what is at stake overall, there have not been many attempts to come up with aggregate evaluations for all ecosystem services to be found in forests. Fortunately, there have been a few exploratory efforts with respect to individual tropical-forest countries, and the "total economic value" assessed seems to be made up for the most part of what are construed in this chapter to be ecosystem services, whether actual or potential, whether present or future.

The analytic methodology spans direct-use values such as timber, non-timber forest products, medicinal plants, plant genetic resources, hunting and fishing, recreation and tourism, also education and human habitat; while indirect use values include soil conservation, nutrient cycling, watershed protection, flood control, microclimatic effects and carbon sequestration. In addition, there is existence value, being the value conferred by assuring the survival of a resource. On top of all these, and perhaps the most important in the indefinitely long run, there are what are called option values, including potential values of future use, whether direct or indirect. These values, also known as "passive use" values, exist where individuals who do not intend to make use of environmental resources would nevertheless feel a loss if these resources were to disappear; they may wish to see species conserved "in their own right," or they may be interested in retaining use options for other humans now and in the future. The concept is considered to include bequest value, option value, and future "information" value.

All in all, this analytic approach confirms what one might intuitively suppose, viz. that the bulk of the values in question constitute, or at least reflect, ecosystem services values.

Using this conceptual construct, the net present value of Korup Park in Cameroon (discounted at 8 percent) is estimated to be $7.5 million (Ruitenbeek 1989). The Khao Yai Park in Thailand is estimated to be worth $4.8 million for biodiversity habitat and $0.4–$1.0 million for ecotourism per year. (The park's watershed and carbon sequestration values, plus non-use values, are not included; nor does the calculation reflect the opportunity costs of logging forgone.) The joint annual total (for biodiversity and tourism)of $5.2–$5.8 million is to be contrasted with the almost $7 million of income lost to local villagers through traditional uses forgone by virtue of the park's existence (Dixon and Sherman 1990).

In Mexico's forests, a suite of environmental values have been calculated to be worth some $4 billion per year, or $80 per hectare (Adger et al. 1995). In Panama's forests, the total economic value, including both use and non-use values, has been calculated to be $500 per hectare per year (de Groot 1994). The country with the most detailed evaluation is probably Costa Rica. The total economic value of the country's thirteen thousand square kilometers of wildlands, the great majority of them being tropical forests, is estimated (with a discount rate of 8 percent) to be $102–$214 per hectare per year, with a net present value of $1,278–$2,871 per hectare. This translates into an annual value for all wildlands of between $133 million and $278 million, and a net present value of between $1.7 billion and $3.7 billion. Note that because of externality effects, only 34 percent of the total economic value accrues to Costa Rica, with the remaining 66 percent going to the global community (Castro 1994; Constantino and Kishor 1993).

A summary review of ecosystem services in several dozen tropical forests indicates that the hypothetical overall value of sustainable use of one hectare of forest is about $220 per year, made up of: minor forest products $69, recreation $12, watershed functions $10, hunting and fishing $5, option and existence values $16, and timber $110 (49 percent) (Pimentel et al. 1996).

All the figures adduced above are preliminary and exploratory. They need to be firmed up with due dispatch—and the same for whatever other material values are inherent in the world's forests and are amenable to economic analysis. Then, and only then, shall we be in a position to give "real world" regard to the immediate costs of saving forests, putatively put at $30 billion per year for tropical forests alone (United Nations 1992). Along the way we should bear in mind two further points. First is that as the incomes of many if not most people rise, it seems they are likely to give greater recognition to the value of ecosystem services and to be willing to pay more for them. The second and more important point is that as forests continue to disappear, taking many of their ecosystem services with them, the value of remaining services will tend to become more significant (which implies a reduced discount rate for future evaluation efforts).

Conclusion

The biggest calculation is specially imponderable. If we carry on with a business-as-usual approach and with no conservation measures of scope and scale to match the destruction threat, we may well find in fifty years' time that the earth will be largely bereft of its forests. How shall we respond to those descendants who ask how we could afford to watch the terminal reduction of what has been the predominant type of vegetation on the earth for hundreds of millions of years? Will they not rather ask, "How could you not afford to save the forests in light of all that has been ultimately and irretrievably lost in environmental terms alone?" As this chapter has demonstrated, the forests' ecosystem services are manifold and abundant, and they are crucial to both the ecological stability and the economic viability of societies worldwide—no less.

Acknowledgments

It is a pleasure to acknowledge the helpful comments on an early draft of this chapter, from Michael Apps, David Duthie, Jim MacNeill, and Sten Nilssen. In addition, I especially appreciate the many supports—data searching, statistics checking, and iterative critiques—from my research associate, Jennifer Kent. All errors, misinterpretations, and the like remain the responsibility of the one who perpetrated them in the first place, viz. the author. The chapter has been written with financial support from my Pew Fellowship in Conservation and Environment.

References

Adamowicz, W.L., W. White and W.E. Phillips, eds. 1993. *Forestry and the Environment.* Wallingford, England: Commonwealth Agricultural Bureau International.

Adger, W.N., K. Brown, R. Cervigni and D. Moran. 1995. "Total economic value of forests in Mexico." *Ambio* 24: 286–296.

Alexeyev, V. 1991. *Human and Natural Impacts on the Health of Russian Forests.* Moscow, Russia: Institute of Forest and Timber Research, Siberian Branch, USSR Academy of Sciences.

Anderson, A.B., and E.M. Ioris. 1992. "The Logic of Extraction: Resource Management and Income Generation by Extractive Producers in the Amazon Estuary." In *Conservation of Neotropical Forests: Working from Traditional Resource Use*, K. H. Redford and C. Padoch, eds. New York: Columbia University Press.

Apps, M., and D. Price, eds. 1996. *Forest Ecosystems, Forest Management and Global Climate Change*. New York: Springer-Verlag.

Auclair, A.N.D., and T.B. Carter. 1993. "Forest wildfires as a recent source of CO_2 at mid-high northern latitudes." *Canadian Journal of Forest Research* 23: 1528–1536.

Ba, L.K. 1977. *Bio-Economics of Trees in Native Malayan Forest*. Kuala Lumpur, Malaysia: Department of Botany, University of Malaya.

Bochuan, H. 1991. *China on the Edge: The Crisis of Ecology and Development*. San Francisco: China Books and Periodicals Inc.

Brown, K., and W. N. Adger. 1994. "Economic and political feasibility of international carbon offsets." *Forest Ecology and Management* 68: 217–229.

Brown, K., and D. W. Pearce. 1994. "The Economic Value of Non-Marketed Benefits of Tropical Forests: Carbon Storage." In *The Economics of Project Appraisal and the Environment*, J. Weiss, ed., pp. 102–123. London: Edward Elgar.

Bruijnzeel, L. A. 1990. *Hydrology of Moist Tropical Forests and Effects of Conservation: A State of Knowledge Review*. Paris: International Hydrological Programme, UNESCO.

Castro, R. 1994. *The Economics Opportunity Costs of Wildlands Conservation Areas: The Case of Costa Rica*. Cambridge, Mass.: Department of Economics, Harvard University.

Chan, N.W. 1986. "Drought trends in northwestern peninsular Malaysia: Is less rain falling?" *Wallaceana* 44: 8–9.

Chopra, K. 1993. "The value of non-timber forest products: An estimation for tropical deciduous forests in India." *Economic Botany* 47: 251–257.

Constantino, L., and N. Kishor. 1993. *Forest Management and Competing Land Uses: An Economic Analysis for Costa Rica*. Washington, D.C.: The World Bank.

Cool, J.C. 1980. *Stability and Survival: The Himalayan Challenge*. New York: Ford Foundation.

de Groot, R. S. 1994. "Environmental Functions and the Economic Value of Natural Ecosystems." In *Investing in Natural Capital: The Ecological Economics Approach to Sustainability* A. N. Jansson, M. Hammer and R. A. Costanza, eds., pp. 151–168. Washington, D.C.: Island Press.

Dixon, J. A., and P. B. Sherman. 1990. *Economics of Protected Areas: A New Look at Benefits and Costs*. Washington, D.C.: Island Press.

Dixon, R. K., S. Brown, R. A. Houghton, A. M. Solomon, M. C. Trexler and J. Wisniewski, 1994. "Carbon pools and flux of global forest ecosystems." *Science* 263: 185–190.

Dosso, H., et al. 1981. "The Tai Project: Land use problems in a tropical forest." *Ambio* 10, nos. 2–3: 120–125.

Durning, A. T. 1993. *Saving the Forests: What Will it Take?* Washington, D.C.: Worldwatch Institute.

Ehrlich, P. R., and A. H. Ehrlich. 1992. "The value of biodiversity." *Ambio* 21: 219–226.

Fankhauser, S. 1995. *Valuing Climate Change: The Economics of the Greenhouse*. London: Earthscan Publications Ltd.

FAO (Food and Agriculture Organization). 1992. *Third Interim Report on the State of Tropical Forests*. Rome: Food and Agriculture Organization (Forest Resources Assessment Project).

FAO (Food and Agriculture Organization). 1993a. *Forest Products Yearbook 1993*. Rome: Food and Agriculture Organization.

FAO (Food and Agriculture Organization). 1993b. *More than Wood: The Major Significance of "Minor" Forest Products*. Rome: Food and Agriculture Organization.

FAO (Food and Agriculture Organization). 1995. *Forest Resources Assessment 1990: Global Synthesis*. Rome: Food and Agriculture Organization.

Fleming, W. M. 1979. *Environmental and Economic Impacts of Watershed Conservation on a Major Reservoir Project in Ecuador*. Santa Fe, Mexico: Environmental Improvement Division.

Gash, J. H. C., and W. J. Shuttleworth. 1992. "Tropical Deforestation: Albedo and the Surface-Energy Balance." In *Tropical Forests and Climate* (special issue of *Climatic Change* N. Myers, ed., 19, nos. 1–2), 123–134. Dordrecht, Netherlands: Kluwer Academic Publishers.

Godoy, R., R. Lubowski, and A. Markandya. 1993. "A method for the economic valuation of non-timber tropical forest products." *Economic Botany* 47: 220–233.

Grimes, A., and 10 others. 1993. *Valuing the Rain Forest: The Economic Value of Non-Timber Forest Products in Ecuador*. New Haven, Conn.: School of Forestry and Environment, Yale University.

Grubb, M.J. 1993. "The politics and economics of climate change after Rio." *International Journal of Environment and Pollution* 3: 179–186.

Henderson-Sellers, A., and V. Gornitz. 1984. "Possible climatic impacts of land cover transformations, with particular emphasis on tropical deforestation." *Climatic Change* 6: 231–257.

Hodgson, G., and J. A. Dixon. 1988. *Logging Versus Fisheries and Tourism in Palawan*. Honolulu: East-West Environment and Policy Institute.

Houghton, J. T., G. J. Jenkins, and J. J. Ephramus, eds. 1990. *Climate Change: The IPCC Scientific Assessment (Final Report of Working Group 1)*. New York: Cambridge University Press.

Houghton, J. T., L. G. M. Filho, B. A. Callandar, N. Harris, A. Kattenburg, and K. Maskell, eds., 1995. *The Science of Climate Change: The Second Assessment Report of the Intergovernmental Panel on Climate Change*, Contribution of Working Group I. Cambridge, England: Cambridge University Press.

Houghton, R. A. 1993. "Role of Forests in Global Warming." In *World Forests for the Future: Their Use and Conservation,* K. Ramakrishna and G. M. Woodwell, eds., pp. 21–58. New Haven, Conn.: Yale University Press.

Houghton, R. A., and G. M. Woodwell. 1989. "Global Climate Change." *Scientific American* 260: 36–44.

Isaev, A. S., G. N. Korovin, A. I. Utkin, A. A. Pryazhanikov, and E. G. Zamolodchikov. 1993. "Estimation of stores and annual deposition of carbon in phytomass of forest ecosystems of Russia." *Russian Forest Sciences* 5: 3–10. New York: Allerton Press.

Isaev, A., G. Korovin, D. Zamolodchikov, A. Utkin, and A. Pryaznikov. 1995. "Carbon stock and deposition in phytomass of the Russian forests." *Water, Air and Soil Pollution* 82: 247–256.

Jardine, K. 1994. *The Carbon Bomb: Climate Change and the Fate of the Northern Boreal Forests.* Amsterdam, Netherlands: Stichting Greenpeace Council.

Kolchugina, T. P., and T. S. Vinson. 1995. "Role of Russian Forests in the Global Carbon Balance." *Ambio* 24: 258–264.

Kurz, W. A., and M. J. Apps. 1995. "An analysis of future carbon budgets of Canadian boreal forests." *Water, Air and Soil Pollution* 82: 321–331.

Magrath, W., and P. Arens. 1989. *The Costs of Soil Erosion on Java: A Natural Resource Accounting Approach.* Washington, D.C.: The World Bank.

Mahmood, K. 1987. *Reservoir Sedimentation: Impact, Extent and Mitigation.* Washington, D.C.: The World Bank.

McNeely, J. A., and K. R. Miller. 1984. *National Parks, Conservation and Development: The Role of Protected Areas in Sustaining Society.* Washington, D.C.: Smithsonian Institution Press.

Meher-Homji, V.M. 1992. "Probable Impact of Deforestation on Hydrological Processes." In *Tropical Forests and Climate* (special issue of *Climatic Change* 19, nos. 1–2), N. Myers, ed., pp. 163–174. Dordrecht, Netherlands.

Mori, S. A. 1992. "The Brazil Nut Industry: Past, Present and Future." In *Sustainable Harvesting and Marketing of Rain Forest Products*, M. J. Plotkin and L. M. Famolare, eds. Washington D.C.: Island Press

Myers, N. 1988. "Environmental degradation and some economic consequences in the Philippines." *Environmental Conservation* 15, no. 3: 205–214.

Myers, N. 1989. *Deforestation Rates in Tropical Forests and Their Climatic Implications.* London: Friends of the Earth.

Myers, N. 1992a. *The Primary Source: Tropical Forests and Our Future.* New York: W. W. Norton.

Myers, N., ed. 1992b. *Tropical Forests and Climate* (special issue of *Climatic Change* 19, nos. 1–2). Dordrecht, Netherlands: Kluwer Academic Publishers.

Myers, N. 1996. "The world's forests: Problems and potentials." *Environmental Conservation* 23.

Myers, N., and T. J. Goreau. 1992. "Tropical Forests and the Greenhouse Effect: A Management Response." In *Tropical Forests and Climate* (special issue of *Climatic Change* 19, nos. 1–2), N. Myers, ed., pp. 215–226. Dordrecht, Netherlands: Kluwer Academic Publishers.

Nilsson, S. 1993. "Sustainability of Siberian Forests." *Environmental Conservation* 20: 177.

Nilsson, S. 1994. "Air Pollution and European Forests." In *Acid Rain: Current Situation and Remedies*, J. Rose, ed. Amsterdam, Netherlands: Gordon and Breach Science Publishers.

Nilsson, S., ed. 1995. *Boreal Forests: The Role of Research*. Laxenburg, Austria: International Institute for Applied Systems Analysis.

Panayotou, T., and P. S. Ashton. 1992. *Not by Timber Alone*. Washington, D.C.: Island Press.

Paucar, A., and A. Gardner. 1981. *Establishment of a Scientific Research Station in the Yasuni National Park of the Republic of Ecuador*. Washington, D.C.: The National Zoo.

Peters, C. M., A. H. Gentry and R. O. Mendelsohn. 1989. "Valuation of an Amazonian rain forest." *Nature* 339: 655–656.

Phisphumvidhi, P. 1981. *Water Resources for Water Supply in Bangkok, Thailand*. Bangkok, Thailand: Metropolitan Water Works Authority.

Pimentel, D., M. McNair, L. Buck, M. Pimentel, and J. Kamil. 1996. "The value of forests to world food security." *Human Ecology*, in press.

Postel, S. 1992. *Last Oasis: Facing Water Scarcity*. New York: W. W. Norton. Reiger, H.C., ed. 1977. *Himalayan Mountain Ecosystems*. New Delhi, India: Max Mueller Bhavan.

Riley, L.F. 1995. "Criteria and indicators of sustainable forest management in Canada." *Water, Air and Soil Pollution* 82: 67–70.

Ruitenbeek, J. 1989. *Social Cost-Benefit Analysis of the Korup Project, Cameroon*. Godalming, England: World Wide Fund for Nature–U.K.

Ruitenbeek, J. 1992. "The rainforest supply price: A tool for evaluating rain forest conservation expenditures." *Ecological Economics* 6, no. 1: 57–78.

Ryan, M., and C. Flavin. 1995. "Facing China's Limits." In *State of the World 1995*, L. R. Brown, D. Denniston, C. Flavin, H. French, H. Kane, N. Lenssen, M. Renner, D. Roodman, M. Ryan, A. Sachs, L. Starlie, P. Weber, and J. Young, eds., pp. 113–131. New York: W. W. Norton.

Salati, E., and C. A. Nobre. 1992. "Possible Climatic Impacts of Tropical Deforestation." In *Tropical Forests and Climate* (special issue of *Climatic Change* 19, nos. 1–2), N. Myers, ed., pp. 177–196. Dordrecht, Netherlands: Kluwer Academic Publishers.

Sale, J. B. 1983. *The Importance and Values of Wild Plants and Animals in Africa*. Gland, Switzerland: International Union for Conservation of Nature and Natural Resources.

Sfeir-Younis, A. 1986. *Soil Conservation in Developing Countries: A Background Report*. Washington, D.C.: The World Bank.

Smith, T. M., and H. H. Shugart. 1993. "The Transient Response of Terrestrial Carbon Storage to a Perturbed Climate." *Nature* 361: 523–526.

Stocks, B. J. 1991. "The Extent and Impact of Forest Fires in Northern Circumpolar Countries." In *Global Biomass Burning: Atmospheric, Climate and Biospheric Implications*, J. S. Levine, ed., pp. 197–202. Cambridge, Mass.: MIT Press.

Trexler, M. C. 1991. *Minding the Carbon Store: Weighing U.S. Forestry Strategies to Slow Global Warming.* Washington, D.C.: World Resources Institute.

United Nations (Conference on Environment and Development). 1992. *Agenda 21: Action Program.* New York: United Nations.

U.S. Forests Service. 1993. *Viability Assessments and Management Considerations for Species Associated with Late-Successional and Old-Growth Forests of the Pacific Northwest.* Washington, D.C.: U.S. Forest Service.

Vitousek, P. M., and P. A. Matson. 1992. "A Commentary On: Effects of Tropical Deforestation on Global and Regional Atmospheric Chemistry." In *Tropical Forests and Climate* (special issue of *Climatic Change* 19, nos. 1–2) N. Myers, ed. pp. 159–162. Dordrecht, Netherlands: Kluwer Academic Publishers.

Wiens, T. 1989. *Philippines: Environmental and Natural Resource Management Study.* Washington, D.C.: The World Bank.

Woodwell, G. M. 1993. "Forests: What in the World are They For?" In *World Forests for the Future: Their Use and Conservation,* K. Ramakrishna and G. M. Woodwell, eds., pp. 1–20. New Haven, Conn.: Yale University Press.

Woodwell, G. M., 1995. "Biotic Feedbacks from the Warming of the Earth." In *Biotic Feedbacks in the Global Climatic System: Will the Warming Feed the Warming?* G. M. Woodwell and F. T. Mackenzie, eds., pp. 3–21. New York: Oxford University Press.

Woodwell, G.M. and F.T. Mackenzie, eds. 1995. *Biotic Feedbacks in the Global Climatic System: Will the Warming Feed the Warming?* New York: Oxford University Press.

Woodwell, G. M., F. T. Mackenzie, R. A. Houghton, M. J. Apps, E. Gorham, and E. A. Davidson. 1995. "Will the Warming Speed the Warming?" In *Biotic Feedbacks in the Global Climatic System: Will the Warming Feed the Warming?* G. M. Woodwell and F. T. Mackenzie, eds., pp. 393–412. New York: Oxford University Press.

World Health Organization. 1994. *Health for All in 2000.* Geneva: World Health Organization.

Wotton, B. M., and M. D. Flannigan. 1993. "Length of the fire season in a changing climate." *Forestry Chronicle* 69: 187–192.

Zak, D. 1995. "Response of Terrestrial Ecosystems to Carbon Dioxide Fertilization." In *Elements of Change 1994,* S. J Hassol and J. Katzenberger, eds., pp. 202–204. Aspen, Colorado: Aspen Global Change Institute.

Chapter 13

Ecosystem Services in Grasslands

Osvaldo E. Sala and José M. Paruelo

The grassland biome covers an enormous fraction of the surface of the earth. Grasslands are the potential natural vegetation of approximately 25 percent of the land surface of the earth, or 35×10^6 km^2 (Shantz 1954, Graetz 1994). These are systems mostly limited by water, which are dominated by grasses and have a variable woody component. Humans utilize these areas as grazing lands or transform them into croplands depending mostly on water availability and the amount of subsidies received by agriculture in each individual country. Most of the mesic grasslands have been converted into agricultural land, whereas a large fraction of the arid and semi-arid grasslands remain as such. Subsidies to agriculture make transformation of grasslands into croplands economically feasible in regions that otherwise would remain as native grasslands, such as the western portion of the North American Great Plains (Hannah et al. 1995).

Grasslands produce an array of goods and services for humankind, but only a few of them have market value. Meat, milk, wool, and leather are the most important products currently produced in grasslands that have a market value. Simultaneously, grassland ecosystems confer to humans many other vital and often unrecognized services such as maintenance of the composition of the atmosphere, maintenance of the genetic library, amelioration of weather, and conservation of soils. The fact that humans take for granted the provision of these grassland services is not an indication of their value. In many cases, the value of services provided by grasslands in terms of production inputs and sustenance of plant and animal life (see chapter 3 by

Goulder and Kennedy in this book) may be larger than the sum of the products with current market value.

In this chapter we will focus on those services that currently have no market price and for which society has difficulty in assessing value. We will discuss the role of natural grasslands in maintaining the composition of the atmosphere and the genetic library, as well as ameliorating the weather and conserving the soil. Our approach will be to compare natural grasslands under moderate grazing with alternative land uses, which include drastic changes such as transformation into croplands and more subtle changes that grasslands undergo when grazed with different intensities. We will evaluate the ecological effect and the economic value of these changes in land-use practices.

Maintenance of the Composition of the Atmosphere

Grasslands sequester in the soil large quantities of carbon (C) as soil organic matter, which are rapidly transferred into the atmosphere when plowed and converted into agricultural land. In comparison with other ecosystems such as forests, grasslands store most of their C belowground (Burke et al. 1989, Moraes et al. 1995). Carbon stocks in grasslands are largely determined by abiotic factors; they increased with precipitation mainly as a result of increased primary production (input) and decrease with increasing temperature as a result of increased decomposition (output) (Burke et al. 1989).

Tillage associated with the transformation of grasslands into croplands increases soil organic matter decomposition and decreases carbon stocks mainly as a result of breaking soil aggregates and exposing residues to decomposers (Elliot 1986). Carbon losses as a result of cultivation are very large. Results of a study comparing native and cultivated soils in the Great Plains of the United States indicated that cultivation resulted in C losses ranging between 0.8 and 2 kg m^{-2} when the average C content of soils for the region ranges between 2 and 5 kg m^{-2} (Burke et al. 1989) (figure 13.1).

Carbon losses as a result of cultivation vary according to climate and site characteristics, increasing with precipitation and silt content and decreasing with temperature (Burke et al. 1989). In general, C losses as a result of cultivation track C stocks, with larger losses occurring in soils with larger C stocks (figure 13.1). The loss of carbon as a result of cultivation of grasslands occurs very rapidly, but recovery after abandonment occurs at a slower rate. For example, in the Great Plains of North America C stocks after plowing decreased significantly and very rapidly (Cole et al. 1989), but after fifty years of abandonment stocks had not yet reached the levels of native soils (Burke et al. 1995, Ihori et al. 1995).

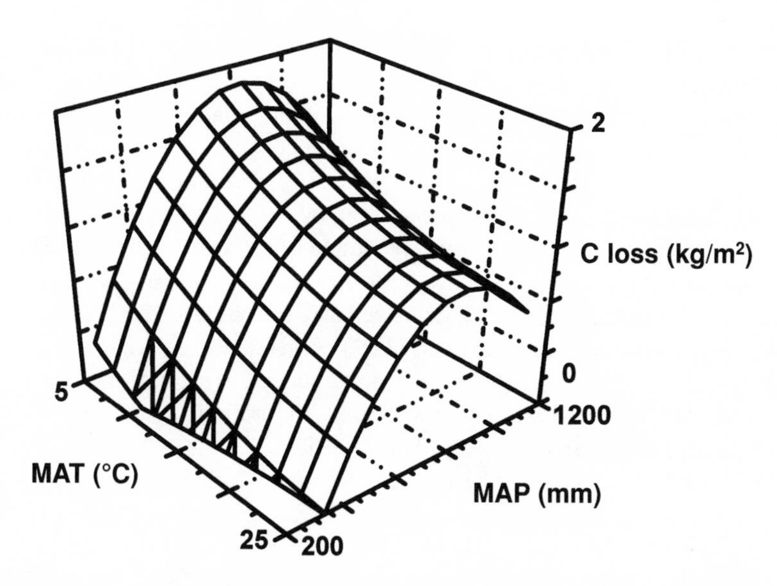

Figure 13.1. Abiotic controls on the amount of carbon lost as a result of cultivation in the Great Plains of North America. MAT is mean annual temperature and MAP is mean annual precipitation. Carbon loss is the difference in C stocks between cultivated and native soils for different locations in the Great Plains of North America.
Source: Redrawn from Burke et al. 1989.

At a global scale, agriculture has made a significant contribution to the observed increase in atmospheric CO_2. Analysis of tree rings, which are indicative of the CO_2 concentration of the atmosphere in the past, showed that the rapid transformation of native ecosystems into croplands that occurred between 1860 and 1890 contributed one and a half times the amount of CO_2 produced by all the fossil fuel emissions through 1950 (Wilson 1978).

The transformation of grasslands into croplands significantly contributes to the global increase of the atmospheric concentration of CO_2. Changes in the concentration of CO_2 have multiple direct effects on the functioning of plants and animals (for a review see Bazzaz 1990). Here we will focus on the very important indirect effect of increasing CO_2 on climate. There is general agreement that increases in the atmospheric concentration of trace gases such as CO_2, methane, nitrous oxide, and CFC's will result in disruptions of global climate systems (Mitchell et al. 1990).

Carbon loss from grasslands contributes to the global CO_2 increase and to climate change. The effects of increasing CO_2 on climate are expected to be major. For example, doubling of the concentration of CO_2 in the atmosphere will result in an increase of the temperature of the earth ranging between 1.5 and 4.5 °C and in an increase of global precipitation parallel with an increase in evaporation ranging between 3 and 15 percent (Mitchell et al. 1990). Most of the uncertainties are now reduced to the timing and the geographical distribution of those changes. Climate change results from several factors changing simultaneously. Most recent projections suggest an increase between 2.0 and 2.4 °C by the year 2100 (Kattenberg et al. 1995).

These disruptions of the climate system will have a negative impact on a majority of countries, mostly through impacts on agricultural production (Paruelo and Sala 1993, Rosenzweig and Parry 1994). The size of the impact varies according to the climate scenario chosen and the geographical location of the country (Kane et al. 1992). Based upon ecological and economic information briefly described above, scientists have been able to develop models that estimate the costs of adding carbon to the atmosphere (Nordhaus 1991, Fankhauser and Pearce 1994).

The costs of CO_2 emissions have been estimated based on the negative effects that increasing CO_2 has on climate: $20.4 per tonne of C for the period 1991–2000, $22.9 for 2001–2010, $25.4 for 2011–2020, and $27.8 for 2021–2030 (Fankhauser and Pearce 1994). The costs of CO_2 emissions increase through time because an extra tonne of CO_2 added to an already large stock of atmospheric CO_2 will result in more damage than a tonne emitted when CO_2 was low. Based on the estimates of the effects of cultivation on C stocks and the estimated costs of CO_2 emissions described above, we calculated the value of carbon sequestration in grasslands to be $200 per ha with a range between $160 and $400/ha (table 13.1).

The value of carbon sequestration by grasslands is large compared to the value of land and the annual production of goods with market value such as meat, wool, and milk. The market value of land for counties in eastern Colorado (U.S.) ranges between $311 and $1,633/ha with a direct average of $798/ha (U.S. Department of Commerce 1995). The net cash return for farms in the same region ranges between $5 and $144/ha/yr with an average of $47/ha/yr. This comparison of the services with nonmarket value, such as carbon sequestration, against the market value of goods and services is valid since both are based on data from the same region. Our estimates of the impact of cultivation on CO_2 emissions, the value of carbon sequestration, as

Table 13.1. The value of maintaining native grasslands: Carbon sequestration

Carbon loss	10×10^3 kg/ha[a]
Cost	$0.02 per kg of C
TOTAL	$200/ha

[a]Values observed ranged between 8×10^3 kg/ha and 20×10^3 kg/ha.

well as the value of land and production were based on data from eastern Colorado.

We want to stress the hysteresis of this process; while transformation of grasslands into croplands yields large amounts of C to the atmosphere in a relatively short period of time, the reverse process of abandonment of croplands and their slow transformation into native grasslands sequesters only modest amounts of carbon over relatively long periods of time (Ihori et al. 1995). A study of C accumulation showed that after fifty years of abandonment, C stocks increased 3,000 kg/ha, which results in a value of $60/ha or $1.20 $ha^{-1}yr^{-1}$ (Burke et al. 1995, Ihori et al. 1995).

Agriculture and the transformation of grasslands also affect the dynamics of other trace gases such as methane and nitrous oxide. Both are active greenhouse gases in that they are transparent to the radiation of the sun but absorb radiation emitted by the earth. Therefore, increases in the atmospheric concentration of trace gases lead to increases in the temperature of the earth and severe disruptions of the climate system.

Field experiments comparing native grasslands and adjacent cultivated plots have shown that cultivation decreases the uptake of methane and increases the emissions of nitrous oxide, contributing to the increasing concentrations of these gases in the atmosphere (Mosier et al. 1991). The absolute quantities of carbon that are emitted as methane and its concentration in the atmosphere are quite small compared to CO_2. However, methane has a greenhouse effect that is twenty to fifty times larger than CO_2 (Shine et al. 1990). The energy emitted by the earth comprises a range of wavelengths, and different greenhouse gases absorb in different wavelengths. Methane absorbs in a range of wavelengths, where current absorption is quite low, and therefore small additions of this gas into the atmosphere will result in large changes in the greenhouse effect and the temperature of the earth.

Field experiments showed that native grasslands take up 2.6 g C ha^{-1} d^{-1} as methane, while adjacent wheat fields uptake only half of this magnitude (Mosier et al. 1991). The cost of methane emissions has been calculated in a similar manner as that described for the cost of CO_2 emissions (Fankhauser and Pearce 1994). Combining information about the effect of cultivation and the cost of methane emissions, we calculated the current annual costs of cultivation associated with methane emissions (table 13.2). As in the case of CO_2, the cost of methane emissions increases with time. Therefore, the cost for the forty-year period 1991–2030 is forty times larger than the cost of emissions in 1991.

Nitrous oxide is also a trace gas with greenhouse effect; its greenhouse effect is two orders of magnitude larger than CO_2. Nitrous oxide is emitted by grasslands and croplands, but croplands emit at a higher rate than native grasslands, and this rate is even larger in fertilized than in unfertilized crops

Table 13.2. The value of maintaining native grasslands: Methane uptake

Methane uptake	0.474 kg C ha^{-1} yr^{-1}
Methane cost	\$0.11/kg CH_4
Current annual cost	\$0.05/ha
Projected cost for 1991 to 2030	\$2.70/ha

Table 13.3. The value of maintaining native grasslands: Nitrous oxide emissions

Nitrous oxide emissions	0.191 kg N ha^{-1} yr^{-1}
Nitrous oxide cost	\$2.94/kg of N
Current annual cost	\$0.60/ha
Projected cost for 1991 to 2030	\$28.50/ha

(Mosier et al. 1991). The damage caused by nitrous oxide emissions is even greater than the cost of methane and CO_2 emissions because of its larger greenhouse effect (Fankhauser and Pearce 1994). We estimated the annual and the accumulated nitrous oxide emission costs for the period 1991–2030 based upon the difference in emissions between grasslands and adjacent wheat fields and the cost per unit of nitrogen emitted as nitrous oxide (table 13.3).

The costs of emissions of methane and nitrous oxide associated with cultivation are small when compared with the costs of CO_2 associated with transforming grasslands into croplands. We need to take into account that cultivation releases huge amounts of carbon as CO_2 after cultivation only once (during a relatively short period of time). The annual benefits of capturing CO_2 after returning croplands into grasslands are much smaller, and they are comparable to the annual benefits associated with reductions in methane and nitrous oxide emissions from maintaining grasslands as such.

Genetic Library

Grasslands provide an important service to humans by maintaining a large storehouse of genetic material referred as the genetic library. Norman Myers thoroughly describes the ecosystem services associated with the maintenance of the genetic library (see chapter 14 of this book). In this section we instead recognize the uniqueness of grasslands for their contribution to maintaining a global genetic library. Semi-arid systems are particularly important in terms of their biological diversity. For example, drylands in South America are richer in number of mammal species, and have more endemic taxa, than lowland Amazonian rainforest (Mares 1992). This is partially the result of the huge area covered by arid and semi-arid ecosystems.

Another important aspect of grasslands is that the majority of the centers of origin of domesticated plants and animals are located within water-limited systems, primarily composed of grasslands (Vavilov 1951, McNeely et al. 1995). These are the ecosystems where annual grasses and legumes are most abundant. Wheat, barley, onions, and peas all share the same center of origin in the grasslands of the Mediterranean region, in an area known as the Fertile Crescent, which extends from Greece eastward. This area also is the center of origin of many domesticated animals such as goats, sheep, and cattle.

Therefore, the genetic resources of grasslands have a disproportionately large conservation value for humans, who depend on a limited number of grassland species for nutrition, medicine, fiber, and shelter. Grasslands represent the natural ecosystem from where a large fraction of domesticated species originated, and where wild populations related to the domesticated species and their associated pests and pathogens still thrive. These areas are most likely to provide new strains that are resistant to diseases or contain new features important for humankind.

Amelioration of Weather

Changes in the utilization of grasslands, such as those resulting from differential grazing intensity and ultimately overgrazing, as well as the more drastic transformation of grasslands into croplands, have important effects on climate at different scales. We will attempt to demonstrate here that moderately grazed natural grasslands provide a valuable service to humans by ameliorating climate.

Grazing results in changes in the structure of the community and in the composition of plant species in the Patagonian steppe (León and Aguiar 1985). Cover of the grasses preferred by sheep decreases and bare soil groundcover increases along a gradient of increasing grazing intensity. After a threshold in grass cover is crossed, a largely unpalatable shrub (*Mulinum spinosum*) invades, and its dominance continues to increase. These changes in community structure and composition result in changes in albedo, which is the amount of energy reflected by the land surface (Aguiar et al. 1996) (figure 13.2). From light to moderate grazing intensity, there is an increase in the amount of energy reflected, which is related to a decrease in plant cover and an increase in bare soil. Further increases in grazing intensity result in the invasion of the shrub *Mulinum spinosum*, with the resulting increase in cover and decrease in albedo. Besides changes in albedo, grazing also modifies vegetation roughness length, another parameter affecting climate, which varies from 0.02 m in the grass-dominated portion of the gradient to 0.09 m in the shrub-dominated portion of the gradient. These changes in roughness length have the potential to alter local circulation patterns and regional climate.

Similar changes in community structure and climate have been observed along the U.S.-Mexico border (Balling 1988, Bryant et al. 1990). As a result of differences in land use between the two countries, there is a sharp difference in community structure along the border in what was once the same community with the same climate. The Mexican side has lower grass cover and correspondingly more bare ground. The lower plant cover results in higher albedo, as in the Patagonian case. The increase in reflectance has been suggested to decrease temperature and convective precipitation, leading to a positive feedback toward desertification, where overgrazing leads to lower precipitation, lower primary production, and (if stocking rate remains constant) further overgrazing (Charney 1975). Comparison of long-term climate data sets of the Sonoran Desert on both sides of the U.S.-Mexico border showed that the Mexican side was 2.3°C warmer than the U.S. side.

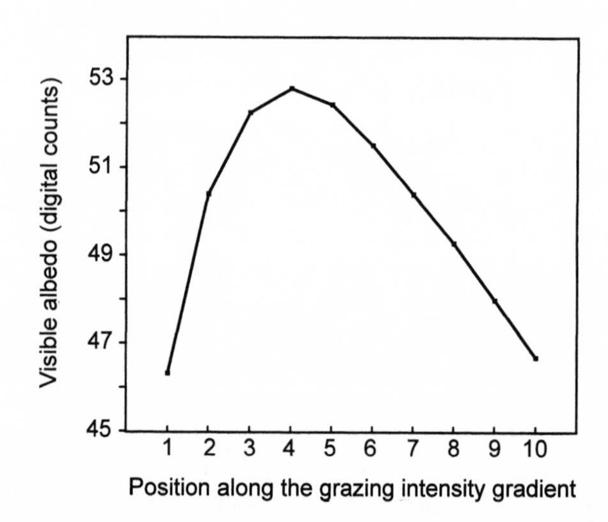

Figure 13.2. Changes in albedo, which is the fraction of incoming solar radiation reflected back into the atmosphere, along a gradient of grazing intensity from light to heavy grazing.
Source: Redrawn from (Aguiar et al. 1996).

These data do not support the Charney (1975) hypothesis, suggesting that the decrease in plant cover reduces transpiration and the energy loss by means of latent heat. Reduced transpiration seems to be more important in altering climate than does the decrease in energy absorbed as a result of increased reflectance.

The drastic transformation of grasslands into agricultural land modifies the energy balance of a region (figure 13.3). The effect of a shift from a grassland into a wheat field or into wheat-soybean relay double cropping system on the energy balance is documented by the changes in the Normalized Vegetation Difference Index (NDVI). The NDVI is an index derived from the reflectance in the red and infrared bands measured by satellites, which shows strong correlation with vegetation attributes such as biomass and production (Running 1990). The three land cover types differ in the seasonal dynamics of the NDVI, reflecting seasonal change in leaf area, albedo, and evapotranspiration. Pielke et al. (in press) showed that these types of changes in land use may affect significantly the mesoscale climate.

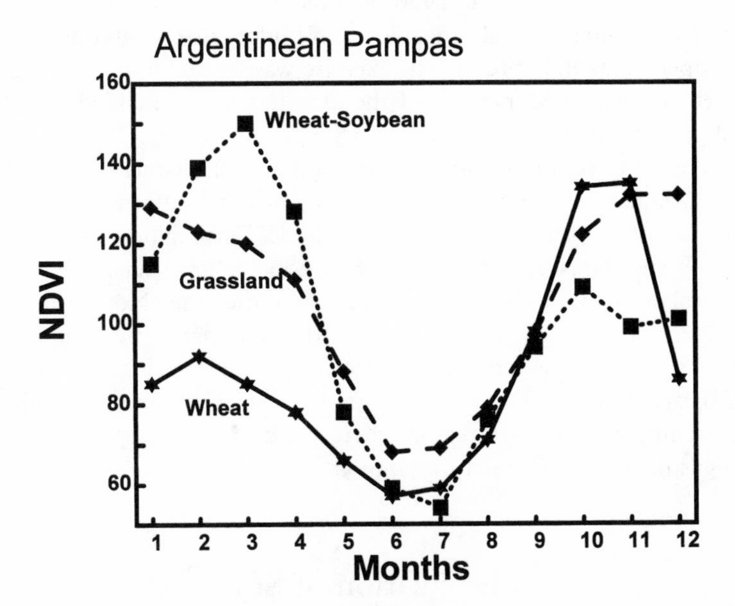

Figure 13.3. Changes in the quality and the amount of energy reflected as a result of transforming native grassland into different kinds of agricultural land. Annual pattern of the Normalized Vegetation Difference Index (NDVI) for a native grassland, a wheat field, and double cropping wheat-soybean in the Argentinean Pampas. The NDVI is an index derived from the reflectance in the red and infrared bands measured by satellites, which is strongly correlated with leaf area, biomass, and primary production.

Table 13.4. Regional climate modeling exercise for North America: The Central Plains case

	Current	Natural	Difference
Temperature (K)	296.15	295.99	0.16
Precipitation (mm d^{-1})	3.83	3.65	0.18

Source: Copeland et al., *Journal of Geophysical Research* (in press).

We have described examples in which variations in grazing intensity or agricultural use have resulted in changes to the local climate by means of changes in albedo, roughness length, and evaporation. Do these effects scale up, and can they be seen at the regional level? This question cannot be answered experimentally because there are very few regions identical with respect to climate but with different land-use patterns. This large-scale question needs to be answered using simulation models.

An exercise that used a climate model that operates at regional scales (Pielke et al. 1992, Pielke et al. 1996) allowed a detailed comparison of climate under potential natural vegetation and under current land-use conditions (Copeland et al. in press). The exercise was limited to the continental United States, where 60 percent of the area has been modified from the original potential natural vegetation. For the purpose of this chapter, we focused exclusively on the Central Plains region of the United States. This area was originally entirely covered by grasslands and currently contains a combination of croplands and native grasslands. This region still maintains a relatively large area as native grasslands because of water limitation on agricultural production. Current changes in land use in the North American Great Plains are estimated to have already caused warmer conditions, mainly as a result of the reduction of green cover and transpiration during part of the year (table 13.4). Precipitation has increased slightly, which still is of great importance for a region where the average precipitation is low and ranges between 300 and 1,000 mm yr^{-1}.

Conservation of Soil

Increases in grazing intensity result in profound changes in the functioning of ecosystems. We have already discussed examples indicating how grazing modifies plant and bare soil cover. More subtle changes occur as a result of grazing before changes in cover, which include changes in plant species composition and soil conditions, are evident (Sala et al. 1986, Chaneton and Lavado 1996). The range science literature has abundant examples demonstrating that heavy grazing and overgrazing have negative impacts on soil erosion (Branson et al. 1981). Most of the effects of grazing on soil erosion are related to the reduction in plant biomass and cover, as well as to the increase in bare ground. Animals have also a direct effect on grasslands by trampling and compacting the soil surface, which in some cases decreases water infiltration and consequently increases runoff and soil erosion. Heavily grazed plots, in Colorado in the United States, showed double the erosion rate of moderately grazed or ungrazed plots (Dunford 1949). Similarly, in grasslands of western Texas it has been shown that a clear relationship exists between plant biomass (which is controlled by grazing) and sediment yield (Bedunah and Sosebee 1986).

Humans have a more drastic impact on erosion when they plow grasslands and transform them into croplands. A comparison of two crops, wheat and sorghum, versus a native grassland in Texas, shows almost negligible soil losses in the native grassland and huge losses (on the order of tons per ha) in any of the crop systems (figure 13.4) (Jones et al. 1985). It is clear that grasslands provide an important service by controlling soil erosion.

Erosion results in multiple on-site and off-site costs. On-site costs are those occurring within the piece of land under consideration and are those that ranchers and farmers are usually most concerned about. This kind of cost accounts for losses in production potential, infiltration, water availability, and nutrient availability. Off-site erosion costs include expenditures such

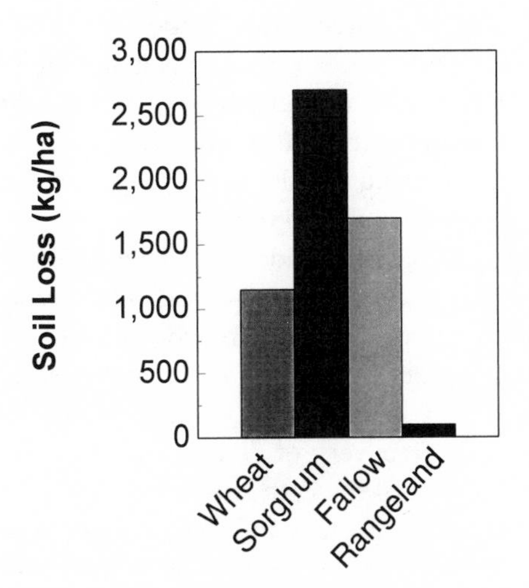

Figure 13.4. Soil loss as a function of land use, wheat, sorghum, fallow, and rangeland. Results are six-year averages.
Source: Redrawn from Jones et al. 1985.

as the increased costs of obtaining a suitable water supply, maintaining navigable channels and harbors, increased drainage problems, increases in flood damage, increased costs of maintaining roads, and a decreased potential for water power. In economic terminology, off-site erosion costs are "externalities" to the production process.

The off-site costs of erosion to society are huge. In the United States, the off-site erosion costs are $17 billion per year (using 1992 dollars) (Pimentel et al. 1995). This enormous cost occurs in a country that has a moderate erosion rate of seventeen tons ha^{-1} yr^{-1} as a result of investment in technology and erosion control mechanisms. Poorer countries in Asia, Africa, and South America average much larger erosion rates of forty tons ha^{-1} yr^{-1}. The on-site costs for the United States are also quite high, $27 billion per year. The magnitude of the erosion problem can be appreciated when costs are scaled up to national or global levels. In the United States the total cost of soil erosion is $44 billion per year, or $100 per hectare of cropland or pasture. At the global level, soil erosion is enormous (75×10^9 tons of soil), which results in costs of $400 billion per year or $70 per person per year (Pimentel et al. 1995).

Conclusions

Grasslands provide humans with many services, most of which currently have no market value. Native grasslands contribute to maintaining the composition of the atmosphere by sequestering carbon, absorbing methane, and reducing emissions of nitrous oxide. Grasslands maintain a large genetic library, ameliorate regional climate, and preserve the soil from devastating erosion. Our estimates suggest that, in many cases, the value of these services are comparable to the value of the services that have a market value, such as production of meat, wool, and milk.

Hysteresis in the ability of grasslands to provide services is a pervasive phenomenon. Grasslands contain large quantities of carbon in their soils that are rapidly released into the atmosphere when plowed. However, the reverse process of accruing carbon is very slow. Similarly, native grasslands represent a reservoir of biological diversity, which is rapidly depleted after cultivation or overgrazing. Recovery of diversity is very slow, or may never occur, depending on the size of the disturbed area.

The underestimated value of grasslands has consequences for decision makers, researchers, and society as a whole. Errors in valuation may lead to inappropriate decisions on the fate and best use of natural resources for society. The ability to provide goods and services with market value is not necessarily related to the ability to provide other services that currently may not have a market value. Ignoring the value of services with no market value may be seriously misleading.

Research efforts are guided either by scientific curiosity or by problem-solving needs. Scientific curiosity is the most important driving force accounting for the major accomplishments of humankind in understanding the functioning of nature. However, in many instances, scientific curiosity also has led to major applications. The search for solving problems is the most important motivation for applied research, but in many instances, it has also illuminated basic issues. Not recognizing all the services provided by grasslands misled scientists since all of the applied management questions are aimed at maximizing the production of goods and services with market value. Large numbers of studies of grazing systems trying to maximize meat production contrast with scarce or nonexistent studies of management techniques aiming at maximizing biological diversity, carbon sequestration, or soil preservation.

Acknowledgments

We thank A. T. Austin, D. Hooper, M. Keller, and L. Pierce for helpful discussions and suggestions. This work was partially supported by the University of Buenos Aires, Consejo Nacional de Investigaciones Cientificas of Argentina, and the U.S. Natiuonal Science Foundation through the Cross-Site LTER DEB-9416815. José M. Paruelo was partially supported by NSF LTER program BSR 90-11659.

References

Aguiar, M. R., J. M. Paruelo, O. E. Sala, and W. K. Lauenroth. 1996. "Ecosystem responses to changes in Plant Functional Type composition: An example from the Patagonian steppe." *Journal of Vegetation Science* 7: (in press).

Balling, R. C. 1988. "The climatic impact of a Sonoran vegetation discontinuity." *Climatic Change* 13: 99–109.

Bazzaz, F. A. 1990. "The response of natural ecosystems to rising global CO_2 levels." *Annual Review of Ecology and Systematics* 21: 408–414.

Bedunah, D. J., and R. E. Sosebee. 1986. "Influence of mesquite control on soil erosion on a depleted range site." *Journal of Soil and Water Conservation* 41:131–135.

Branson, F., G. Gifford, K. Renard, and R. Hadley. 1981. "Runoff and water harvesting." In: *Rangeland Hydrology*, E. H. Reid, ed., pp. 73–110. Dubuque, Iowa: Kendall/Hunt Publishing Company.

Bryant, N. A., L. F. Johnson, A. J. Brazel, R. C. Balling, C. F. Hutchinson, and L. R.

Beck. 1990. "Measuring the effect of overgrazing in the Sonoran Desert." *Climatic Change* 17: 243–264.

Burke, I. C., W. K. Lauenroth, and D. P. Coffin. 1995. "Soil organic matter recovery in semiarid grasslands: Implications for the conservation reserve program." *Ecological Applications* 5: 793–801.

Burke, I. C., C. M. Yonker, W. J. Parton, C. V. Cole, K. Flach, and D. S. Schimel. 1989. "Texture, climate, and cultivation effects on soil organic matter content in U.S. grassland soils." *Soil Science Society of America Journal* 53: 800–805.

Chaneton, E. J., and R. S. Lavado. 1996. "Soil nutrients and salinity after long-term grazing exclusion in a flooding Pampa grassland." *Journal of Range Management* 49: 182–187.

Charney, J. G. 1975. "Dynamics of deserts and droughts in the Sahel." *Quarterly Journal of the Royal Meteorological Society* 101: 193–202.

Cole, C. V., I. C. Burke, W. J. Parton, D. S. Schimel, and J. W. B. Stewart. 1989. "Analysis of historical changes in soil fertility and organic matter levels of the North American Great Plains." In: *Proceedings of the International Conference on Dryland Farming*, P. W. Unger, T. V. Sneed, and R. W. Jensens, eds., pp. 436–438. College Station: Texas A&M University.

Copeland, J. H., R. A. Pilke, and T. G. F. Kittel. "Potential climatic impacts of vegetation change : A regional modelling study." *Journal of Geophysical Research* (in press).

Dunford, E. G. 1949. "Relation of grazing to runoff and erosion on bunchgrass ranges." *Rocky Mountain Forest and Range and Experimental Station Notes* 7: 1–2.

Elliot, E. T. 1986. "Aggregate structure and carbon, nitrogen, and phosphorus in native and cultivated soils." *Soil Science Society of America Journal* 50: 627–632.

Fankhauser, S., and D. W. Pearce. 1994. "The Social Costs of Greenhouse Gas Emissions." In: *The Economics of Climate Change*, Paris: Organisation for Economic Cooperation and Development—International Energy Agency.

Graetz, D. 1994. "Grasslands." In *Changes in Land Use and Land Cover: A Global Perspective*, W. B. Meyer and B. L. Turner II, eds., pp. 125–147. Cambridge, England: Cambridge University Press.

Hannah, L., J. L. Carr, and A. Lankerani. 1995. "Human disturbance and natural habitat: A biome level analysis of a global data set." *Biodiversity and Conservation* 4: 128–155.

Ihori, T., I. C. Burke, W. K. Lauenroth, and D. P. Coffin. 1995. "Effects of cultivation and abandonment on soil organic matter in northeastern Colorado." *Soil Science Society of America Journal* 59: 1112–1119.

Jones, O. R., H. B. Eck, S. J. Smith, G. A. Coleman, and V. L. Hauser. 1985. "Runoff, soil and nutrient losses from a rangeland and dry-farmed cropland in the Southern High Plains." *Journal of Soil and Water Conservation* 40: 161–164.

Kane, S., J. Reilly, and J. Tobey. 1992. "An empirical study of the economic effects of climatic change on world agriculture." *Climatic change* 21: 17–35.

Kattenberg, A., F. Giorgi, H. Grassl, G. A. Meehl, J. F. B. Mitchell, R. J. Stouffer, T.

Tokioka, A. J. Weaver, and T. M. L. Wigley. 1995. "Climate Models—Projections of Future Climate." In *Climate Change: The IPCC Scientific Assessment*, J. T. Houghton, G. J. Jenkins, and J. J. Ephraums, eds. Cambridge, England: Cambridge University Press.

León, R. J. C., and M. R. Aguiar. 1985. "El deterioro por uso pasturil en estepas herbáceas patagónicas." *Phytocoenologia* 13: 181–196.

Mares, M. A. 1992. "Neotropical mammals and the myth of Amazonian biodiversity." *Science* 225: 976–979.

McNeely, J. A., M. Gadgil, C. Leveque, C. Padoch, and K. Redford. 1995. "Human Influence on Biodiversity." In *Global Biodiversity Assessment*, V. H. Heywood, exec. ed., pp. 715–821. Cambridge, England: Cambridge University Press.

Mitchell, J. F. B., S. Manabe, V. Meleshko, and T. Tokioka. 1990. "Equilibrium Climate Change and its Implications for the Future." In *Climate Change: The IPCC Scientific Assessment*, J. T. Houghton, G. J. Jenkins, and J. J. Ephraums, eds. 135–164. Cambridge, England: Cambridge University Press.

Moraes, J. L., C. C. Cerri, J. M. Melillo, D. Kicklighter, C. Neill, D. L. Skole, and P. A. Steudler. 1995. "Soil carbon stocks of the Brazilian Amazon Basin." *Soil Science Society of America Journal* 59: 244–247.

Mosier, A., D. Schimel, D. Valentine, K. Bronson, and W. J. Parton. 1991. "Methane and nitrous oxide fluxes in native, fertilized and cultivated grasslands." *Nature* 350: 330–332.

Nordhaus, W. D. 1991. "To slow or not to slow: The economics of the greenhouse effect." *Economic Journal* 101: 920–937.

Paruelo, J. M., and O. E. Sala. 1993. "Effect of global change on maize production in the Argentinean pampas." *Climate Research* 3: 161–167.

Pielke, R. A., W. R. Cotton, R. L. Walko, C. J. Tremback, W. A. Lyons, L. C. Grasso, M. E. Nicholls, M. D. Moran, D. A. Wesley, T. J. Lee, and J. H. Copeland. 1992. "A comprehensive meterological modeling system—RAMS." *Meteorology and Atmospheric Physics* 49: 69–91.

Pielke, R. A., T. J. Lee, J. H. Copeland, J. L. Eastman, C. L. Ziegler, and C. A. Finley. 1996. "Use of USGS-provided data to improve weather and climate simulations." *Ecological Applications* (in press).

Pimentel, D., C. Harvey, P. Resosudarmo, K. Sinclair, D. Kurz, M. McNair, S. Crist, L. Shpritz, L. Fitton, R. Saffouri, and R. Blair. 1995. "Environmental and economic costs of soil erosion and conservation benefits." *Science*. 267: 1117-1123.

Rosenzweig, C., and M. L. Parry. 1994. "Potential impact of climate change on world food supply." *Nature* 367: 133–138.

Running, S. W. 1990. "Estimating Terrestrial Primary Productivity by Combining Remote Sensing and Ecosystem Simulation." In *Remote Sensing of Biosphere Functioning*, R. J. Hobbs and H. A. Mooney eds., pp. 65–86. New York: Springer Verlag.

Sala, O. E., M. Oesterheld, R. J. C. León, and A. Soriano. 1986. "Grazing effects upon plant community structure in subhumid grasslands of Argentina." *Vegetatio* 67: 27–32.

Shantz, H. L. 1954. "The place of grasslands in the earth's cover of vegetation." *Ecology* 35: 142–145.

Shine, K. P., R. G. Derwent, D. J. Wuebbles, and J. J. Morcrette. 1990. "Radiative Forcing of Climate." In *Climate Change: The IPCC Scientific Assessment*, J. T. Houghton, G. J. Jenkins, and J. J. Ephraums, eds., pp. 41–68. Cambridge, England: Cambridge University Press.

U.S. Department of Commerce. 1995. *Census of Agriculture 1992.* Washington, D.C.: Bureau of the Census.

Vavilov, N. I. 1951. *The Origin, Variation, Immunity and Breeding of Cultivated Plants.* New York: Ronald Press.

Wilson, A. T. 1978. "Pioneer agriculture explosion and CO_2 levels in the atmosphere." *Nature* 273: 40–41.

Part IV

Case Studies

Chapter 14

Biodiversity's Genetic Library

Norman Myers

This chapter reviews the manifold contributions of biodiversity and its genetic resources to modern agriculture, medicine, and industry. It also looks at wild species as models for research and reviews a few further applications of biodiversity in support of human welfare (Myers 1983, Oldfield and Alcorn 1991). The chapter assesses the scope and scale of current contributions to these sectors; the prospect for future contributions, notably via biotechnology; and some of the financial, commercial, and economic values of these contributions, both present and prospective. The chapter emphasizes the role of populations and races as well as species, with their differing degrees of genetic variability.

Let us note that all forms of biodiversity are both generated and maintained by natural ecosystems. Axiomatic as this may seem to some readers, it is well to remind ourselves that no one can exist without the life-support basis supplied by natural ecosystems.

There are solid reasons of biology, ecology, genetics, evolution, aesthetics, and ethics for us to regret the loss of any species, or even a species' population. But since the world is run in many people's eyes by the wisdom of the marketplace, we shall make that the focus for evaluation. This is not to say that the author does not appreciate the many other values, which for the most part are more valuable than those reviewed here.

From morning coffee to evening nightcap we benefit in our daily lifestyles from our fellow species. Without recognizing it, we utilize hundreds of products each day that owe their origin to wild plants and animals. Our welfare is intimately tied up with the welfare of biodiversity. Well may conservationists proclaim that by saving the lives of wild species, we may be saving our own. We enjoy the manifold benefits of biodiversity's genetic library after scientists have conducted intensive investigation of only 1 in 100 of earth's 250,000 plant species, and a far smaller proportion of the millions of animal species.

Plants, being stationary, have a hard time warding off threats from their environment in the form of plant eaters, too much heat or too much cold, too much water or not enough, and so forth. An animal can cope with these problems by simply walking, crawling, flying, or swimming away from the scene. A plant must fight its battle where it grows. So a plant's survival strategy lies in the chemical compounds it produces in its tissues in order to cause indigestion for herbivorous mammals, insects, and the like, and in order to help it through times of stress from heat, drought, and other adverse conditions.

Indeed a plant's life is one long struggle against difficult circumstances. As a result, the plant kingdom presents an extraordinary array of biocompounds of an abundance and diversity that can hardly be visualized until one starts to dabble in biochemistry. Especially productive of these biocompounds are plant communities that live in areas of "biological warfare," notably in areas with climatic extremes such as deserts and rainforests. In deserts, plants have to deal with a hostile environment, and in rainforests they have to deal with exceptional competition from huge numbers of species crowded into confined localities. When scientists go out to search for specialized chemicals in the plant kingdom, they often head for arid lands and jungles.

Much the same applies to coral reefs, where there is a plethora of biodiversity in a confined area and hence acute competition for living space. As a response, coral reef animals have evolved all manner of compounds to ward off neighbors, many of these compounds being toxins with multiple applications in medicine.

In short, biodiversity represents some of the most valuable natural resources with which we can confront the unknown challenges of the future. As the demand for biodiversity grows and the supply dwindles, moreover, biodiversity's value will steadily increase. Slowly but steadily we are coming to learn that birds and reptiles, insects and sea slugs, fishes and mammals, flowers and mosses, even fungi and bacteria, hold all manner of goodies in store for us if only we can ensure their survival.

Improved Forms of Existing Crops

Nothing is more basic to our material needs than our daily bread. The wheat and corn crops of North America, like those of Europe and other major grain-growing regions, have been made bountiful principally through the efforts of crop breeders rather than through huge amounts of fertilizers and pesticides—and crop breeders are increasingly dependent on genetic materials from wild relatives of wheat and corn. In common with all agricultural crops, the productivity of modern wheat and corn is sustained through constant infusions of fresh germplasm with its hereditary characteristics (FAO 1993; Potter et al. 1993). Thanks to this regular "topping up" of the genetic or hereditary constitution of the United States' main crops, the Department of Agriculture estimates that germplasm contributions lead to increases in productivity that average around 1 percent annually, with a farm-gate value that now tops $1 billion (U.S. National Research Council 1992).

And "we" means each and every one of us. Whether we realize it or not, we enjoy the exceptional productivity of modern corn not only when we eat our daily bread. It appears in our diets as cornflakes and popcorn and each time we enjoy a soft drink, a beer, or a whisky, all of which contain corn sugar. Since corn is widely used to feed cattle, pigs, and poultry, we effectively consume it each time we enjoy a breakfast omelette, a lunchtime escalope, or a dinner-table steak. It turns up, moreover, each time we read a magazine; because cornstarch is used in the manufacture of sizing for paper, the reader of this book is enjoying corn by virtue of the "finish" of the page he or she is looking at right now. The same cornstarch contributes to our lifestyles whenever we put on a shirt or a blouse. Cornstarch likewise contributes to glue, so we benefit from corn each time we mail a letter. And the same applies, through different application of corn products, whenever we wash our face, apply cosmetics, take aspirin or penicillin, chew gum, eat ice cream (or jams, jellies, catsup, pie fillings, salad dressings, marshmallows, or chocolates), and whenever we take a photograph, draw with crayons, or utilize explosives. Corn products also turn up in the manufacture of tires, in the molding of plastics, in the drilling for oil, in the electroplating of iron, and in the preservation of human blood plasma. Millions of motorists now utilize corn whenever they put gasohol instead of gasoline into their car fuel tanks.

Regrettably, wild gene pools are being rapidly depleted, as is illustrated by the case of wheat. In 1995 wheat flourished across an expanse of more than 230 million hectares, featuring a rough average of two million stalks per hectare. This means that the total number of individuals in 1995 exceeded 460 trillion, probably a record (Myers 1995). As a species, then, wheat is the opposite of endangered. But because of a protracted breeding trend toward genetic uniformity, the crop has lost the great bulk of its populations and most of its genetic variability. In extensive sectors of wheat's original range where wild strains have all but disappeared, there is virtual "wipeout" of endemic genetic diversity. Of Greece's native wheats, 95 percent have become extinct; and in Turkey and extensive sectors of the Middle East, wild progenitors find sanctuary from grazing animals only in graveyards and castle ruins. As for wheat germplasm collections, they were described more than one dozen years ago as "completely inadequate"—and that was without considering future broadscale problems such as acid rain and enhanced UV-B radiation (Damania 1993).

The commercial value of crop plant germplasm is significant. Wheat and corn germplasm collected in developing countries by the International Maize and Wheat Improvement Center near Mexico City benefits industrialized countries to the tune of $2.7 billion a year. In Italy, wheat germplasm contributes $300 million a year to the pasta industry. In Australia, grain varieties have boosted harvests by as much as $2.2 billion between 1974 and 1990. One-fifth of the value of the billion-dollar U.S. rice crop is attributed to genetic infusions. In all these instances, it is a case of germplasm from the South assisting agriculture in the North. For a South-South linkage, note that in India the introduction of new rice strains from wild relatives and primitive cultivars has increased yields by about $75 million a year—and the cost of the two seed banks that have made the increase possible is about $1 million a year (Evanson 1991).

For a detailed assessment, consider the case of corn. In 1970, 70 percent of the seed corn grown by U.S. farmers owed its ancestry to six inbred lines. When a leaf fungus blighted cornfields from the Great Lakes to the Gulf of Mexico, America's great corn belt almost came unbuckled. The disease eliminated 15 percent of the entire crop and as much as half of the crop in several states of the South, pushing up corn prices by 20 percent and causing losses to farmers, plus increased costs to consumers, worth more than $2 billion. The damage was halted with the aid of blight-resistant germplasm of various kinds, with a genetic ancestry that originally derived from Mexico. Not that the critical genetic material was worth $2 billion. Various other factors contributed to the turnaround, including the professional expertise of plant breeders and their research infrastructures, plus management systems, extension and educational services, marketing networks, and many other contributory factors. Nevertheless, it is apparent that the new wild-source genes played a major part in confronting the 1970 disaster: plant geneticists are no better than the genetic material they have to work with.

Hence the significance of a late-1970s discovery of an obscure plant, *Zea diploperennis*, in a montane forest of southwestern Mexico, being the most primitive known relative of modern corn (Iltis et al. 1979). It was surviving only in three tiny patches, covering a mere four hectares. Equally to the point, this last relict habitat in Mexico was acutely threatened by settlement schemes, squatter cultivators, and timber-cutting enterprises. The wild corn is a perennial, i.e., it resprouts every year. The few weedy stalks have been crossbred with conventional corn, making a perennial hybrid. Corn growers could be spared the annual expense of ploughing and sowing, since the crop would come up by itself year after year, just like grass. This advance in itself could offer very large savings, worth at least $4 billion a year.

Still more gains could lie with this wild corn's capacity to grow in unusual environments and to resist several prominent diseases. The plant has been discovered at elevations between 2,500 and 3,250 meters, where its cool montane habitats are often damp. This feature offers the prospect of growing crossbred corn in wet soils beyond the survival capacities of conventional corn and thereby expanding the cultivation range around the earth by as much as one-tenth. Even if cornfields in these marginal environments were to average only half the harvest of those in more hospitable areas, the additional output could be worth at least $1 billion per year. Furthermore, the Mexican variety is immune or tolerant to at least four of eight major viruses and mycoplasmas that are significant to corn growers around the world. These diseases now cause at least a 1 percent loss to the world's corn harvest each year, worth at least $500 million. All in all, the wild corn could eventually have an annual value of $6.8 billion (Fischer and Hanemann 1984).

We can tell a related story with respect to rice. In the early 1970s Asia's rice fields were hit by a "grassy stunt" virus that threatened to devastate rice production across more than thirty million hectares from India to Indonesia. Fortunately, a single gene from a wild rice offered resistance against the virus. Then in 1976 another virus, known as "ragged stunt" disease, emerged; and again, the most potent source of resistance proved to be a wild rice. The economic returns from these wild rices would be more than enough to pay for all the expenses of preserving the collection of rice germplasm at the International Rice Research Institute in the Philippines. Several other diseases could impose widespread losses on Asia's rice crop, but at least one hundred wild rices appear to harbor resistance (Chang 1987).

As for less important crops, several, ranging from banana and coffee to oilpalm and rubber, owe their productivity in major measure to wild relatives found in tropical forests. Just their export value was worth more than $20 billion in 1991 (Panayotou and Ashton 1992).

New Foods

Next, consider the scope for new foods. During the course of human history we have utilized around 3,000 plant species for food. Yet the earth contains at least another 75,000 edible plants. Of this cornucopia of plant foods, only about 150 have ever been cultivated on a large scale, and a mere 20 produce 90 percent of our food (Pimentel et al. 1992). We are essentially using the same limited number of plant species that have served humankind for millennia. All our modern crops are confined to particular environments, meaning that many potential food-producing areas are left unexploited. What if we could draw on additional plant species that could serve as crops in environments that are too dry or too saline, for example, for conventional agriculture?

There are numerous instances of underexploited food plants with proven potential (Wickens et al. 1989). For instance, Aborigines in Australia have used scores of plants, especially fruits and bulbs, as food. They favor certain yams that are well adapted to dry conditions, opening up the possibility that crossbreeds with established forms of yams could allow this important crop to be extended to several further regions. Another dryland plant, the yeheb-nut bush of Somalia, grows prolific bunches of pods that contain seeds the size of peanuts (though they taste more like cashew nuts), making a nutritious food that Somalis prefer to staples such as corn and sorghum. In addition, the yeheb's foliage supplies tasty fodder for livestock. Being adapted to arid environments, the yeheb could assist desert dwellers in many parts of the tropics. It is being brought back from the very verge of extinction in the wild through domestication efforts in Somalia (Myers 1989).

Similarly, a marine plant from the west coast of Mexico, known as eel-grass, produces grain that the Seri Indians grind into flour. This plant opens up the prospect that we could use the seas to grow bread. Many other little-known crops have exceptional potential, such as the amaranth, a grain crop of the Andes.

Perhaps the most promising category of wild foods comprises vegetables. The main center investigated to date is Southeast Asia, where at least three hundred vegetable species have been used in native cultures, about eighty of them still growing only in the wild in forest habitats (Vietmeyer 1986). A second center is the highlands of Ethiopia, where leafy grass vegetables prove a promising source of plant protein, yielding as much as alfalfa or soybean. When considering the potential of these Ethiopian vegetables, we should recall that a single wild species of the same genus has provided us, through plant breeding, with cabbage, kale, broccoli, cauliflower, and brussels sprouts.

Especially pertinent is the winged bean, a vegetable native to the island of New Guinea. The vinelike plant contains far more protein than potato, cassava, or several other crops that serve as principal sources of food for millions of people in the tropics. In fact, it offers a nutritional value equivalent to soybean, with 40 percent protein and 17 percent edible oil, plus vitamins and other nutrients. Its capacity to match the soybean might remind us that the United States grew sporadic patches of the soybean for at least a century before the plant was finally upgraded into a widespread crop, until today it is the premier protein crop in the world, flourishing in dozens of temperate-zone countries. As a result of genetic improvement, the winged bean is now upgrading the diets of hundreds of millions of people in more than fifty countries of the developing tropics.

Many other leafy food plants are important on a local scale, while remaining unknown elsewhere. At least 1,650 of them in tropical forests are reputed to contain roughly as much protein as legumes. They also feature some five to ten times more calcium than legumes and fruits, from two to six times as much iron, and ten to one hundred times more carotene (a yellow pigment in the green chlorophyll). In addition, these leafy vegetables often contain as much vitamin C as the best fruits, together with an abundance of vitamin A (Myers 1992).

Next, some wild fruits. Temperate-zone plants have given us only about 10 fruit species altogether, whereas the tropics have supplied almost 200 species, and another 3,000 species are available in unexploited form (Smith et al. 1990). The main tropical source is the rainforests, particularly in Southeast Asia, where around 125 fruit species are cultivated, and more than 100 other fruit trees grow wild in the forests, several of them producing edible fruits (others offer potential for crossbreeding with established crop species). A notable instance is the durian, with delectable taste and execrable smell: consuming a durian is like eating an almond-flavored custard in a public toilet. Also from Southeast Asia comes the rambutan, a bright red table fruit. Perhaps tastiest of all fruits from Southeast Asia is the mangosteen, though regrettably the plant appears to offer little genetic variability, placing all the greater value on its wild gene reservoirs. For those people who favor citrus fruits such as oranges and tangerines, the pummelo offers a suitably stimulating taste; it also yields a larger harvest than most citrus crops, and it grows in saline conditions.

These vegetables and fruits are but a few examples of new foods awaiting us in the wild. Indeed, a number of them have already made their way into our supermarkets. North American stores now feature all manner of new vegetables and fruits: from 1970 to 1985, the number of items available doubled to more than 130, and in certain instances to as many as 250. This specialty produce, mostly from Asia and Latin America, had become a $200-million-a-year business ten years ago (Vietmeyer 1986).

Among some leading entrants into the market are exotic items such as boniato (a dry-fleshed sweet potato), caladaza (pumpkinlike in appearance), dasheen (a form of taro root), tindora (a cucumberlike vegetable), jicama (a sweet-tasting root), calamondin (a sour, limelike citrus used as seasoning), jack fruit (the world's largest tree fruit, weighing 27 kilograms or more, featuring a yellow filling with a musky flavor), longan (a small-sized relative of the lychee), carambola (a sweet fruit resembling a bright yellow star when cut), manzano (a banana with a pink skin), cherimoya (described by Mark Twain as "deliciousness itself"), chayote (a type of squash with a taste halfway between apple and cucumber), and malanga (a starchy root). In addition there are lemongrass, kabocha squash, yucca, pigeon pea, Barbados cherries, sapodilla, bitter melons, and a new Philippines variety of sweet potato, among more than two hundred other little-known products.

All these items could become everyday items in our supermarkets around the world. True, certain persons may suppose these out-of-the-way items will never become accepted on a large scale by the general public. But they could reflect that we are today familiar with many crops our parents scarcely heard of: avocados, artichokes, bean sprouts, snow peas, shiitake mushrooms, chili peppers, tofu, nori, daikon, adjuki beans, and garbanzos. Recall too our experience with the kiwi fruit, introduced into American supermarkets as recently as 1961, and within twenty years enjoying a market of more than 10,000 tonnes worth $22 million a year.

Next, and as a category of environmentally adapted crop varieties, consider salt-tolerant plants. Because many natural environments are too saline for conventional crops, a major strategy for innovative agriculture lies with salt-tolerant plants. Saline soils around the world amount to 9.5 million square kilometers (the size of the United States), to be compared with total world croplands of 14.4 million square kilometers. The U.S. Department of Agriculture believes that more than 600,000 square kilometers, or one-fifteenth of the United States, overlie aquifers with a salinity of three thousand parts per million, which, while only one-tenth as salty as the oceans, is three times as salty as most conventional crops will tolerate. If American farmers could mobilize those saline underground water stocks that are readily pumpable, they would have access to as much water as would fill Lake Michigan six times over. Right now the nonsaline groundwater stocks of the United States are being depleted so rapidly that American agriculture could face a crisis of water supplies early next century.

Salt-dominated environments also include those irrigated croplands that, through misuse and overuse, have become too salty for further cultivation. Salinization is a major problem of some 350,000 square kilometers of croplands, and it claims a further 2,000 square kilometers each year. One-eighth of California's croplands are affected by gross salinization, and as much as one-quarter in the lower Rio Grande Valley. To rehabilitate salinized lands through conventional methods costs as much as $1,000 a hectare.

Fortunately, we may soon be able to look on the salinity problem as an opportunity, thanks to salt-tolerant plants—otherwise known as halophytes, from the Greek *halo* meaning salt and *phyte* meaning plant. As many as fifteen hundred plant species could qualify, some being wild relatives of commercial barley, wheat, sorghum, rice, millet, sugarbeet, tomato, date palm, and pistachio, plus several kinds of forage plants for livestock such as alfalfa, ladino clover, creeping bentgrass, Bermuda grass, and various reeds and rushes. For example, a strain of barley has been discovered that, deriving all its moisture from seawater, produces almost 1.2 tonnes of grain per hectare. A similar prospect appears in store with a number of wheat strains. In Israel, salt water irrigates adapted types of sorghum, soybean, and avocado (Squires 1994).

Finally, insects as food. Creatures with more than four legs are part of home cooking in many parts of Africa, Asia, and Latin America. No wonder: your typical creepy-crawlie contains 60–70 percent protein, more calories than soya beans or meat, and abundant minerals and vitamins. In Mexico, people consume more than three hundred species of insects, and in southern Africa insects can comprise as much as two-thirds of the animal protein diet. Crickets convert plants into biomass five times faster than cows, while a swarm of African locusts can weigh as much as thirty thousand tonnes (de Foliart 1992).

Medicines and Pharmaceuticals

One in four medicines and pharmaceuticals owes its origin to germplasm materials or other vital products of plant species, and another one in four to animals and microorganisms. These products include antibiotics, analgesics, diuretics, and tranquilizers, among a host of similar items (Joyce 1992). The commercial value of these medicines and pharmaceuticals in developed nations, including both prescription and nonprescription materials, topped $40 billion a year in the late 1980s (McNeely et al. 1993).

The contraceptive pill stems from a tropical forest plant. Three promising responses to AIDS derive from plant materials (Djerassi 1992). The bark of a yew tree in the U.S. Pacific Northwest contains a biocompound, taxol, that damages cancer cells unaffected by other drugs. Taxol could help at least 100,000 Americans with breast, lung, and ovarian cancers (regrettably, it takes nine thousand kilograms of bark from two thousand to four thousand trees to produce one kilogram of taxol; fortunately, it may shortly become synthesizable in the laboratory) (Kingston 1993). A child suffering from leukemia in 1960 faced only one chance in ten of remission, but today such a child enjoys nineteen chances in twenty thanks to two potent drugs derived from alkaloids of Madagascar's rosy periwinkle. These two drugs, also used against Hodgkin's disease and a number of other cancers, generate commercial sales totaling more than $200 million per year in the United States alone. According to the National Cancer Institute and the Economic Botany Laboratory outside Washington, D.C., tropical forests alone could well contain twenty plants with materials for several further anti-cancer "superstar" drugs (Douros and Suffness 1980).

Plant-derived anti-cancer drugs now save around thirty thousand lives in the United States each year, with annual economic benefits already amounting to an estimated $370 billion (1990 dollars) in terms of lives saved, suffering relieved, morbidity reduced, and worker productivity maintained. We can at least double these figures to determine values for all developed nations (Principe 1991 and 1996). Many anti-cancer plants are to be found in the tropics, meaning for the most part the developing nations—and this too is where the great majority of extinctions are occurring.

As indicated, tropical forest plants are specially important to medicine. A number of analysts have attempted an economic assessment of the plants' potential worth overall, not just for anti-cancer purposes. Early estimates for total present value range from $420 billion (Pearce and Puroshothaman 1993) to $900 billion (Gentry 1993). The latest estimate (Mendelsohn and Balick 1995) proposes that each new plant-derived drug is worth an average of $94 million to the private pharmaceutical company that discovers and develops it and $449 million to society as a whole.

There are around 125,000 flowering plants in tropical forests. In many species five separate parts can be important, viz. roots, stems, leaves, flowers, and fruit, yielding an average of six different extracts. This means that about 750,000 potential extracts are available. Given that each sample can be screened in some 500 different ways to test for new drugs, the total possible individual tests amount to 375 million. Between one in fifty thousand and one in one million tests result in a commercial drug, which means that at least 375 potential drugs are available from flowering plants in tropical forests. To date, 48 drugs have been discovered (including vincristine, vinblastine, curare, quinine, codeine, and pilocarpine), or only one-eighth of all potential drugs. So we can reasonably expect that another 328 drugs await discovery, with a putative total value of $3–$4 billion to individual pharmaceutical companies and $147 billion to society as a whole (Mendelsohn and Balick 1995). (For an alternative set of calculations that are broadly in the same ballpark, see Principe 1991 and 1996.)

Allowing for plant-derived drugs and pharmaceuticals from other parts of the world (mainly from developing nations of the tropics), the cumulative commercial value of these products (combined prescription and over-the-counter sales) to developed nations alone during the 1990s is estimated to amount to $500 billion (1987 value) (Principe 1991 and 1996). Moreover, the calculation applies to just the commercial value as reflected through the sales of these products. The economic value, including the benefits of reduced morbidity and mortality, could well be several times larger.

Suppose that until the year 2050 we witness the extinction every two years of one plant species with medicinal or pharmaceutical potential. The cumulative retail-market loss from each such extinction would amount to $12 billion for the United States alone (Principe 1996). At the same time we should note that there is an uncertain amount of overlap and substitutability among drugs; we probably would not need all 375 of the ones from tropical forests. In addition, the plant component of a new medicinal or pharmaceutical is only a small part of the final product (including research and development, manufacture, and marketing); in money terms it may amount to no more than 5–10 percent of the cross-counter price. But the plant component bears an absolute value in the sense that without it there would be no product. In addition again, the consumer surplus of the drug purchaser may be substantial insofar as many drugs, especially life-saving drugs, tend to be price inelastic.

A good number of medicinals and pharmaceuticals come from animals too. Notable sources are amphibians, which tend to be beset by all manner of predators and diseases. Three Australian tree frogs are endowed by evolution with the ability to secrete a host of chemical weapons from glands in their skin. Some of these biocompounds are toxins to ward off other animals, while others such as caerin 1.1 protect them against infection, this being an antimicrobial compound with potential for antibiotic and antiviral drugs. In the rainforests of Ecuador is a frog that secretes a painkiller with two hundred times the potency of morphine. In similar style, a number of insects secrete substances including hormone and birth-control analogs, cardiotonic factors, wound-healing promoters, and antiviral agents (Eisner 1992). So much for land and freshwater creatures. Still more bounty is to be found in the marine realm. The sea urchin yields holothurin, a substance that may help with treatment of coronary disorders, even cancer. The octopus yields an extract that relieves hypertension; the seasnake produces an anticoagulant; and the menhaden harbors an oil that helps with atherosclerosis. A Caribbean sponge produces a compound that acts against diseases caused by viruses, much as penicillin did for diseases caused by bacteria; this widely hailed discovery offers us the prospect of curing a wide range of viral diseases from the common cold upwards. The skeletons of shrimps, crabs, and lobsters feature a material known as chitin, which contains an enzyme, scientifically termed chitosanase, that serves as a preventive medicine against fungal infections and helps to heal other wounds more quickly and to inhibit certain categories of malignant cells. Even the lowly barnacle, that bane of sailors, could soon make life easier for us when we sit in the dentist's chair. The adhesive that enables the animal to cling to ship bottoms could be adapted into a cement for tooth fillings—and it could even replace the binding pins now used to set bone fractures.

Let us end this section by looking at a creative initiative that bodes well for conservation of biodiversity. Until recently, source nations of medicinal materials have not received any financial return from those pharmaceutical companies that have benefited. Eli Lilly, exploiter of the rosy periwinkle for two anti-cancer drugs, has enjoyed sales that may well have averaged $100 million per year since the drugs were first marketed in the early 1960s. By contrast, the original home of the plant, Madagascar, has not received a single cent. To this extent, Madagascar has sensed less incentive to safeguard the thousands of other plant species that are endemic to the country and face imminent loss of habitat through deforestation—even though certain of them may contain startpoint materials for other wonder drugs.

Fortunately, a deal has been struck between Costa Rica and Merck, the world's largest pharmaceutical company, with 1991 sales of $6.6 billion. Merck has agreed to pay Costa Rica's National Institute of Biodiversity a sum of $1 million in exchange for access to the country's plant species and other biodiversity. True, this upfront payment amounts to a mere 0.1 percent of the company's R&D budget and a still smaller proportion of its sales turnover. But if the Merck bioprospectors make a discovery that results in a commercial product, Merck will pay the Institute a share of the royalties, probably in the order of 1–10 percent. One hundred thousand dollars of the upfront money and 50 percent of Costa Rica's royalty income will go to on-ground conservation. This pioneering initiative may provide a model for replication in other countries.

Industry

Both plants and animals serve many needs of modern industry. As technology advances in a world growing short of many things except shortages, industry's need for new raw materials expands with every tick of the clock. To illustrate the manifold purposes served by biodiversity materials, consider just a single example, natural rubber. In its many shapes and forms, rubber supports us, literally and otherwise, in numerous ways during our daily round.

Other specialized materials contribute to industry by way of gums and exudates, essential oils and ethereal oils, resins and oleoresins, dyes, tannins, vegetable fats and waxes, insecticides, and multitudes of other biodynamic compounds. Many wild plants bear oil-rich seeds with potential for the manufacture of fibers, detergents, starch, and general edibles—even for an improved form of golf ball (Mann et al. 1994).

Biodiversity serves our needs in still further and unusual ways. Certain plants can sequester heavy metals and other mineral contaminants (Baker et al. 1988). Other species act as pollution indicators; notable instances include lichens, dragonflies, tiger beetles, and amphibians (Ahmadjian 1994). The larvae of several aquatic flies—notably mayflies, stone flies, caddis flies, and true flies—can be used to identify point sources of chemical contaminants in water bodies, especially with respect to molybdenum, manganese, and copper (Root 1990). Other animals, such as earthworms and certain fish, birds, and mammals, serve as biological monitors of various kinds of widespread pollution. A number of plants, for instance the water hyacinth, act as first-rate depolluting agents in sewage lagoons. A few plants can even register radiation, some of them more sensitively than a dosimeter. Certain animals, such as dogs, cats, horses, chimpanzees, and snakes, appear able to anticipate slight earth tremors and thus to warn of impending earthquakes (Kerr 1980).

A number of tree species—especially beech, elm, oak, sycamore, willow, and elder—serve to clean up city pollutants, notably sulfur dioxide. Trees also act as air coolants. A twenty-meter shade tree can mitigate 900,000 BTUs of heat, worth three tonnes of air conditioning costing $20 a day in the United States (Cairns and Niederlehner 1994).

Also important are plant species such as euphorbias that contain hydrocarbons rather than carbohydrates. Hydrocarbons are what make petroleum petroleum—and plant hydrocarbons, while very similar to those in fossil petroleum, are practically free of sulfur and other petroleum contaminants. Of plant species that appear to be candidates for "petroleum plantations," several can grow in areas that have been rendered useless through, for example, strip mining. Hence the prospect that land degraded by extraction of hydrocarbons from beneath the surface can be rehabilitated by growing hydrocarbons above the surface. Remember too that a petroleum plantation need never run dry like an oil well (Hinman and Hinman 1992).

Finally, recall that the bulk of the world's species are insects. Some readers may ask what creepy-crawlies have ever done for industry, let alone any other economic sector. Consider the oilpalm plantations of Malaysia. Until the early 1980s the pollination of millions of oilpalm trees was done by human hand, an inefficient and expensive way of performing the task. Then the plantation owners asked themselves how the oilpalm got itself pollinated in its native habitats of West Africa's forests. Researchers went off to Cameroon, where they found the job was undertaken by a tiny weevil. Start-up stocks of the weevil were taken back to Malaysia, where they were released into the plantations. (There was no problem of ecological complications with other species, since the weevil confined its attentions to the oilpalm.) The pollination is now entirely accomplished by the weevil, with savings that already amounted to $150 million per year in the early 1980s

(Greathead 1983). We can reflect on that the next time we utilize cosmetics or other products that may contain palm oil—and the next time we hear of dozens of insect species becoming extinct every day.

Research Models

Not only does biodiversity supply materials for our direct use, but it also provides research models for many medicines and industrial products. Scientists would have had a hard time devising synthetic rubber if they had not had a "blueprint" in the molecular structure of natural rubber. Each time we enter a high-rise office block or apartment building, we might reflect that the edifice has probably been constructed on the principles of metal-beam architecture, a building design that owes its constructional inspiration to the physical makeup of the giant water lily in the heart of Amazonia.

Similarly, the polar bear's hairs are not white but, being tiny hollow tubes with no pigment, merely appear white by virtue of their rough inner surfaces that, like transparent snowflakes, reflect invisible light. This feature allows the polar bear to funnel ultraviolet light inwards to warm its body; it also enables manufacturers to do a better job with cold-weather clothing for humans and may even lead to incorporation of solar-energy "light pipe" collectors for heating houses and offices.

We can likewise learn from biodiversity models for better health. Animal physiology affords many clues to the origins and nature of human ailments. Diseases of the heart and circulatory system, for example, can be studied by comparative investigation of long-flying birds such as the stormy petrel and the albatross. These birds feature highly developed hearts, and their circulatory systems are in top-notch state too, in order for them to accomplish their immense annual migrations. It is the superb physical equipment of these birds that has advanced our understanding of cardiomyopathy, a failing in humans caused by overdevelopment of the heart muscle, obstructing blood overflow. Similar health clues are becoming available from research into hummingbirds, creatures that spend much of their entire waking life in flight. The same applies to those butterflies that wander four thousand kilometers each year, and to locusts and other insects that sustain high levels of activity for weeks if not months on end.

The desert pupfishes of Nevada and California rank among the most threatened creatures on earth. They reveal a remarkable tolerance to extremes of temperature and salinity, an evolved attribute that could assist research into human kidney diseases. The Florida manatee, another endangered species, features poorly clotting blood, offering scope for research into hemophilia. Primates, more closely related to humans, are especially valuable for medical research: the cotton-topped marmoset, a species of monkey susceptible to cancer of the lymphatic system, is helping to produce an anticancer vaccine. Squids, with nerve fibers one thousand times larger in cross section than humans', supply neuroscientists with crucial insights into our own nervous system.

As for industrial applications, the woodpecker has a neck built to withstand severe whiplash, which has offered a blueprint for crash helmets. A species of chalcid wasp with unusual capacity for hovering has aided with the design of an improved helicopter. For a lengthy listing of further industrial applications, see Myers (1983) and Oldfield (1989).

Biotechnology

Biotechnology places a still greater premium on biodiversity with its genetic resources (Frederick and Egan 1994, Moore et al. 1992). The best biotechnologist is no better than the basic materials he or she has to work with. Genetic variability in species and populations opens up abundant opportunities in agriculture, for instance, where the famous Green Revolution is being superseded by a still more revolutionary departure, the Gene Revolution. This breakthrough in agrotechnology may soon enable us to harvest crops from deserts, farm tomatoes in seawater, grow superpotatoes in many new localities, and enjoy entirely new crops such as a "pomato." The sophisticated techniques of genetic engineering, isolating, and manipulating the hereditary materials of each species' makeup may eventually bring us closer to the day when we can send more people to bed with a full stomach (Hobbelink 1991).

Biotechnology can also assist with new forms of pesticides. The Monsanto Company in St. Louis, Missouri, has genetically engineered several leading crops, including corn, potatoes, and cotton, to produce a potent insecticide known as cholesterol oxidase. Plants with the gene kill a wide variety of insects, including caterpillars that attack corn, caterpillars and boll weevils that attack cotton, and the Colorado beetle that attacks potatoes. This form of biotechnology-based insect control would have no effect on beneficial insects that are now killed inadvertently by pesticide spraying (Allen 1995).

It is microorganisms that offer most promise to date for biotechnology, notably by helping us to maintain a pollution- and waste-free environment (Bull et al. 1992). They can be used to generate products ranging from biodegradable plastics to hydrocarbons. We can also counter global warming by utilizing those microorganisms that trap and recycle carbon dioxide from the atmosphere, e.g., marine planktonic algae that convert carbon dioxide into scales of calcium carbonate, whereupon the scales sink to the ocean floor and ultimately form chalk (Branden and Schneider 1994).

The species most useful for biotechnology thus far tend to be algae, fungi, nematodes, viruses, and bacteria. Bacteria, being microscopic unicellular organisms, are found almost everywhere and in exceptional numbers. One milliliter of water can contain 10 million bacteria as well as 10,000 protozoa. Bacteria comprise 10 percent of a human's dry body weight. (But not all of them are super small: a bacterium living in the intestine of a Red Sea surgeonfish is a full five millimeters long, large enough to be seen with the naked eye and one million times as massive as a typical bacterium.) In just a single hectare of temperate-region pasture, there can be 3,000 kilograms of bacteria, together with 4,000 kilograms of fungi, 380 of protozoa, 200 of algae, and 120 of nematodes, for a total of 7,700 kilograms—whereas mammals make up only 1.2 kilograms and birds 0.3 of a kilogram (Pimentel et al. 1992). A single gram of temperate-forest soil can contain billions of bacteria, many of them from species yet to be identified. Altogether, microorganisms make up one-fifth of global biomass, the same proportion as for animals, the rest being plants (Wilson 1994).

Bacteria are specially helpful by undertaking biological reactions that were once thought impossible, e.g, in cleaning up toxic chemicals. A notable instance is the removal of chlorine from aromatic compounds, this being a critical step in breaking down compounds including such major pollutants as PCBs, dioxins, chlorinated phenols, and chlorinated benzenes.

Despite their extravagant abundance as a category of organisms, however, certain microorganism species can be prone to extinction (Cairns 1993). Fungi, being exceptionally sensitive to atmospheric pollutants, are undergoing a "catastrophic decline" in many industrial parts of Europe and the United States (Jaenike 1991). This is all the more regrettable since the basidomycete fungi, for example, with about thirty thousand species identified to date, reveal novel metabolites with antibiotic, antiviral, phytotoxic, and cytostatic activities.

Moreover, certain bacteria contrast with the frequent trend toward cosmopolitan distribution among microorganisms and are confined to ultra-localized habitats. An example is *Thermus aquaticus* or taq, thriving only in the boiling water of hot springs. By virtue of its highly specialized lifestyle and elevated temperatures, it is the source of an enzyme that serves as a catalyst for the polymerase chain reaction, a method of producing millions of copies of any DNA sequence. Its confined habitat could leave it vulnerable to summary extinction were the boiling water to be diverted for energy or industrial purposes.

Conclusion

There is ample evidence to demonstrate that:

- biodiversity contains abundant stocks of genetic materials to support several leading sectors of human welfare;
- these stocks, while almost entirely unexploited, often possess exceptional economic value; and
- the stocks are nonetheless being depleted at a rate that will grossly reduce our scope for dealing with many basic economic problems of the future.

Acknowledgments

This chapter has been written with financial support from my Pew Fellowship in Conservation and Environment. I thank my research associate, Jennifer Kent, for her many incisive suggestions and her other hyper-helpful contributions.

References

Ahmadjian, V. 1994. "Lichens are more important than you think." *BioScience* 45: 124.

Allen, W. H. 1995. "Second insecticide gene for crops." *BioScience* 45: 387.

Baker, A., R. Brooks, and R. Reeves. 1988. "Growing for gold, copper and zinc." *New Scientist* (March 10): 44–48.

Branden, C. I., and G. Schneider, eds. 1994. *Carbon Dioxide Fixation and Reduction in Biological and Model Systems*. Oxford: Oxford University Press.

Bull, A. T., M. Goodfellow, and J. H. Slater. 1992. "Biodiversity as a source of innovation in biotechnology." *Annual Review of Microbiology* 46: 219–252.

Cairns, J. 1993. "Can microbial species with a cosmopolitan distribution become extinct?" *Speculations in Science and Technology* 16: 69–73.

Cairns, J., and B. R. Niederlehner. 1994. "Estimating the effects of toxicants on ecosystem services." *Environmental Health Perspectives* 102: 936–939.

Chang, T. T. 1987. "The impact of rice on human civilization and population expansion." *Interdisciplinary Science Reviews* 12(1): 63–69.

Damania, A. B., ed. 1993. *Biodiversity and Wheat Improvement*. Chichester, England: John Wiley.

Djerassi, C. 1992. "Drugs from third world plants: The future." *Science* 258: 203–204.

Douros, J.D., and M. Suffness. 1980. "The National Cancer Institute's Natural Products Antineoplastic Development Program." *Recent Results in Cancer Research*, S. K. Carter and Y. Sakurai, eds. 70: 21–44.

Eisner, T. 1992. "The hidden value of species diversity." *BioScience* 42: 578.

Evanson, R. E. 1991. "Genetic Resources: Assessing Economic Value." In *Valuing Environmental Benefits in Developing Economies*, J. R. Vincent, E. W. Crawford, and J. Hoehn, eds., pp. 169–181. East Lansing, Michigan: Michigan State University Press.

FAO (Food and Agriculture Organization). 1993. *Harvesting Nature's Diversity*. Rome: Food and Agriculture Organization.

Fischer, A. C., and W. M. Hanemann, 1984. *Option Values and the Extinction of Species*. Berkeley, California: Department of Agricultural Economics, University of California.

Frederick, R. J., and M. Egan. 1994. "Environmentally compatible applications of biotechnology." *BioScience* 44:529–535.

Gentry, A. 1993. "Tropical Forest Biodiversity and the Potential for New Medicinal Plants." In *Human Medicinal Agents from Plants*, A. D. Kinghorn and M. F. Balandrin, eds., pp. 13–24. Washington, D.C.: American Chemical Society.

Greathead, D.J. 1983. "The multi-million dollar weevil that pollinates oil palm." *Antenna* (Royal Entomological Society of London) 7: 105–107.

Hinman, C. W., and J.W. Hinman. 1992. *The Plight and Promise of Arid Land Agriculture*. New York: Columbia University Press.

Hobbelink, H. 1991. *Biotechnology and the Future of World Agriculture*. London: Zed Books.

Iltis, H. H., J. F. Doebley, R. M. Guzman, and B. Pazy. 1979. "*Zea diploperennis* (Gramineae), a New Teosinte from Mexico." *Science* 203: 198–201.

Jaenike, J. 1991. "Mass extinction of European fungi." *Trends in Ecology and Evolution* 6: 174–175.

Joyce, C. 1992. "Western medicine men return to the field." *BioScience* 42: 399–403.

Kerr, R. A. 1980. "Quake prediction by animals gaining respect." *Science* 208: 695–696.

Kingston, D. G. I., 1993. "Taxol: An Exciting Anticancer Drug from *Taxus brevifolia*." In *Human Medicinal Agents from Plants*, A. D. Kinghorn and M. F. Balandrin, eds., pp. 138–148. Washington, D.C.: American Chemical Society.

Mann, J., R. S. Davidson, J. B. Hoggs, D. V. Banthorbe and J. B. Harborne. 1994. *Natural Products: Their Chemistry and Biological Significance*. London: Longman.

McNeely, J., S. Laird, C. Meyer, R. Gomez, A. Sittenfeld, D. Janzen, M. Gollin, and C. Juma, eds. 1993. *Biodiversity Prospecting: Using Genetic Resources for Sustainable Development*. Washington, D.C.: World Resources Institute.

Mendelsohn, R., and M. J. Balick. 1995. "The value of undiscovered pharmaceuticals in tropical forests." *Economic Botany* 49: 223–228.

Moore, H. D. M., W. V. Holt, and G. M. Mace, eds. 1992. *Biotechnology and the Conservation of Genetic Diversity*. Oxford: Oxford University Press.

Myers, N. 1983. *A Wealth of Wild Species: Storehouse for Human Welfare*. Boulder, Colo.: Westview Press.

Myers, N. 1989. "Loss of Biological Diversity and Its Potential Impact on Agriculture and Food Production." In *Food and Natural Resources*, D. Pimentel and C. W. Hall, eds., pp. 49–68. San Diego, Calif.: Academic Press.

Myers, N. 1992. *The Primary Source: Tropical Forests and Our Future*. New York: Norton.

Myers, N. 1995. "Population and biodiversity." *Ambio* 24(1): 56–57.

Oldfield, M. L. 1989. *The Value of Conserving Genetic Resources*. Sunderland, Mass.: Sinauer Associates.

Oldfield, M. L., and J.B. Alcorn, editors, 1991. *Biodiversity: Culture, Conservation, and Ecodevelopment*. Boulder, Colo.: Westview Press.

Panayotou, T., and P. S. Ashton. 1992. *Not by Timber Alone*. Washington, D.C.: Island Press.

Pearce, D., and S. Puroshothaman, 1993. *Protecting Biological Diversity: The Economic Value of Pharmaceutical Plants*. London: Center for Social and Economic Research into the Global Environment, University College London.

Pimentel, D., U. Stachow, D. A. Takaes, H. W. Brubaker, A. R. Dumas, J. J. Meaney, J. A. S. O'Neil, D. E. Onsi, and D. B. Corzilius. 1992. "Conserving biological diversity in agricultural/forestry systems." *BioScience* 42: 354–362.

Potter, C. S., J.I. Cahen, and D. Janczewski, eds. 1993. *Perspectives on Biodiversity: Case Studies of Genetic Resource Conservation and Development*. Washington, D.C.: American Association for the Advancement of Science.

Principe, P. 1991. "Valuing Diversity of Medicinal Plants." In *Conservation of Medicinal Plants*, O. Akerele, V. Heywood, and H. Synge, eds., pp. 70–124. Cambridge: Cambridge University Press.

Principe, P. 1996. "Monetizing the Pharmocological Benefits of Plants." In *Tropical Forest Medical Resources and the Conservation of Biodiversity*, M. J. Balick, W. Elisabetsky and S. Laird, eds. New York: Columbia University Press.

Root, M. 1990. "Biological monitors of pollution." *BioScience* 40: 83–86.

Smith, N. J. H., D. L. Plucknett, J. T. Williams and P. Greening. 1990. *Tropical Forests and Crop Genetic Resources*. Washington, D.C.: International Fund for Agricultural Research.

Squires, V. 1994. "Overcoming salinity with salt water: Salt bushes as a useful crop." *Search* 25(1): 9–12.

U.S. National Research Council. 1992. *Managing Global Genetic Resources: The U.S. National Plant Germplasm System*. Washington, D.C.: National Academy Press.

Vietmeyer, N. D. 1986. "Lesser-known plants of potential use in agriculture and forestry." *Science* 232: 1379–1384.

Wickens, G. E., M. Haq, and P. Day, eds. 1989. *New Crops for Food and Industry*. London: Chapman and Hall.

Wilson, E.O. 1994. *Naturalist*. Washington, D.C.: Island Press.

Chapter 15

IMPACTS OF MARINE RESOURCE EXTRACTION ON ECOSYSTEM SERVICES AND SUSTAINABILITY

Les Kaufman and Paul Dayton

The few large islands that pass for continents on this planet supply only a portion of humanity's needs. The rest comes from marine ecosystems, which offer a huge and often unappreciated source of food, recreation, and other types of wealth dependent upon a large array of marine species and ecological services. There may be as many as three thousand marine species subject to commercial extraction, with perhaps a third of these also taken through recreational fisheries. Marine fisheries are thought to account for some 20 percent of total animal protein consumed by humans, at a value of between $50 billion and $100 billion annually. Ecotourism and other living resources probably add a few billion more; the exact amount has not been tabulated. This is a massive ecosystem service representing vitally important nutritional and economic benefits to billions of people, especially poor people.

Although frequently referred to as "harvesting," marine resource extraction is accompanied by very little "planting" and in some cases severely damages the host system. There are other striking differences between fisheries and agronomy. Where land harvests focus on a few domesticated species, all wild species in the sea are fair game, an impact superimposed on already formidable natural mortality and uncertainty. Terrestrial harvests are much more selective than marine fisheries, which are destructive to a wide array of species and components of the habitat. Population structure of domestic animals is optimized by the farmer; the population structure of marine living resources is degraded, not improved, by fishing pressure. In these ways the removal of goods from the sea is more similar to mining than farming. The difference—and this is crucial—is that most of these marine populations could be extracted in a sustainable manner. What is required is that the integrity of the ecosystem that supports this bounty be respected and maintained. Unfortunately, this simple and obvious objective has rarely been met. Among the most serious problems is the burden of proof about who is damaging what. Many stakeholders behave as if harvesting wild resources is a right rather than a privilege: if their extraction methods are destructive, it is somebody else who must prove this to be so. This will not work. Sustainability is possible only if the privilege of extraction is limited to those interested in the long-term welfare of the resource. The purpose of this chapter is to suggest that we move in that direction and to explain why it is so urgent and important to do so now. Marine systems are large, but they can be depleted. Marine systems are interconnected but not enough to ensure the regeneration of devastated areas (Vermeij 1993, Roberts 1995a, b). These simple empirical truths must be effectively incorporated into marine policy.

The ocean is certainly huge, but the part that is relevant to human economy is relatively small. It is mostly found within the upper thin skin of 100 meters of depth. Even this is concentrated within a threadlike band ringing the world's coastlines, plus a few important hot spots scattered about in regions of upwelling (e.g., Haney 1986) and points of aggregation (such as cod migratory routes and group spawning grounds (Rose 1993). Such areas are highly sensitive from the standpoint of marine species survival (Vermeij 1993). Logically, the world's fisheries are directed toward these focal areas (Dayton et al. 1995), precisely where the risks to marine systems, endangered species, and resource sustainability are the highest (Mangel 1993). Widespread fishery collapse has shown the oceans to be alarmingly responsive to the cumulative effects of human intervention.

Kinds of Goods

The sea provides three kinds of goods: (1) bulk raw materials that have low unit value, such as seaweed (Doty et al. 1986) and clupeid fishes (menhaden, herring, etc.), minerals, building materials, as well as the targets of subsistence fisheries; (2) high-value species such as shellfishes and top carnivores (e.g., billfishes, tunas, and sharks); and (3) materials and live organisms with a very high nonfood value, removed from the wild in relatively small quantities. The targets of recreational fishing and hunting fit mostly in this third category, as do precious corals, nacre and pearls, ornamental shells and corals, ornamental species for the home aquarium trade, and the genes that code for valuable natural products as raw material for biotechnology.

Bulk Raw Materials

A familiar example of a nonliving bulk raw material from the sea is manganese from deep-sea manganese nodules, whose extraction could ultimately become a problem for some deepwater communities. Currently, however, the most serious extraction costs are localized to nearshore marine habitats. For example, the mining of coral reef limestone for cement manufacture poses a serious threat to some coral reef habitats (White 1987, Salvat 1987, Clark and Edwards 1995). Every spring, the young of both anadromous and catadromous fishes mass in regions of low salinity and high productivity in coastal estuaries. These areas correspond roughly to the intersection of maximum cross-sectional area or flow volume and minimum salinity, exactly the sorts of places where cooling facilities for large power plants are often located. Consequently, billions of eggs and larvae are consumed by power plants as they extract immense volumes of freshwater as coolant for condensation coils. Offshore drilling for oil, and the associated activities of oil transport, storage, and processing in coastal facilities, also threatens water quality, ecosystem integrity, and endangered species.

Most important, however, is the process of biomass removal, which has two principal effects on ecosystem services: habitat destruction and the diversion of productivity. Some of the biomass removed is cycled into terrestrial communities, but most is redirected into portions of the marine food web it normally does not nourish. Deforestation of coastal watersheds, both to remove wood and to open land for agriculture and other forms of development, poses one obvious type of threat to adjacent marine systems. Nutrient runoff and increased turbidity alter littoral ecology and destabilize fragile systems such as coral reefs so that when next hit with a major disturbance, natural or otherwise, their ability to regenerate is severely impaired (Kaufman 1986, Hughes 1994). Mangrove forests, an important source of wood for building and charcoal for cooking in the developing world, are especially under siege (Hellier 1988). Mangals are also nursery areas for a wide array of fishes and are crucial to a variety of narrow-habitat endemics (e.g., yellow warbler, mangrove cuckoo, proboscis monkey).

The amount of marine biomass extracted in fisheries is sobering: 8 percent of global aquatic primary production, focused so as to account for between 24 and 35 percent of upwelling and continental shelf productivity (Pauly and Christensen 1995). Fishery collapse can result through either the failure of target species themselves (and often their associates: Polovina et al. 1994, Lowry et al. 1989, Parsons 1992, Ainley et al. 1994) or the failure of the systems on which the target species depend. Ecosystem failure is manifested in a variety of ways. Food web changes and their cascading ecological effects brought about by direct human actions (including bycatch, ghost fishing, habitat destruction, and so forth) are unfolding on a grand scale and have far-reaching ecosystem impacts. Such effects are graphically exhibited in the clear, shallow waters of coastal coral reef communities (Roberts 1995a, b). Human exploitation draws from all points in the food web, especially in poor countries where human population is so high (and rapidly growing) and the demand for protein so intense that coastal foraging leaves literally no stone unturned and no edible tidbit overlooked (Dayton et al. 1995).

High-Unit-Price Food Resources

Many seafood commodities command a sufficiently high price to justify fisheries in which much or even most of what is caught is thrown away. Groundfishes (mostly cods, flatfishes, and their relatives) are on the bottom of a luxury seafood market that currently tops out with such delicacies as New Zealand green-lipped mussels, sea scallops, tropical and temperate shrimps of many species, virtually all lobsters and crayfishes, all salmonids, Alaskan king crab, and various caviars. As the price of a particular item goes up, the amount of wastage that can be tolerated in its pursuit likewise increases, as does the amount of deep ecosystem damage inflicted during extraction. Foreign investment in this process has retracked local fisheries in poor nations away from providing local protein at affordable rates, to supporting massive export fisheries to garner foreign revenue (e.g., Kaufman 1992).

Taxonomic and Temporal Nonspecificity of Fishing Gear

Marine catches are highly diverse, so if one or two species are worth much more than the others, it makes economic sense to dump all the rest overboard as bycatch. Bycatch is a major factor in world fisheries: most fishing gear is highly nonspecific, and bycatch constitutes 40 percent or more of the catch (Androkovich and Stollery 1984, Pauly 1988, Dayton et al. 1995). Depending on locality and fishing method, there may be anywhere from 4 to 100 pounds of animals wasted for every pound of shrimp landed (e.g.,

Safina 1994). In the North Sea, discards and offal constitute five to ten times the catch of German sole. This amounts to 71,000 tons of offal and 109,000 tons of discards that are either consumed by seabirds or settle to the bottom and decompose. In the North Sea, offal now contributes 50 percent of the diet of fulmar, 40 percent for the herring gull, 60 to 70 percent for great black-backed gull, 50 percent for kittiwake and 10 percent for gannet. Experimental studies found that 90 percent of the offal is consumed in the Shetland Islands, 15 to 35 percent in France, and 83 percent in Germany, where 50 to 80 percent of the discards are also consumed. This boosts seabird populations, disrupts seabird communities, and could become the major force driving seabird dynamics, a situation reminiscent of garbage dumps in urban areas.

Another form of discard consists of the millions of benthic invertebrates crushed or dislodged by trawls each year. *Arctica islandica* assumed prominence in the diets of cod and flatfishes only after trawling appeared on the scene (Dayton et al. 1995). Dredge-associated mortality of sea scallops is high, especially on hard bottoms (Caddy 1973, Shepard and Auster 1991). Those scallops not killed outright are damaged in ways that increase vulnerability to predation by crabs and starfishes (Shepard and Auster 1991). Scallop dragging is also nonselective, killing or reducing habitat quality for lobsters (Jamieson and Campbell 1985). Harvesting of Irish moss (an alga) has been shown to negatively impact both lobsters and scallops (Scarratt 1973, Pringle and Jones 1980). The discards that are not immediately consumed act as organic pollution and can deplete benthic oxygen (Wassenberg and Hill 1990, Oug et al. 1991) and alter benthic patch dynamics. In one case decomposing discards were associated with a disease that brought down a scallop fishery (Jones 1992).

That most of the bycatch from a trawlfishery dies is obvious to the fishermen who shovel or pitchfork it over the side. Nonetheless, it was recently deemed necessary to generate hard data on this for a northwest Atlantic groundfishery. The hypothesis that the bulk of the bycatch is lethally traumatized was confirmed, with the perverse caveat that certain naturally compressed fishes (i.e., pleuronectiform, or "flat" fishes) survived at a higher rate than "round" fishes (e.g., cod) just recently flattened, or at least damaged, by the weight of the rest of the catch (Carr et al. 1995).

Longlines are often considered to be relatively clean, but in just a few years they have killed over 90 percent of some wandering albatross colonies (see Croxall 1990, Brothers 1991). Used in conjunction with glow sticks, they can also be devastating to leatherback turtle populations (Dayton et al. 1995). The problem of bycatch is not merely one of nonspecificity of gear. Predators aggregate around aggregations of their prey, and virtually any gear dragged through their midst will kill a great many through drowning and entanglement. Examples of bycatch are far too numerous, and the phenomenon too pervasive, for a detailed review here. The reader is referred to Alverson et al. 1994 and Dayton et al. 1995.

Obviously, it is advantageous to reduce the amount of bycatch, and many gear improvements have been developed to help do so. By and large they have helped, but these improvements have been partly offset by twin-beam trawls, paired trawling, and other new ways to increase the total size of the catch, and with it, the absolute volume of bycatch taken. How much effort should we devote to further bycatch reduction? There are really three separate issues here: economic, ethical, and ecological. The economic issue is simple: the amount of bycatch a fisherman can tolerate depends simply on the total value of the catch; if bycatch reduces this too much, then it must in turn be reduced. Ethically, any fishery that threatens either to drive species to extinction or to harm individual animals held in special favor is abhorrent to much of the world. Thus, the value of reducing the bycatch of birds, marine mammals, and sea turtles is set very high. In actuality, the ethical issues may be subsumed within the ecological effects of reducing species diversity, eliminating key functional groups, and rerouting biomass through the food web in novel ways. In terms of long-term human welfare, these lattermost are the truly important issues.

One of the most insidious hidden ecosystem costs of extracting food from the sea relates to the loss of fishing gear, which then continues to kill fish (Anderson 1988, Segawa 1990). Unfortunately data are limited, and it is difficult to derive quantitative estimates of lost gear, but Jaist (1994) reports many observations and some hard data that suggest this is a massive worldwide problem. It would appear that perhaps 10 to 30 percent of the world gear inventory is lost annually, and gill nets and traps may continue to fish and damage fragile benthic habitats for months or even years (e.g., Bohnsack 1992). Even recreational fishing gear can leave behind a killer trace. Recent attention has been drawn to the consumption of lead shot and fishing weights by waterfowl, leading to their poisoning and death in large numbers. Campaigns are under way to encourage the use of substitute, lead-free gear (Anon. 1986, 1993).

Habitat Alteration

The analogy of overfishing to deforestation (e.g., Hagler 1995) is oddly appropriate in ways not originally intended. Sea bottom habitats derive their characteristics in large part from sessile biota that sculpt both the habitat itself and feeding regimes for all the other organisms in the community, including the juveniles of commercially important species. Coral reef habitats are exquisitely sensitive to the kinds of massive disturbance that accompany intensive fishing (see Roberts 1995a for a thorough review). The use of poison, dynamite, or line weights to extract fish biomass from the labyrinth of coral reef habitats is very effective the first time but cannot be repeated productively in the same location more than once every century or so without serious threat to the physical structure of the reef itself (Gomez et al. 1987, Eldredge 1987, Alcala and Gomez 1987). In East Africa, overfishing is correlated to explosions in sea urchin populations, which can then destroy the reef framework (McClanahan 1995; McClanahan and Muthiga 1988, 1989). Too few herbivorous urchins and fishes can have equally devastating effects on coral reef growth. Hurricane damage to Caribbean coral reefs and their herbivore populations, on top of decades of overfishing of herbivorous fishes (Levitan 1992), and then followed by the near-extinction of hard-grazing urchins by an epizootic (Bauer and Ageter 1987) resulted in reefs in the tropical west Atlantic being overgrown by fleshy algae (Woodley et al. 1981; Kaufman 1983, 1986; Hughes 1994). Fishing pressure can also indirectly alter the abundance of coral predators, an effect amplified by interaction with natural disturbance (McClanahan 1995b, Kaufman 1983). In temperate waters, where algae are the principal source of biogenic habitat structure, high grazing pressure can flip sublittoral assemblages from a high-relief kelp forest to a low-relief, coralline-dominated "urchin barrens." This is not so much the elimination of structural habitat as the substitution of one type for another—it is not well known that the urchins themselves offer an important refugium for several important species beneath their canopy of spines (Tegner and Dayton 1991). Conversely, removal of the urchins (as, for example, by the sushi fishery for urchin roe) can result in regrowth of dense kelp forest in a few years.

Biogenic structure is equally important, though muted in scale, on offshore banks and shelf edge communities in the temperate zone (Auster et al. 1991). Here, hard bottoms are encrusted with sessile invertebrates. On the Georges Bank off the New England coast, transition juvenile (*sensu* Kaufman et al. 1992) groundfishes are closely associated with hard bottoms with mature fouling communities (Valentine, pers. comm., Lough et al. 1989). On soft bottoms juvenile groundfishes are associated with such relief as does exist: i.e., isolated sponge colonies, amphipod tubes, shell lag, and feeding depressions created by crabs, scallops, and rays (Lough et al. 1989; Langton and Robinson 1990; Auster et al. 1991, 1994). Redfish are strongly associated with tube-building anemones (Shepard et al. 1987). Such habitats, which nurture the groundfishery by providing both food and shelter from predators, are wiped clean by the action of trawls. Trawling destroys the benthic infauna, spreading a moving banquet before all the sea's scavengers and opportunists. This food glut could inflate populations of small, motile predators, thus increasing mortality of juvenile groundfishes during times when such handouts are not as readily available. In any event, it is clear that by several processes demersal trawls can have an enormous impact on juvenile fishes. Even for scallops, it is the young, affixed to shell lag by strong byssal fibers, that are most at risk from exposure to trawling (Shepard and Auster 1991).

Prior to the advent of sophisticated underwater reconaissance devices, the idea that fishing with mobile gear destroys fish habitat was controversial. Early studies purporting to demonstrate minimal or no damage by trawls were flawed by having been conducted in environments already severely disturbed by trawls; later studies demonstrated clear relationships between trawling and habitat degradation (Auster et al. 1996) and between trawling damage to habitat and actual reductions in fishery landings (e.g., Peterson et al. 1987, Sainsbury 1988). The loss of a meaningful baseline is a very serious problem that will continue to plague future studies.

Skeptics still point to heavily trawled fisheries that appear to be thriving (though rarely for long), or else cite the effects of natural disturbance as being far greater than that caused by the fishery. Sea bottoms are indeed subject to various forms of natural disturbance but only in shallower waters (i.e., >50 m depth), and even there it is much more episodic and less ubiquitous than are the effects of mobile gear dragged across the benthos. In the northwest Altantic (e.g., George Bank) trawling has been so intensive since the 1960s that the area of sea bottom trawled each year prior to 1991 (for which area data could be obtained) has averaged 3 to 4 times the total available area (Auster et al. 1996)! Most likely, any one site within in a prime fishing area has been disturbed much more often than this. In effect, trawling has extended the zone of frequent physical disturbance, as well as the overall intensity of this disturbance, into depths far beyond those usually affected by even the most severe and unusual storms. The true intensity and frequency of this disturbance is much higher than these numbers indicate, since trawling is not evenly distributed over the Georges and other banks, but instead is concentrated in relatively small areas where fishing is best. By and large these coincide with the highly structured bottoms so critical to the recruitment process on which the entire future of the fishery depends.

Aquaculture

The logical solution to declining fishery yields is to restore fish populations and the ecosystem that supports them so that desired food species are again produced in abundance. There are two ways to do this: The first is ecological reconstruction; the other is aquaculture. There is much that is good about it, but in the end analysis, aquaculture is not magic. Like any fishery it is still a process of resource extraction. Instead of extracting the target species, however, aquaculturists mine biomass from the environment, usually one trophic level down from their charges, to fuel production. The coopted food chain is more reliable than leaving the job of stock survivorship and extraction to nature, however. Primary and secondary production are much more efficiently funneled into desired product than in nature, and the animals are shielded from natural sources of mortality. This prevents some leakage of biomass out of the desired track and back into other parts of the ecosystem. On the other hand, fishing is still taking place, and there is still bycatch, waste, and habitat destruction (Larsson et al. 1994). The wastes generated by most aquaculture are returned to the coastal environment, burdening and sometimes overtaxing the detritivore food chain (Folke and Kautsky 1989, Folke and Kautsky 1992, Folke et al. 1994). Today's aquaculture operations are best thought of as oblique wild fisheries, a sort of plumbing to reroute biomass from natural systems into artificial ones. Like other forms of plumbing, expensive plumbers are needed to keep it going.

Many forms of aquaculture have been tried, but the prevalent one today is intensive monoculture. Various species of shrimps, crayfishes, oysters, mussels, clams, conchs, urchins, sea cucumbers, and other invertebrates have been brought into successful intensive monoculture. The list of marine target fishes, while heavily weighted by salmonids in the temperate zone, also includes sturgeons, paddlefish, temperate marine basses, porgies, seawater-acclimated tilapias, and grow-out of wild-spawned eels. One curious thing about intensive monoculture is that it so often is focused on species and genetic stocks alien to the sites where they are being cultured. This leads to a host of problems including: introduction of alien fishes and invertebrates; deleterious competition between planted or escaped cultured stock and native species and stocks; increased predation on wild fish stocks (Washington and Koziol 1993); outbreaks of epizootics, often exotic ones (Hindar et al. 1991); and reduced genetic diversity in wild stocks (Reisenbichler and Phelps 1988, Reisenbichler et al. 1992, Gall et al. 1992). Removal of running-ripe wild brood stock to be stripped for hatchery production can threaten weak natural runs of anadromous fishes like salmonids (Waples 1991). Large quantities of organic matter are produced at high concentration by aquaculture operations, and the benthos beneath floating cages is usually dead and often anoxic as a result. Some forms of monoculture, such as tropical shrimp production, can result in extensive destruction of coastal marine habitats such as mangrove forests, which themselves have substantial value and are already quite threatened (Larsson et al. 1994). Ironically, intensive shrimp culture has sometimes replaced indigenous polyculture systems that are far less damaging to the environment (Folke and Kautsky 1992).

Information Goods

At the very high end of the value spectrum for marine goods comes a broad array of living materials that are so rare and desirable that they are treated as precious commodities. These include skeletal material from marine organisms, chemical extracts and the genes that can enable other organisms to produce them on demand, and living marine organisms themselves, prized as pets, ornaments, curios, and public exhibits. Among the more conventional precious substances from the sea are antipatharian skeleton ("black coral"), red coral, nacre, and pearls. Pricey curios include virtually all seashells, egg cases, and whole scelaractinian coral colonies, whose extraction exacts a host of both direct and indirect effects on marine ecosystems (Wells and Alcala 1987). Some aquarium hobbyists brighten dark northern and southern winters with, literally, a piece of the tropics: a chunk of living reef framework ("live rock") complete with epifaunal and inquiline invertebrates. The demand for live rock, which is both fascinating to observe and of value in maintaining water quality in an aquarium, is so great that particularly accessible portions of coral reef systems are in danger of being dismantled and sold, piece by piece.

Pound for pound, live marine organisms are among the most valuable of all commonly exchanged goods on earth. Marine aquarium fishes range in value from several hundred to several thousand dollars a pound. Of course, these animals are very small and one does not plan on selling them by the ton. Even so, current demand far exceeds supply. Nearly six million fishes swim through more than a million marine aquariums in U.S. households alone (Hoff 1993). The hobby is also very popular in several European and Asian countries. Rather like houseplants, turnover is high. Few aquarium fishes survive longer than a year.

The trade in live marine aquarium fishes and invertebrates has been praised by some as a largely untapped resource for some of the world's very poorest nations (Andrews 1990, Emmanuel et al. 1990, Hoff 1993) and damned by others as wasteful and destructive (Wood 1985, Randall 1987, Andrews 1990). Skilled aquarium fish collectors can extract specimens from the reef framework with little damage to the reef habitat. Alternatively, unskilled collectors can drag nets, spread clouds of highly toxic anesthetics, and pulverize coral colonies to overcome their diminutive quarry. Experience in places like Australia suggests that an aquarium fish export industry can be a sustainable and desirable enterprise when properly conducted

(Couchman and Beumer 1992). Indeed, the good that can come from broadened awareness for and valuation of marine biodiversity by aquarium enthusiasts could well be worth the minimal environmental impacts of a well-managed aquarium fish trade. Some of the greatest consumers of exotic marine specimens are huge public aquariums, now all the rage in urban centers around the world. These institutions supposedly exist to promote aquatic conservation through public education (McCormick-Ray 1993, Kaufman and Zaremba 1995, Dean 1995, Atkinson et al. 1995).

Recent attention drawn to animal rights extremism has perversely boosted the search for marine models as one alternative to the use of birds and mammals as laboratory subjects. Though growing in importance, marine species are still not well utilized as research models, and few biologists fully appreciate the vast potential marine organisms hold in modern biological and biomedical research. Part of the reason for this is that molecular biologists and biomedical researchers are rarely trained in organism biology, much less in natural history or the comparative method. At the Marine Biological Laboratory in Woods Hole, Massachusetts, the juxtaposition of molecular, cellular, and organism biologists has led to recent acceptance of several new marine animal models, of which one of the most exciting is the reef squid, a culturable replacement for wild-caught northern squid species used in neurobiological research (Lee et al. 1994). Certain fishes offer an attractive new model for the study of regulatory processes in human bone (Smith-Vaniz et al. 1995). Natural products chemistry is another underappreciated class of marine goods. Here the potential is vast and the impacts potentially modest (Pietra 1990, Vogel 1994).

Conclusions

Nearly all extractive practices in use to date are damaging to marine systems and unlikely to be sustainable at a satisfactory level of yield. This is due to the direct effects of the removal of organisms and biomass; the destruction of marine habitats; and complex interactions that sweep through food webs with profound and often unpredictable consequences. The threats to ecological integrity in marine systems are worse in combined severity and geographical scope than anything on land, and at times seem *designed* to breed frustration and despair. Consider comments gleaned from early reviews of this manuscript: "Does this mean that we can't fish at all?"; "Are we to conclude that fishing should only be permitted with bare hands?"; and ". . . maybe marine people have seen more of life's darkest side than others." Nevertheless, marine conservationists probably could do well to focus a bit less energy on spreading gloom and doom and more on problem solving.

One approach with particular promise is the development of a co-management community. This is essentially a group of civic leaders who oversee the relationship between a resource-based community, and the ecosystem that allows the community to exist. The idea is as old as human society, but our notion of how it might work for marine resources is a bit different. If it works, this approach could in part address the fundamental issues of responsibility and burden of proof (Dayton et al. 1995) raised at the outset of this chapter. The critical steps are as follows.

1. *Create Skilled, Powerful Co-management Communities:* To be effective, stewardship responsibility must sit with those most affected by the welfare of the resource. A diversity of skills is required of the community's leaders, and it is mandatory that the community extend beyond fisherpeople and bureaucrats to include resident conservationist biologists, ecological economists, teachers, socioeconomists, and investors. The residency requirement is important. Not only must these academics, bureaucrats, and businesspeople share a personal stake in the outcome, but they must also earn the trust of the often highly independent-minded fisherfolk. For the sake of continuity, training apprenticeships within such communities are essential.

2. *Agree on Ecosystem Baselines and Goals:* The magnitude of past disturbance to marine communities has made it difficult to locate pristine localities against which we can measure current and future impacts. Marine reserves, strategically located, can serve this purpose while in some cases also providing ecotourism, recreation, and sources of larvae to supply grounds open to extraction; the sizes and locations of the reserves must allow for both sources and sinks of larvae. Expectations of yield for open areas must be based on constant comparison between these areas and adjacent reserves, through a monitoring program led by co-management community scientists.

3. *Increase Flexibility Concerning When, Where, and How to Extract the Resource:* Co-management community leaders must manage their relationship with the natural trust like an investment portfolio, distributing and minimizing risk from year to year, or at least taking calculated risks (overfishing or habitat disturbance) with full knowledge of the estimated loss if the system is damaged. This requires much greater flexibility than is currently typical in marine resource management. Deficits also exist in the underlying science. Vermeij (1993) highlighted the special importance and vulnerability of areas of high productivity, advising extreme caution not to disturb such places. Roberts (1995a) advised the opposite: i.e., that fisheries should be concentrated in high-productivity, low-diversity areas to help spare precious low-productivity, high-diversity habitats such as coral reefs. The unfortunate truth is that there is nowhere left to run and no place left to ruin.

Impacts must be managed in accordance with unique local perspectives. Over time there will emerge a coherent, guiding wisdom specific to and appropriate for each region. In this way, something of the culture of conservation achieved over millennia by aboriginal societies can flower again within the context of our modern, technological world.

4. *Engage Government Assistance:* While it is important that co-management communities have sufficient authority and autonomy, it is incumbent on government to create a climate supportive to their work. Political and economic instruments are needed to facilitate interstate and international cooperation on transborder issues, such as straddling stocks (highly migratory species) and trade. Various forms of incentive will be required to encourage compliance with conservation measures, to allow the local co-management communities to achieve the right balance of professional skills, and to provide relief to fishermen and their families during what will undoubtedly be an error-prone exercise in its early phases. Government must also provide adequate, dedicated funds for problem solving. Recent proposals to form an Institute for the Environment charged with these responsibilities are a very positive example. The U.S. Biological Survey was another one, but it did not get too far. Though possibly costly, this role for government as a facilitator, midwife, nurturer, and legal umbrella for the emergence of functional co-management communities should ultimately prove far more cost effective than the current system of sweeping regulations and inadequate enforcement.

Acknowledgments

The authors would like to thank many colleagues on the Internet for drawing our attention to materials and references we would otherwise have overlooked, especially Rod Fujita of the Environmental Defense Fund. This work was facilitated by support to both authors from the Pew Charitable Trusts, through the Pew Fellows Program for Conservation and the Environment.

References

Ainley, D. G., W. J. Sydeman, S. A. Hatch, and U. W. Wilson. 1994. "Seabird Population Trends along the West Coast of North America: Causes and the Extent of Regional Concordance." In J. R. Jehl, and N. K. Johnson, eds., *A Century of Avifaunal Change in Western North America. Studies in Avian Biology, No. 15*, pp. 119–133. Cooper Ornithological Society, Camarillo, California.

Alcala, A. C., and E. D. Gomez. 1987. "Dynamiting Coral Reefs for Fish: A Resource-Destructive Method." In B. Salvat, ed., *Human Impacts on Coral Reefs: Facts and Recommendations*, pp. 51–60. Antenne de Tahiti Mus. E.P.H.E. Tahiti, French Polynesia.

Alverson, D. L., M. H. Freeberg, J. G. Pope and S. A. Murawski. 1994. *A Global Assessment of Fisheries Bycatch and Discards*. FAO Fisheries Technical Paper. No 339. Rome, FAO.

Anderson, I. 1988. "'Wall of death' confronts wildlife in the Pacific." *New Scientist* 118:26.

Andrews, C. 1990. "The ornamental fish trade and fish conservation." *Journal of Fish Biology* 37 (Sup.A):53–59.

Andrews, C., and L. Kaufman. 1994. "Captive Breeding Programmes and Their Role in Fish Conservation. In *Creative Conservation*, P. J. S. Olney, G. M. Mace, and A. T. C. Feistnereds, eds., pp. 338–351. Chapman and Hall, London.

Androkovich, R. A., and K. R. Stollery. 1984. "A stochastic dynamic programming model of bycatch control in fisheries." *Mar. Resource Econ.* 9(1):19–30.

Anonymous. 1986. "Government acts to save swans from lead." *New Scientist* 112:9.

Anonymous. 1993. "Getting the lead out of fishing gear." *Environment* 35:21.

Atkinson, M. J., B. Carlson, and G. L. Crow. 1995. "Coral growth in high nutrient, low-pH seawater: A case study of corals cultured at the Waikiki Aquarium, Honolulu, Hawaii." *Coral Reefs* 14: 215–223.

Auster, P. J., R. J. Malatesta, S. C. LaRosa, R. A. Cooper, and L. L. Stewart. 1991. "Microhabitat utilization by the megafaunal assemblage at a low relief outer continental shelf site—Middle Atlantic Bight, USA." *Journal of Northwest Atlantic Fisheries Science* 11:59–69.

Auster, P. J., R. J. Malatesta, and C. L. S. Donaldson. 1994. "Small-Scale Habitat Variability and the Distribution of Postlarval Silver Hake, *Merlucomanagement communityius bilinearis*." In D. Stevenson and E. Braasch, eds., *Gulf of Maine Habitat: Workshop Proceedings*. ME-NH Sea Grant. UNHMP-T/DR-SG-94-18.

Auster, P. J., R. J. Malatesta, R. W. Langton, L. Watling, P. C. Valentine, C. L. S. Donaldson, E. W. Langton, A. N. Shepard, and I. G. Babb. 1996. "The impacts of mobile fishing gear on seafloor habitats in the Gulf of Maine (Northwest Atlantic): Implications for conservation of fish populations. *Reviews in Fisheries Science* 4: 185–202.

Bauer, J. C., and C. J. Ageter. 1987. "Isolation of bacteria pathogenic for the sea urchin *Diadema antillarum* (Echinodermata: Echinoidea)." *Bulletin of Marine Science* 40:161–165.

Bohnsack, J. 1992. "Reef Resource Habitat Protection: The Forgotten Factor." In R. Stroud, ed., *Stemming the Tide of Coastal Fish Habitat Loss*, pp. 117–129. National Coalition for Marine Conservation, Savannah.

Brothers, N. 1991. "Albatross mortality and associated bait loss in the Japanese longline fishery in the Southern Ocean." *Biological Conservation* 55:255–268.

Caddy, J. F. 1973. "Underwater observations on tracks of dredges and trawls and some effects of dredging on a scallop ground." *Journal of the Fisheries Research Board of Canada* 30:173–180.

Carr, H. A., M. Farrington, J. Harris, and M. Lutcavage. 1995. "Juvenile bycatch and codend escape survival in the Northeast groundfish industry—Assessment and mitigation." A report of the New England Aquarium to the National Oceanic and Atmospheric Administration pursuant to NOAA Award No. NA36FD0091. 95 pp.

Clark, S, and A. J. Edwards. 1995. "Use of artificial reef structures to rehabilitate reef flats degraded by coral mining in the Maldives." *Bulletin of Marine Science* 55:724–744.

Couchman, D., and J. P. Beumer. 1992. *The Commercial Fishery for the Collection of Marine Aquarium Fishes in Queensland: Status and Management Plan.* Department of Primary Industries, Brisbane.

Croxall, J. P. 1990. "Impact of incidental mortality on Antarctic marine vertebrates." *Antarctic Science* 21(1):1.

Dayton, P. K., S. F. Thrush, M. T. Agardy, and R. J. Hoffman. 1995. "Environmental effects of marine fishing." *Aquatic Conservation of Marine and Freshwater Ecosystems* 5:153.1–153.28.

Dean, S. 1995. "Aquariums: Emerging Leaders in the Environmental Conservation Movement." In C. M. Wemmer, ed., *The Ark Evolving.* Smithsonian Institution, Front Royal.

Doty, M. S., J. F. Caddy, and B. Santelices. 1986. *Case Studies of Seven Commercial Seaweed Resources.* FAO Fisheries Technical Paper No. 281. FAO, Rome.

Eldredge, L. G. 1987. "Poisons for Fishing on Coral Reefs. In B. Salvat, ed., *Human Impacts on Coral Reefs: Facts and Recommendations.* Antenne de Tahiti Mus. E.P.H.E. Tahiti, French Polynesia.

Emmanuel, P., V. Anand, and T. J. Varghese. 1990. "Notes on marine ornamental fishes of Lakshadweep." *Seafood Export Journal* 22(4):13–18.

Folke, C., and N. Kautsky. 1989. "The role of ecosystems for a sustainable development of aquaculture." *Ambio* 18(4):234–243.

Folke, C., and N. Kautsky. 1992. *Aquaculture with its environment: Prospects for sustainability. Ocean & Coastal Management* 17:5–24.

Folke, C., N. Kautsky, and M. Troell. 1994. "The costs of eutrophication from salmon farming: Implications for policy." *Journal of Environmental Management* 40:173–182.

Gall, G., D. Bartley, B. Bentley, J. Brodziak, R. Gomulkiewicz, and M. Mangel. 1992. "Geographic variation in population genetic structure of chinook salmon from California and Oregon." Fishery Bulletin 90:77–100.

Gomez, E. D., A. C. Alcala and H. T. Yap. 1987. "Other Fishing Methods Destructive to Coral." In B. Salvat, ed., *Human Impacts on Coral Reefs: Facts and Recommendations.* Antenne de Tahiti Mus. E.P.H.E. Tahiti, French Polynesia.

Hagler, M. 1995. "Deforestation of the deep: Fishing and the state of the oceans." *The Ecologist* 25:74–79.

Haney, J. C. 1986. "Seabird affinities for Gulf Stream frontal eddies: Responses of mobile marine consumers to episodic upwelling." *Journal of Marine Research* 44:361–384.

Hellier, C. 1988. "The mangrove wastelands." *The Ecologist* 18(2):77–79.

Hindar, K., N. Ryman, and F. Utter. 1991. "Genetic effects of cultured fish on natural fish populations." *Canadian Jour. Fish. Aquat. Sci.* 48 (Suppl. 1):124–133.

Hingco, T. G., and R. Rivera. 1991. "Aquarium Fish Industry in the Philippines: Toward Development or Destruction?" In L. M. Chou, T-E. Chua, H. W. Khoo, P. E. Lim, J. N. Paw, G. T. Silvestre, M. J. Valencia, A. T. White, and P. K. Wong (eds)., *Towards an Integrated Management of Tropical Coastal Resources*, pp. 249–253. ICLARM Conf. Proc. No. 22. Manila, Philippines.

Hof, F. H. 1993. Marine Ornamental Fish Culture. In M. Carrillo, L. Dahle, J. Morales, P. Sorgeloos, N. Svennevig, and J. Wyban, eds., *From Discovery to Commercialization*, p. 298. Spec. Publ. Eur. Aquacult. Soc. No. 19. Oostended (Belgium).

Hughes, T. P. 1994. "Catastrophes, phase shifts, and large-scale degradation of a Caribbean coral reef." *Science* 265:1547–1551.

Jamieson, G. S., and A. Campbell. 1985. "Sea scallop fishing impact on American lobsters in the Gulf of St. Lawrence." *Fisheries Bulletin* 83:575–586.

Jones, J. B. 1992. "Environmental impact of trawling on the seabed: A review, New Zealand." *Journal of Marine Freshwater Research* 26:59–67.

Kaufman, L. S. 1983. "Effects of Hurricane Allen on reef fish assemblages near Discovery Bay, Jamaica." *Coral Reefs* 2: 1–5.

Kaufman, L. S. 1986. "Why the Ark is Sinking." In *The Last Extinction*. MIT Press, Cambridge, MA.

Kaufman, L. S., J. Ebersole, J. Beets and C. MacIvor. 1992. "A key phase in the recruitment dynamics of coral reef fishes: Postsettlement transition." *Environmental Biology of Fishes* 34:109–114.

Kaufman, L. S. 1992. "Catastrophic change in species-rich freshwater ecosystems, the lessons of Lake Victoria." *Bioscience* 42: 846.

Kaufman, L. S., and F. Zaremba. 1995. "Conservation Research at Aquariums: The Social Contract." In C. M. Wemmer, ed., *The Ark Evolving*. Smithsonian Institution, Front Royal.

Langton, R. W., and W. E. Robinson. 1990. "Faunal association on scallop grounds in the Western Gulf of Maine." *Journal of Experimental Marine Biology and Ecology* 144:157–171.

Larsson, J., C. Folke, and N. Kautsky. 1994. "Ecological limitations and appropriation of ecosystem support by shrimp farming in Colombia." *Environmental Management* 18(5):663–676.

Lee, P. G., P. E. Turk and W. T. Yang. 1994. "Biological characteristics and biomedical applications of the squid *Sepioteuthis lessoniana* cultured through multiple generations." *Biological Bulletin* 186:328–341.

Levitan, D. 1992. "Community structure in times past: Influence of human fishing pressure on algal-urchin interactions." *Ecology* 73:1597–1605.

Lough, R. G., P. C. Valentine, D. C. Potter, P. J. Auditore, G. R. Bolz, J. D. Neilson, and R. I. Perry. 1989. "Ecology and distribution of juvenile cod and haddock in relation to sediment type and bottom currents on eastern Georges Bank." *Marine Ecology Progress Series* 56:1–12.

Lowry, L. F., K. J. Frost and T. R. Loughlin. 1989. "Importance of Walleye Pollock in the Diets of Marine Mammals in the Gulf of Alaska and Bering Sea, and Implications for Fishery Management. In *Proceedings of the International Symposium on the Biology and Management of Walleye Pollock*, pp. 701–726. University of Alaska Sea Grant Report 89–01.

Mangel, M. 1993. "Effects of high-seas driftnet fisheries on the northern right whale dolphin *Lissodelphis borealis*." *Ecological Application* 3:221–229.

McClanahan, T. R. 1995a. "A coral reef ecosystem-fisheries model: Impacts of fishing intensity and catch selection on reef structure and processes." *Ecological Modelling* 80(1): 1–19.

McClanahan, T. R. 1995b. "Coral-eating snail *Drupella cornus* population increases in Kenyan coral reef lagoons." *Marine Ecology Progress Series* 115:131–137.

McClanahan, T. R., and N. A. Muthiga. 1988. "Changes in Kenyan coral reef community struture and function due to exploitation." *Hydrobiologia* 166:269–276.

McClanahan, T. R., and N. A. Muthiga. 1989. "Patterns of predation on a sea urchin, *Echinometra matthei* (de Blainville), on Kenyan coral reefs." *Journal of Experimental Marine Biology and Ecology* 126:77–94.

McCormick-Ray, M. G. 1993. "Aquarium science: The substance behind an image." *Zoo Biology* 12(5):413–424.

Munro, J.L., J. D. Parrish, and F. H. Talbot. 1987. "The Biological Effects of Intensive Fishing upon Coral Reef Communities." In B. Salvat, ed., *Human Impacts on Coral Reefs: Facts and Recommendations*. Antenne de Tahiti Mus. E.P.H.E. Tahiti, French Polynesia.

Naylor, R. L. 1996. "Invasions in agriculture: Assessing the cost of the golden apple snail in Asia." *AMBIO* 25(7): 442–447.

Nickleson, T. E. 1986. "Influences of upwelling, ocean temperature, and smolt abundance on marine survival of coho salmon (*Oncorhynchus kisutch*) in the Oregon Production Area." *Canadian Journal of Fisheries and Aquatic Science* 43:527–535.

Oug, E., T. E. Lein, R. Kufner, and I. O. B. Falk-Petersen. 1991. "Environmental effects of a herring mass mortality in Northern Norway: Impact on and recovery of rocky shore and soft-bottom biota." *Sarsia* 76:195–287.

Parsons, T. R. 1992. "The removal of marine predators by fisheries and the impact of trophic structure." *Marine Pollution Bulletin* 25:51–53.

Pauly, D., and V. Christensen. 1995. "Primary production required to sustain global fisheries." *Nature* 374:255–257.

Peterson, C. H., H. C. Summerson, and S. R. Fegley. 1987. "Ecological consequences of mechanical harvesting of clams." *Fishery Bulletin* 85:281–298.

Pietra, F. 1990. *A Secret World: Natural Products of Marine Life*. Birkhauser Verlag, Boston.

Polovina, J., G. T. Mitchum, N. E. Graham, M. P. Craig, E. E. DeMartini, and E. M. Flint. 1994. "Physical and biological consequences of a climate event in the Central North Pacific." *Fisheries Oceanography* 3:15–21.

Pringle, J. D., and D. J. Jones. 1980. "The interactions of lobster, scallop, and Irish moss fisheries off Borden, Prince Edward Island." *Canadian Technical Report of Fisheries and Aquatic Science* No. 973.

Randall, J. E. 1987. "Collecting Reef Fishes for Aquaria." In B. Salvat, ed., *Human Impacts on Coral Reefs: Facts and Recommendations*. Antenne de Tahiti Mus. E.P.H.E. Tahiti, French Polynesia.

Reisenbichler, R. R., and S. R. Phelps. 1988. "Genetic variation in steelhead from the North Coast of Washington." *Canadian Journal of Fisheries and Aquatic Science*. 46:66–73.

Reisenbichler, R. R., J. D. McIntyre, M. F. Solazzi, and S. W. Landino. 1992. "Genetic variation in steelhead of Oregon and northern California." *Transactions of the American Fisheries Society* 121:158–169.

Roberts, C. M. 1995a. "Effects of fishing on the ecosystem structure of coral reefs." *Conservation Biology* 9(5):988–995.

Roberts, C. M. 1995b. "Rapid build up of fish biomass in a Caribbean marine reserve." *Conservation Biology* 9:815–826.

Rose, G. A. 1993. "Cod spawning on a migration highway in the north-west Atlantic." *Nature* 366:458–460.

Safina, C. 1994. "Where have all the fishes gone?" *Issues in Science and Technology* 10(3):37–43.

Sainsbury, K. J. 1988. "The Ecological Basis of Multispecies Fisheries and Management of a Demersal Fishery in Tropical Australia." In J. A. Gulland, ed., *Fish Population Dynamics,* 2nd edn. Wiley, New York.

Salvat, B. 1987. "Dredging in Coral Reefs." In B. Salvat, ed., *Human Impacts on Coral Reefs: Facts and Recommendations*. Antenne de Tahiti Mus. E.P.H.E. Tahiti, French Polynesia.

Scarratt, D. J. 1973. "The effects of raking Irish moss (*Chondrus crispus*) on lobsters in Prince Edward Island." *Helgol. wiss. Meer*. 24:415–424.

Segawa, S. 1990. "Driftnet fishing: Net disagreement." *Nature* 345:567.

Shepard, A. N., R. B. Theroux, R. A. Cooper, and J. R. Uzmann. 1987. "Ecology of Ceriantharia (Coelenterata, Anthozoa) of the northwest Atlantic." *Fisheries Bulletin* 84:625–646.

Shepard, A. N., and P. J. Auster. 1991. "Incidental (Non-Capture)Damage to Scallops Caused by Dragging on Rock and Sand Substrates. In S. E. Shumway and P. A. Sandifer, eds., *An International Compendium of Scallop Biology and Culture*, pp. 219–230. World Aquaculture Society, Baton Rouge, LA.

Smith-Vaniz, W., L. S. Kaufman, and J. Glowacki. 1995. "The natural history of fish hyperostosis: Cellular bone within an acellular skeleton." *Marine Biology* 121: 573–580.

Tegner, M. J., and P. K. Dayton, 1991. "Sea urchins, El Niños, and the long-term stability of southern California kelp forest communities." *Marine Ecology Progress Series*. 77:49–63.

Vermeij, G. J. 1993. "Biogeography of recently extinct marine species: Implications for conservation." *Conservation Biology* 7(2):391–397.

Vogel, J. H. 1994. *Genes for Sale: Privatization as a Conservation Policy*. Oxford University Press, New York.

Waples, R. S. 1991. "Genetic interactions between hatchery and wild salmonids: Lessons from the Pacific Northwest." *Canadian Journal of Fisheries and Aquatic Science* 48(Supl. 1):124–133.

Washington, P. M., and A. M. Koziol. 1993. "Review of the interactions and environmental impacts of hatchery practices on natural and artificial stocks of salmonids." *Fisheries Research* 18:105–122.

Wassenberg, T. J., and B. J. Hill. 1990. "Partitioning of material discarded from prawn trawlers in Moreton Bay." *Australian Journal of Marine and Freshwater Research* 41:27–36.

Wells, S. M., and A. C. Alcala. 1987. "Collecting of Corals and Shells." In B. Salvat, ed., *Human Impacts on Coral Reefs: Facts and Recommendations*. Antenne de Tahiti Mus. E.P.H.E. Tahiti, French Polynesia.

White, A. T. 1987. "Effects of Construction Activity on Coral Reef and Lagoon Systems." In B. Salvat, ed., *Human Impacts on Coral Reefs: Facts and Recommendations*. Antenne de Tahiti Mus. E.P.H.E. Tahiti, French Polynesia.

Wood, E. 1985. *Exploitation of Coral Reef Fishes for the Aquarium Trade*. Marine Conservation Society, Ross-on-Wye, England.

Woodley, J. D., E.A. Chornesky, P. A. Clifford, J. B. C. Jackson, L. S. Kaufman, N. Knowlton, J. C. Lang, M. P. Pearson, J. W. Porter, M. C. Rooney, K. W. Rylaarsdam, V. J. Tunnicliffe, C. M. Wahle, J. L. Wulff, A. S. G. Curtis, M. D. Dallmeyer, B. P. Jupp, M. A. R. Koehl, J. Neigel, and E. M. Sides. 1981. "Hurricane Allen: Initial impact on the coral reef communities of Discovery Bay, Jamaica." *Science* 214: 749–755.

Chapter 16

Ecosystem Services in Subsistence Economies and Conservation of Biodiversity

Kamaljit S. Bawa and Madhav Gadgil

Natural and human-impacted ecosystems provide a range of goods and services to human societies and play a vital role in sustaining many, if not ultimately all, human endeavors and enterprises. The goods, in the form of drinking water, fish and shellfish, wood and nonwood products, constitute significant components of local and national economies and sustain livelihoods of millions of people living in and around ecosystems. Dasmann (1988) has appropriately termed such inhabitants as the ecosystem people, to distinguish them from the biosphere people. Ecosystem people include forest dwellers, herders, fishers, and peasants, who rely on biological resources of local ecosystems to fulfill most of their needs. Biosphere people include urban dwellers of the industrialized societies and people engaged in high-input agriculture and animal husbandry. They do not depend on local ecosystems for their basic needs; the catchment area for their resource needs is the whole biosphere. For the ecosystem people, natural communities have been an integral part of their lives for millennia. In many ways, the ecosystem people behave as integral components of the ecosystems they inhabit (Gadgil 1995).

Environmental economists often make the distinction between goods and services provided by ecosystems. Ecosystem goods harvested by people are, however, often products of ecosystem processes such as biotic interactions, energy flow, and nutrient cycling. In the case of ecosystem people there is an additional justification in considering all goods as ecosystem services. Although the ecosystem people harvest many products from ecosystems, these products are used primarily to sustain their own livelihoods, which are embedded in the ecosystem in which they live. In contrast, the biosphere people extract ecosystem products for commercial purposes. Of course, with monetization of subsistence economies, and with increased commercial exploitation of ecosystem products, the ecosystem people also become engaged in extraction of resources far beyond the levels necessary to meet their own needs, for use by the biosphere people. Nevertheless, even when products are harvested commercially, the ecosystem people often derive only subsistence-level wages for the time spent in harvesting products. These wages, together with products gathered for their own use, may fall more appropriately under ecosystem services than under ecosystem goods.

Resource managers and conservation biologists have until recently ignored the importance of ecosystem services in supporting the livelihoods of ecosystem people and the potential contributions that such people can make to conservation of biodiversity. Livelihoods based on ecosystem services provide a direct link between people and biodiversity in much of the developing world. The reliance of local communities on ecosystem services to sustain their livelihoods can form the basis of grassroots support for conservation efforts. Yet attempts to involve ecosystem people in conservation activities have been sporadic and limited (Western and Henry 1994).

Here we assess the importance of ecosystems in sustaining the livelihoods of millions of ecosystem people. We first define ecosystem services of interest and identify problems in valuation of these services in the context of sustaining livelihoods of ecosystem people. Despite difficulties in valuation, we provide some estimates of the amounts and importance of services. We show that technological changes are leading to the disruption of these services and the unsustainable use of the resources. We conclude with a discussion of policy changes that should enhance the probability of sustainable use of services and goods, and incorporate ecosystem people in efforts to conserve biodiversity. Specifically, this chapter has two goals: the first is to demonstrate the importance of ecosystem services in the livelihoods of ecosystem people and rural poor, and the second is to show how the ecosystem people can be and should be involved in conservation efforts.

Problems in Valuation of Services

Environmental goods and services are often valued by making an initial distinction between use and non-use values (Pearce and Moran 1994). The use value may be subdivided into direct use value and indirect use value. The former refers to the value of goods and the latter to services. Goulder and Kennedy (chapter 3, this volume) distinguish a third type of service, the provision of production inputs, that includes many types of services associated with ecosystem processes and generally included under indirect use value. The non-use value may be distinguished into bequest and existence values. The bequest value simply refers to the value an individual ascribes to the knowledge that others may benefit from a resource or service in the future. The existence value is derived from the knowledge that an environmental resource or service exists. Option value or the price an individual is willing to pay to retain the options of using services at a future date is generally included under the non-use value but can also be applied to goods.

Traditional valuation techniques pose at least five major problems in quantifying ecosystem services with respect to subsistence economies. First, valuation techniques, particularly with respect to indirect use value and non-use values, involve subjective value judgments of people living in modern urbanized societies. Application of such value judgments to societies with radically different social and economic structures not only poses methodological difficulties but also raises moral and ethical issues. If indeed many of the societies that rely on ecosystems for subsistence livelihoods are an integral part of the ecosystem they live in, then valuation exercises tend to place monetary value on a particular lifestyle and culture by those who do not share these lifestyles and cultural values.

Second, the valuation techniques are relatively easy to apply in fully monetized economies that are homogeneous and involve individuals who are well informed about choices and in a position to exercise various options. In contrast, subsistence economies in various parts of the world are very heterogeneous, with different value systems. Moreover, economies often are not fully monetized and individuals are not well informed about choices and are economically and socially constrained to exercise various options.

Third, despite the availability of sophisticated techniques, certain benefits remain difficult to quantify in monetary forms. For example, as is well known, ecosystem people use a wide variety of medicinal plants for their health care. However, contributions of medicinal plants to the local economy are hard to quantify. The efforts to value plant medicines emphasize the option value for pharmaceutical companies but ignore the uses of herbal medicines by local people (Brown 1994). For local communities, surrogate values derived from health care costs of comparable income groups in rural or urban areas may not be used because people in such areas spend less than they would like to because of the high costs in relation to their ability to pay. In contrast, the ecosystem people are free of such constraints because herbal medicines collected from the forest do not impose heavy costs.

The productivity of ecosystem goods, which form the basis of subsistence economies, depends on the functioning of the ecosystems. For example, a multitude of processes are involved in the production of ecosystem goods; the formation of such products as fruits and seeds alone requires a wide array of biotic interactions. Processes such as pollination, seed dispersal, and even nutrient uptake require interaction among plants, animals, and microorganisms. In theory, the functional processes can be valued, but current valuation methods for products do not assign value to functional processes (Vatn and Bromley 1995), implying that such processes are free.

Valuation of cultural, religious, and spiritual services is even more difficult than that of goods and ecological services from natural ecosystems. Interestingly, such services, at least in partly monetized economies and in areas experiencing rapid depletion of natural habitats and ecosystems, may be more vital to the well-being than the goods from the ecosystem. In many countries, sacred groves or ponds persist even in areas where landscape has been modified for centuries to eliminate natural forest, as for example in the Western Ghats and the Meghalaya state of India. These sacred sites in many cases may fulfill only religious, spiritual, and cultural needs. Existence value can perhaps cover cultural, religious, and spiritual services, but apart from the appropriateness of assigning monetary value to such services, there is little experience in applying valuation techniques to such services.

Another problematic area in valuation is the contribution of ecosystems to human knowledge. As is well known, ecosystem people have accumulated a large body of practical knowledge about the uses of organisms and their interactions with the environment for medicinal and other purposes. This practical knowledge provides biosphere people with a springboard for new developments and innovations in medicine, agriculture, forestry, horticulture, animal husbandry, toxicology, and other endeavors. This practical knowledge is lost with the destruction of ecosystems and extinction of species. The value of such knowledge has been increasing steadily with the growing number of environmental challenges and with increasing potential of technology to transform practical knowledge to new goods and services.

The fourth hurdle in valuation is that costs and benefits, actual and perceived, of conserving a given area are different for different sectors of the society (Wells 1992). However, valuation methodologies do not take into account variation among different sectors of society in assigning value to ecosystem services. For example, ecosystem services such as pure air and water and biodiversity may be perceived to be more valuable by a person living in an urban environment, devoid of biodiversity, and full of polluted air and water, than by a person living in or around pristine ecosystems. Thus, for many ecosystem services, actual as well as potential benefits may be less valued by the ecosystem people than by the biosphere people.

Finally, subsistence-level benefits from ecosystems are also undervalued when the human costs associated with the destruction of the ecosystem are not taken into account, as is often the case. The livelihood strategies of ecosystem people tend to be tightly linked to extraction and utilization of goods and services of local ecosystems, for their own use as well as for exchange on the market in an essentially unprocessed form. Such human groups have few skills that can be exchanged for a livelihood when deprived of access to the ecosystem goods and services on which they have long depended. Displacement of ecosystem people is therefore always accompanied by great human suffering, as when tribals are forced to migrate to urban shantytowns when their forest habitat is taken over by mining or river valley projects. The costs of creating impoverished ecological refugees are manifested as expenditures in poverty alleviation programs.

Quantification of Services

Despite the problems in valuing ecosystem services as they relate to subsistence economies, the importance of such services can be assessed in a wide variety of ways. We may estimate the number of people who derive their livelihoods from harvesting ecosystem productivity, the value of particular crops, the contribution to cash income of households, the proportion of households that rely on ecosystem products, the total GDP derived from ecosystem goods, and the value of services on a per hectare basis. The various attributes of these parameters in terms of advantages and disadvantages are listed in table 16.1. It is important to note that none of these parameters incorporates marginal costs of extraction or marginal benefits of biodiversity (see chapter 3, this volume).

Number of persons. In India alone, approximately fifty million people (5 percent of the total population) are assumed to live in and around forests and presumably derive a subsistence level of their livelihood from forest products (NCHSE 1987). In Brazil, 1.5 million people, or 20 percent of the economically active persons in the Amazon region, derive a significant portion of their livelihood from extraction of natural products (Browder 1992). These numbers do not include the people involved in the preserving and marketing of forest-based products, nor the people employed in forest-based industries. For both India and Brazil, the number of people who rely on harvesting of ecosystem products for their livelihoods could easily double when freshwater and marine ecosystems are taken into account, as both countries have huge coastlines and many large rivers. However, the exact number of people dependent on freshwater and marine ecosystems is not known.

Value of specific products. The value of particular products in unprocessed form or the revenue to the state generated by the harvest of the products may provide another mechanism to assess the contribution of natural ecosystems to subsistence economies. Rattan is perhaps one of the most celebrated examples of a precious nontimber forest product. The international

Table 16.1. Attributes of various parameters to assess the contribution of ecosystem services to subsistence economies

	Attributes				
Parameter	Direct Measure of Importance to Ecosystem People	Easy to Estimate	Estimates Non-Use Value	Incorporates Marginal Costs of Extraction and Benefits of Biodiversity	Importance to Policy Makers
Number of persons dependent on ecosystem services for livelihood	X	X			
Value of specific products	X	X			X
Contribution to cash income	X	X			
Proportion of households dependent on ecosystem services for livelihood		X			
Contribution to GDP			X		X
Value per hectare			X		X

and domestic trade in rattan was valued at US$4.0 billion and US$2.5 billion, respectively, in the 1980s (Manokaran 1990). The number of people employed in the rattan furniture industry in Indonesia alone is estimated to be 150,000 (Manokaran 1990). Another example is the tendu leaves from the Indian subcontinent. Tendu leaves from trees of *Diospyros melanoxylon* are used to wrap tobacco to produce bidis, a form of inexpensive cigarettes. The tendu leaves generate an annual revenue of US$160 million for the state of Madhya Pradesh in India. The leaves are just one among the hundreds of various types of nontimber forest products harvested from forests of India.

Contribution to cash income of households. The contribution of ecosystem products to cash income is difficult to ascertain because of limited information about the quantities of goods extracted and their disposal at the household level. The harvested products are used by the household members themselves, made available to others in the community, and sold or bartered for cash or other commodities and services. Cash income is generally reported from the last component pertaining to sale and exchange. However, even for this fractional component, cash income derived from the products can be substantial. For example, indigenous communities in and around the Biligiri Rangan Hills, a protected area in southeast India, derive 48–60 percent of their cash income from ecosystem products (Hegde et al. 1996). In West Bengal, India, nontimber forest products, including fuelwood and fodder from young regenerating forests, contribute 22 percent of the cash income of village households in and around forests (Malhotra et al. 1991). These figures are average figures from all the households surveyed in areas that are heavily forested.

Proportion of households dependent on ecosystem products. The proportion of households that rely on ecosystem products depends largely on the proximity to the ecosystem, size of the catchment area, and economic status of the people. Quantitative data are not available, but we would expect all households of settlements in and around large natural ecosystems in the developing world to derive a substantial portion of their income or livelihoods from ecosystem products. As mentioned earlier, a substantial proportion of the population in countries like India and Brazil is dependent on ecosystem products. With economic growth, the economy should diversify and the proportion of households that derive income from ecosystem services, as well as the extent of reliance on products, might be expected to decline (Godoy and Bawa 1993).

Contribution to GDP. The relative contribution to the gross domestic product could be another indication of the importance of ecosystem goods, even though goods and services from subsistence economies generally are discounted in the calculation of GDP, and GDP does not take into account ecosystem services. Unfortunately, GDP figures do not include all ecosystem products: for terrestrial ecosystems only forest products are included, and timber is often the main entry. However, for India, Lal (1992) has conducted a preliminary analysis of the annual rent from forests for both goods and services and found the rent to be more than 25 percent of the GDP; officially the contribution of forestry to GDP is listed as 1.2 percent. More important, Lal's analysis indicates that ecosystem products, including fuelwood and fodder, which are the basis of subsistence economies, constitute approximately 13 percent of the total value of forest goods and services. We should keep in mind that Lal's preliminary valuation does not include freshwater or coastal ecosystems.

Services per hectare. A number of attempts have been made to quantify the value of ecosystem services on a per hectare basis. Godoy et al. (1993) reviewed a number of studies and estimated the net value of nontimber forest products at approximately US$50 per hectare per year. These estimates include only ecosystem goods. For dry deciduous forests in India, Chopra (1993) estimates the value of nontimber forest products and services such as soil conservation, nutrient cycling, and tourism and recreation to be in the range of US$220–$335 per hectare per year. The use value for nontimber products in certain regions of Mexico has been estimated to be US $330 per hectare per year (Alcorn 1989). Interestingly, the 1991 gross domestic product (GDP) estimates for Mexico and India, respectively, are US$1,501 and US$836 per hectare, and almost certainly do not capture the values for ecosystem services cited above. The various estimates for ecosystem services in the above examples differ with respect to various goods and services included and are based on several untested assumptions. Nevertheless, the estimates are useful in drawing the attention of policy makers to undervaluation of ecosystem goods and services. The refinement of these estimates by incorporation of additional data can provide an assessment of the true importance of ecosystems in sustaining livelihoods of the ecosystem people.

Subsistence Economies, Sustainability, and Conservation of Biodiversity

It is apparent that a very large number of ecosystem people in biodiversity-rich regions of the world are dependent on the harvest of biological resources with their own labor from a limited-resource catchment area. In economic terms, the value of the products extracted by the ecosystem may not be very large because subsistence economies, by definition, involve the most impoverished sectors of society. Inclusion of non-use values into contributions of the ecosystems can provide better estimates of the economic value of the ecosystem, but, although the non-use values are substantial, we lack adequate mechanisms to quantify these values.

However, the real issue is not how valuable these ecosystem services are in relation to subsistence economies, but whether we can build on the ultimate dependence of ecosystem people on ecosystem goods and services of their immediate environments and turn them into stewards of the local living resources and biodiversity, and, in the process, enhance the quality of their lives. We believe the answer to be in the affirmative. There is abundant evidence that ecosystem people, settled for long in a locality and in full control of their own resource base, exhibit a number of cultural practices that promote sustainable use of biological resources and conservation of biodiversity (Gadgil and Berkes 1991; Gadgil et al. 1993). Such practices include limitations on harvest levels, e.g., number of sheep grazed on community pasture or wood harvested from community woodlots; lowering of harvesting pressures when there is evidence of overharvesting, e.g., temporary ban on fishing from coral reef lagoons; total protection in vulnerable life stages, e.g., birds breeding at a heronary; total protection of certain keystone resources, e.g., trees of genus *Ficus* in many parts of India; and the total protection of certain biological communities, e.g., sacred ponds and forests. Such practices, dependent either on a notion of the sacred or taboo or on social conventions, seem to have evolved and persist because they serve long-term interests of a small, well-knit human group in ensuring sustained availability of a diversity of resources.

In recent times, however, the dependence of human societies on diversity of resources from their immediate environments has been greatly reduced by technological progress. People are now capable of moving resources over large distances and transforming them extensively. With access to greatly expanded resource catchments, people may no longer suffer from depletion of resources in their immediate environments. For this category (the biosphere people with access to resources of all the biosphere) there is little motivation to sustainably use and promote persistence of a wide diversity of resources in any particular locality. Often, they do protect environments in their immediate vicinity to ensure healthier, aesthetically more pleasing ambiences for themselves, but this transfers the pressures of resource extraction to localities farther away. These localities tend to be inhabited by people with little economic or political clout, such as the ecosystem people, and results in the loss of control over their own environments to devote these locales to supply resources to biosphere people. At the same time, the ecosystem people have started to receive a trickle of supplies of a diversity of resources through the developing markets. The ecosystem people are thus no longer as completely dependent on a diversity of local resources as before, nor can they regulate unsustainable usage of these local resources. Under the circumstances, they tend to lose their motivation to sustainably use local living resources and conserve local biodiversity, and become suppliers of whatever little they can gather for the larger markets. These resources they tend to gather in an unsustainable fashion, contributing to the degradation of ecosystem goods and services (Gadgil and Guha 1995).

There are several examples of commercialization having deleterious effects on biodiversity. Rattan, mentioned earlier in the chapter, is becoming scarce in many countries (Manokaran 1990). There are numerous other examples of unsustainable extractions of nonwood forest products (Nepstad and Schwartzman 1992, Murali et al. 1996). In general, an increase in trade of exportable ecosystem products by indigenous communities is expected to result in depletion or extinction of populations yielding such products, while species that are nontradeable are likely to increase in abundance (Wilkie et al. 1995).

Many interacting factors are responsible for unsustainable extraction of resources. Commercialization and trade introduce boom and bust cycles and reliance on export markets over which producers have little control. Intrusion of external market forces leads to the loss of control over resources by indigenous groups and to the breakdown of traditional institutions promoting sustainable extraction. Moreover, paucity of resources in degraded environments increases poverty in already impoverished sectors of the society, and poverty, in turn, leads to further deterioration of the environment.

Policy Options

The degradation of ecosystem goods and services affects the local ecosystem people far more directly and adversely than it does any other human group. So, of all people, these local ecosystem people retain the highest levels of motivation for maintaining healthy levels of ecosystem goods and services in their own localities. This motivation cannot, however, be molded into effective action so long as they do not have: (1) control over their resource base; (2) adequate management; (3) incentives to conserve biodiversity; and (4) equity in bearing the cost of conservation. Several policy reforms would therefore be required to integrate the use of ecosystem resources by rural poor and conservation of biodiversity.

Tenurial control. The first and most important prerequisite for turning ecosystem people into stewards for good management of their environments is to restore to them control over the resource base. In the absence of tenurial control over resources or the land or water supporting the resources, the ecosystem people have little incentive to sustainably extract resources to which there is open access.

In India, joint forest management that seeks to partially restore the control of forest resources to local communities has had some success in regeneration of degraded forests. The state forests in India, until the last century, were largely under community control (Gadgil and Guha 1992). Appropriation of the forests by the state during the last one hundred years without addressing the forest-based subsistence requirements of the ecosystem people has created an acute conflict between the people and the government agencies over use and conservation of biodiversity. The joint forest management plan is supposed to resolve some elements of this conflict. The basic concept of the plan is simple: Forest protection committees at the village level safeguard regenerating forests under the control of the state in exchange for access to nontimber forest products as well as a share in timber production. The joint forest management plan apparently has succeeded in regenerating degraded forests, but its contribution to resurrecting original levels of biodiversity and ecosystem functions is not clear. Simple access to state-owned forest resources without an active monitoring program does not eliminate the possibility of resource depletion, nor does it prevent loss of some ecosystem functions. More important, the joint forest management plan circumvents the tenurial issue. In order to be more effective, the joint management plans must address, in addition to property rights, issues related to inventory, productivity, and extraction of ecosystem products (see below).

Management. Tenurial control by itself would not be sufficient to conserve as well as sustainably utilize biodiversity. Without an adequate management plan, modern market forces have the capability to deplete resources even when extraction levels are low. Thus, conservation of biodiversity in many ecosystems is likely to remain an elusive goal without the involvement of the ecosystem people in management plans, regardless of the extent of trade in ecosystem products.

The management plans must be adaptive plans, based on continual monitoring of the abundance and extraction levels of resources being harvested. Extraction should be in proportion to production, which is likely to vary over space and time. Sustainable levels of harvests, particularly for a diverse array of products from terrestrial ecosystems, are difficult to determine because of problems in defining sustainability, and because any large-scale export of materials from the ecosystem is likely to have deleterious consequences on the structure and function of the ecosystem. Given these difficulties, only flexible adaptive management plans could prevent depletion of resources and must be put in place.

Depletion could also be prevented by value addition. Many ecosystem products that form the basis of subsistence economies often leave the point of origin in an unprocessed state. As a result, harvesters realize very low value from extracted products. For example, the Soligas, the indigenous people of the Biligiri Rangan Hills in southwest India, harvest, among other things, amla (*Phyllanthus emblica*) fruits from tropical forests. The fruits, used for pickles, jams, and medicinal products, are exported from the forests in a raw state, and the Soligas secure minimal income based simply on the amount of labor invested in the harvest of fruits. However, the income derived from amla, which averages US$6.60 per capita, could be increased to

US$87.75 per capita if the Soligas were to directly process and market these fruits (Uma Shankar et al. 1996). Amla is only one of the many products harvested by the Soligas. If the economic returns for even five products were to be enhanced by a factor of five for each product, the total income, or value, to the Soligas could be increased by a factor of twenty-five. Another example is rattan. In 1971, the value of rattan exported by Indonesia was US$0.8 million, but in 1988 the value had increased to US$194.6 million, primarily due to value addition (Manokaran 1990). Value addition can enhance the income and reduce the amount extracted in cases where extraction is not sustainable.

There are often several impediments to value addition. Extractors lack the capital or infrastructure for processing the product locally, and the scale of operation is often too small to justify processing at the site of extraction. Moreover, the ecosystem products are not directly marketed by the harvesters. The extractors have little information about the demand, the market channels, and, in many cases, even the eventual fate of the products.

However, value addition by itself cannot promote sustainable harvests or ensure economic benefits to the ecosystem people. Processing and marketing of products will have to be community based and under the full control of local populations. Moreover, the community-based enterprises must have a biological monitoring system that tracks levels of extraction and production and monitors the impact of harvests on ecosystem structure and function. Such community-based enterprises are now beginning to take shape in many ecosystems (Biodiversity Conservation Network 1995). A recent report of FAO further discusses the management, infrastructure, and policy requirements for extractive economics that seek to enhance rural incomes and promote sustainable use of forest products (FAO 1995; see also Murrieta and Rueda 1995).

Incentives. Restoring to ecosystem people full control and management of their own localities may help maintain these ecosystems in better health and provide higher levels of the ecosystem's goods and services, but this alone would not be sufficient to motivate them to maintain high levels of biological diversity. That would require further incentives, and in the modern context these would have to be economic incentives. Thus, if the localities inhabited by the large numbers of ecosystem people in the tropical countries are to maintain or be restored to high levels of biological diversity, we must devise a system of rewards to local communities linked to levels of local biodiversity (Panayotou 1994; Gadgil and Rao 1994, 1995). These rewards should be viewed as service charges to the ecosystem people for helping provide global ecosystem services for the conservation of biological diversity. The ecosystem people ought to provide these services in a highly cost-effective fashion. This is because they automatically acquire the detailed locality and time-specific knowledge of the behavior of local ecosystems, so necessary for effective, adaptive management, in the course of their daily pursuit of obtaining a livelihood. The local ecosystem people are also best situated to monitor all human impacts on the ecosystems, and therefore to control them, provided they have the requisite authority. They also have social structures to minimize exploitation of resources. Finally, being relatively poor, these people would be willing to take on the task of maintaining and restoring local biodiversity for low levels of compensation. There is every reason to believe, therefore, that vesting local ecosystem people with control over their own environments, and paying them service charges to maintain and restore biodiversity would be a very effective way of taking good care of the ecosystems of these parts of the earth. Involving ecosystem people in such a system would enhance their quality of life, as well as confer a measure of dignity on them. That too would be a socially just course of action (Gadgil and Rao 1995).

Equity. A fundamental cause of environmental degradation is the inequity in distribution of benefits and costs of conserving natural resources and biological diversity. As emphasized by Wells (1992), benefits of biodiversity are widely dispersed, whereas costs of conservation are highly localized. In a sense, restoration of tenurial control and provision of economic incentives are designed to relieve inequities in the benefit-cost ratio of conservation. Inequities, however, stem from several socioeconomic factors, and equity can only result from a series of reforms at the local, national, and global levels.

Conclusions

Millions of people depend on natural ecosystems for their livelihoods. It is difficult, however, to quantify the contribution of natural ecosystems to sustain livelihoods because of problems in assigning monetary values to lifestyles, culture, religious beliefs, and many other aspects of people's lives that are intimately associated with their natural surroundings. Subjective and value-laden criteria used in valuation techniques also do not generate much confidence in the figures one may be able to obtain from such methods. Nevertheless, on the basis of the number of people involved in harvesting ecosystem products, the value of particular products, the contribution of cash income to households, the proportion of households that rely on ecosystem products for a substantial portion of the cash income, the GDP derived from ecosystem goods, and the value of products on a per hectare basis, it is obvious that the maintenance of ecosystem services is critical to the well-being of millions, if not billions, of people on earth. The functioning of natural ecosystems is also important for the future increases in incomes of people relying heavily on ecosystem products for their survival.

Currently, commercialization and trade of ecosystem products that primarily benefit people living far away from natural ecosystems, as well as the loss of local control over natural resources, breakdown of traditional regulations and institutions governing the extraction of resources, and inadequate management, are leading to the degradation of many natural ecosystems. The total dependence of ecosystem people on natural ecosystems makes them extremely vulnerable to the disruption of ecosystems. Thus, significant alteration of the structure and function of ecosystems due to land-use changes or uncontrolled commercialization of ecosystem products is often accompanied by great suffering of ecosystem people as well as loss of biodiversity.

The continuous flow of services from ecosystems and conservation of biodiversity require that the ecosystem people be more actively involved in conservation than they have been in the past. For ecosystem people to resume the stewardship of natural resources, tenurial control of resources to local communities must be restored, adaptive management plans fine-tuned to changing resource levels on the basis of continual monitoring resources must be developed, economic incentives to conserve biodiversity must be provided, and inequity in the benefit-cost ratio of conservation at the local level must be reduced. The intimate dependence of a large segment of humanity on ecosystems offers a great opportunity for people-based conservation activities in ecosystems facing degradation.

Acknowledgments

This chapter represents contribution number 41 of the research program in Conservation of Biodiversity and the Environment, jointly coordinated by the Tata Energy Research Institute, New Delhi, and the University of Massachusetts at Boston. The program is supported in part by the MacArthur Foundation. We thank Shaily Menon and Neela DeZoyza for their comments on the manuscript.

References

Alcorn, J. B. 1989. "An economic analysis of Huastec Mayan forest management." In *Fragile Lands of Latin America*, J. O. Browder, ed., pp.182–206. Westview Press, Boulder, Co.

Biodiversity Conservation Network. 1995. *Biodiversity Support Programs, Biodiversity Conservation Network. 1995 Annual Report.* World Wildlife Fund, Washington, D.C.

Browder, J. O. 1992. "Social and economic constraints on the development of market-oriented extractive reserves in Amazon rain forests." *Advances in Economic Botany* 9: 33–42.

Brown, K. 1994. "Approaches to valuing plant medicines: The economics of culture or the culture of economics?" *Biodiversity and Conservation* 3: 734–750.

Chopra, K. 1993. "The value of non-timber forest products: An estimation for tropical deciduous forests in India." *Economic Botany* 47: 251–257.

Dasmann, R. F. 1988. "Toward a biosphere consciousness." In *The Ends of the Earth,* D. Worster, ed., pp. 277–288. Cambridge University Press, New York.

FAO (Food and Agriculture Organization) 1995. *Non-Wood Forest Products for Rural Income and Sustainable Forestry.* Food and Agriculture Organization of the United Nations, Rome.

Gadgil, M. 1995. "Prudence and profligacy: A human ecological perspective." In *The Economics and Ecology of Biodiversity Decline,* T. M. Swanson, ed., pp. 99–110. Cambridge University Press, New York.

Gadgil, M. 1996. "Managing biodiversity." In *Biodiversity: A Biology of Numbers and Difference,* K. J. Gaston, ed., pp. 339–360. Blackwell Scientific Publications, Oxford.

Gadgil, M., and Berkes, F. 1991. "Traditional Resource Management Systems." *Resource Management and Optimization* 18:127–141.

Gadgil, M., F. Berkes, and C. Folke. 1993. "Indigenous knowledge for biodiversity conservation." *Ambio* 22:151–156.

Gadgil, M., and Guha, R. 1992. *This Fissured Land: An Ecological History of India.* University of California Press, Berkeley.

Gadgil, M., and Guha, R. 1995. *Ecology and Equity: The Use and Abuse of Nature in Contemporary India.* Routledge, London.

Gadgil, M., and Rao, P. R. S. 1994. "A system of positive incentives to conserve biodiversity." *Economic and Political Weekly* (Aug. 6):2103–2107.

Gadgil, M., and Rao, P. R. S. 1995. "Designing incentives to conserve India's biodiversity." In *Property Rights in a Social and Ecological Context,* S. Hanna and M. Munasinghe, eds., pp. 53–62. The Beijer International Institute of Ecological Economics and The World Bank. Washington, D.C.

Godoy, R., and Bawa, K. S. 1993. "The economic value and sustainable harvest of plants and animals from the tropical rain forest: Assumptions, hypotheses, and methods." *Economic Botany* 47: 215–219.

Godoy, R., Lubowski, R. and Markandaya, A. 1993. "A method for the economic valuation of non-timber forest products." *Economic Botany* 47: 220–233.

Hall, P., and Bawa, K. S. 1993. "Methods to assess the impact of extraction of non-timber tropical forest products on plant populations." *Economic Botany* 47:234–247.

Hegde, R., Suryaprakash, S., Achot, L., and Bawa, K. S. 1996. "Extraction of non-timber forest product in the forest of Biligiri Rangan Hills, India. 1. Contribution to rural income." *Economic Botany*, in press.

Lal, J. B. 1992. "Economic value of India's forest stock." In *The Price of Forests: Proceedings of a Seminar on the Economics of the Sustainable Use of Forest Resources*, A. Argawal, ed., pp. 43–48. Centre for Science and Environment, New Delhi.

Malhotra, K. C., Deb, D., Dutta, M., Vasulu, T. S., Yadav, G., and Adhikari, M. 1991. "Role of non-timber forest produce in village economy: A household survey in Jamboni Range, Midnapore District, West Bengal." Unpublished paper, Indian Institute of Biosocial Research and Development, 3A Hindustan Road, Calcutta.

Manokaran, N. 1990. *The State of the Rattan and Bamboo Trade*. The Rattan Information Centre, Institute Penylidikan Perhutanan Malaysian (FRIM), Kepong, 52109, Kuala Lumpur.

Murali, K. S., Shankar, U., Shaanker, R.U., Ganeshaiah, K.N. and Bawa, K. S. 1996. "Extraction of non-timber forest product in the forest of Biligiri Rangan Hills, India. 2. Impact of NTFP extraction on regeneration, population structure, and species composition." *Economic Botany*, in press.

Murrieta, J. R., and Rueda, R. P., eds. 1995. *Extractive Reserves*. International Union of Conservation of Nature. Gland, Switzerland.

NCHSE (National Centre for Human Settlements and Environment). 1987. *Documentation on Forest and Rights*, vol. 1. National Centre for Human Settlements and Environment, New Delhi.

Nepstad, D. C., and Schwartzman, S., eds. 1992. *Non-Timber Products from Tropical Forests: Evaluation of a Conservation and Development Strategy*. Special issue of *Advances in Economic Botany* 9:1–164.

Panayotou, T. 1994. *Economic Instruments for Environmental Management and Sustainable Development*. International Environment Program, Harvard Institute for International Development, Harvard University. Manuscript.

Pearce, D., and Moran D. 1994. *The Economic Value of Biodiversity*. Earthscan Publications Limited, London.

Shankar, U., Murali, K. S., Shaanker, R. U., Ganeshaiah, K. N., and Bawa, K. S. 1996. "Extraction of non-timber forest product in the forest of Biligiri Rangan Hills, India. 3. Productivity, extraction and prospects of sustainable harvest of Nelli (*Emblica officinalis*)." *Economic Botany*, in press.

Vatn, A., and Bromley, D. W. (1995). "Choices without prices without apologies." In *The Handbook of Environmental Economics*, D. W. Bromley, ed., pp. 56–81. Basil Blackwell, Oxford, England.

Wells, M. 1992. "Biodiversity conservation, affluence and poverty: Mismatched costs and benefits and efforts to remedy them." *Ambio* 21:237–243.

Western, D., and Henry, H., eds. 1994. *Natural Connections: Perspectives in Community Based Conservation*. Oxford University Press, New York.

Wilkie, D. S., Godoy, R., and Brokaw, N. 1995. "The impacts of trade on indigenous economies and biological diversity: A microeconomic approach." Manuscript.

Chapter 17

ECOSYSTEM SERVICES IN A MODERN ECONOMY: GUNNISON COUNTY, COLORADO

Andrew Wilcox and John Harte

Generic balance sheets showing the economic value of individual ecosystem services and the costs of replacing them are valuable, but they can lack connection to the lives of real people in real socioeconomic settings. Site-specific case studies, on the other hand, can be used to translate the abstract information embodied in more general studies of ecosystem services into tangible and convincing terms that show how ecosystem services affect human economies. The debate over draining swamps in south Florida in the late 1960s offered an excellent example of the importance of site-specificity. In that case, the argument that the Big Cypress Swamp performed the vital service of preventing saltwater intrusion into the underground water supplies of several hundred thousand Gulf Coast residents contributed to the protection of the Cypress Swamp (Harte and Socolow 1971).

This chapter will examine ecosystem services in the context of a case study of Gunnison County, Colorado. We will not attempt to conduct cost-benefit analysis of environmental protection versus development in Gunnison County, or to place a value on environmental goods such as minerals and timber. Rather, we will describe the importance of ecosystem services to the local economy by characterizing local ecosystems, the services they provide, the economic contributions of these services, and the potential for future degradation of ecosystem services. In addition to describing ecosystem-economy interrelations in Gunnison County, we will discuss approaches to quantifying these linkages.

Although we have adopted Gunnison County as our study unit, ecosystem-economy interrelations are considered on multiple scales. Functions performed by county ecosystems provide services to humans on local, regional, and global scales, while locally originated degradation of these services exerts local, regional, and global impacts. Likewise, ecosystem functions and environmental degradation occurring on global and regional scales carry benefits or impacts that may be important locally. Examples of relations between these spatial scales will be elaborated on below.

Gunnison County Case Study

Gunnison County, which lies on the Western Slope of the Rocky Mountains in the upper Gunnison River basin, offers an excellent locale for a study of ecosystem-economy interrelations. The county is sparsely populated, with a population density of approximately three persons per square mile and a total population of approximately eleven thousand. Its economy, which is based on ranching and recreation, is tightly linked to natural resources and ecosystem services. Although the county's ecosystems are for the most part not pristine, functional ecosystems continue to exist and have been relatively well studied. Gunnison County faces the prospect of rapid economic and ecological changes due to plans for transmountain water diversion projects, ski area expansion, molybdenum mining, and recreation and real estate development. These issues, which face many areas in the Rocky Mountain region, highlight the relevance of understanding the role of healthy ecosystems in supporting sustainable local economies.

Gunnison County is characterized by glacially formed terrain with high mountain peaks descending into long ridges and wide river basins. Dominant habitats include Great Basin sagebrush and desert shrub, aspen and conifer forests, and alpine meadows. Seventy percent of the county's 3,237 square miles are publicly owned; this land includes large roadless and wilderness areas. Ecologically important valley bottomlands, which include riparian and grassland habitat and the lower forest ecotone, are dominated by private ownership (Theobald and Riebsame 1995). The county's climate is prohibitive to most forms of agriculture, with long, cold winters, snow accumulations of over 300 inches in some areas, and an average growing season of seventy-seven days (County Information Service 1992). Winter snowpacks melt in late spring and early summer to provide relatively abundant surface water supplies.

Natural resources have been central to Gunnison County's history. Ute Indians seasonally inhabited the area for hundreds of years, subsisting off of fish and wildlife. The area first attracted white settlers in the late 1800s, when towns like Crested Butte, Gothic, and Gunnison boomed as silver and coal mining hubs, and the region's population peaked at over twenty-five thousand in the early 1880s (Duane Vandenbusche, pers. comm. 1996). The influx of settlers to mineral-rich areas of Gunnison County also brought deforestation and air and water pollution, some of the effects of which are still felt. Although the 1893 silver panic drained the area of much of its population, coal mining continued near Crested Butte into the 1950s (Vandenbusche 1980). Other forms of mineral exploitation have continued intermittently in this century, including a uranium mining boom in the 1950s and lead, zinc, copper, silver, and gold extraction (Peckarsky and Cook 1981).

Cattle ranching also developed in the late 1800s and has been an economic mainstay in the county during much of this century. As ranches were established, valley bottomland areas were transformed into hay production and grazing areas through draining of wetlands, irrigation, and removal of willows (Cooper 1993). Since the 1880s, cattle grazing has operated according to a seasonal migration pattern designed to protect forage quality, whereby cattle are moved from lower elevation areas in spring to higher, privately owned valley lands in early summer and on to publicly owned subalpine and alpine areas in late summer and autumn (Theobald et al. 1995). As of 1986, the county supported over thirty-five thousand cattle (HDR Engineering 1991), and ranching contributes $8 million annually to the county economy.

Gunnison County's ranching sector is sustained by the upper Gunnison River basin's water resources. Ranchers are the largest water users, using 81,000 to 108,000 acre-feet per year for irrigated pasture in the upper Gunnison basin (Ken Knox, pers. comm. 1995) (table 17.1). Given the area's climate and soils, viable ranching operations depend on liberal water applications. Although the flooding irrigation practices used by Gunnison County ranchers may be considered inefficient, these methods create substantial return flows and contribute to late-summer baseflow (Tyler Martineau, pers. comm., 1995). Water used to irrigate pasture lands may therefore be reused many times as it moves through the Gunnison River basin. Grazing depends not only on irrigated pasture made possible by water supply, but also on the existence of a healthy, diverse vegetational community characterized by a mixture of grasses and forbs.

Tourism and recreation centered on the area's mountains, rivers, and aesthetic beauty have grown to be the county's largest economic sector in recent decades, contributing $35 million per year to the county economy. The Crested Butte ski area, which opened in the early 1960s and grew rapidly in the 1970s, dominates the county's winter recreation sector and has stimulated a boom in real estate development. Summer recreation, including

Table 17.1. Gunnison County Water Budget (all figures in acre-feet/year)

Water Use Category	Diversion	Consumptive Use
East R. basin Municipal, Domestic & Industrial (1994)[a]	1,155	347
Crested Butte Mountain Resorts snowmaking (1994)[a]	300	60
Irrigation, East R. basin (1994)[a]	88,600	10,500
Irrigation, upper Gunnison R. basin (1994)[b]	—	81,000–108,000
AMAX Mt. Emmons molybdenum mine (projected)[c]	24,000	3,000
Union Park Reservoir Project (projected)[d]	97,000–156,000	97,000–156,000
Natural Flows	**Average Annual Yield**	
Coal Creek near Crested Butte[e]	13,000	
East R. near Crested Butte[f]	165,100	
East R. at Almont[f]	243,500	
Taylor R. below Taylor Park Reservoir[f]	142,300	
Gunnison R. near Gunnison[f]	548,500	

[a]U.S. Bureau of Reclamation, 1995.
[b]K. Knox, personal communication, 1995.
[c]HDR, 1989.
[d]Arapahoe County, 1995.
[e]WRC Engineering, 1991.
[f]U.S. Geological Survey, 1994.

camping, fishing, and biking, has also gained in economic importance. In addition to recreation, the county derives other forms of economic activity from the quality of its ecosystems. The town of Crested Butte, the "Wildflower Capital of Colorado," hosts a Wildflower Festival each summer that attracts hundreds of tourists. The Rocky Mountain Biological Lab, a field research station in the upper East River Valley that depends on the existence of unimpaired alpine ecosystems, adds over $2 million annually to the local economy (Susan Lohr, pers. comm., 1995).

The recreation sector is also dependent on the water supply services provided by the area's riverine systems. Water-based recreation, such as trout fishing and rafting, depends on instream flow and the aesthetics of nondegraded riverine systems. The downhill skiing sector is dependent on both natural snowfall and winter water diversions for snow making. Whereas diversion of water out of its natural channel has historically been considered a prerequisite for economically beneficial use, the growth of instream flow-dependent recreation and increased awareness of ecosystem services has created increased appreciation of instream flow values.

Sparling et al. (1994) estimated the value of instream flows in Gunnison County's Taylor and East River basins in order to assess the economics of transbasin diversion of those waters. Market values for downstream hydropower production, salinity control, and municipal use were estimated at $460 per acre-foot. Nonmarket values, including recreational use values and preservation values, were assessed by a contingent valuation technique, which derived a value of between $350 and $480 per acre-foot. The total value of instream flows summed to between $810 and $940 per acre-foot (Sparling et al. 1994). These figures indicate that instream flow values are substantial.

Economic activity generated by tourism and recreation is an important contributor to the local economy as a whole. A study of fishing at Gunnison County's Blue Mesa Reservoir found that Blue Mesa anglers from outside of the county spent $2.62 million directly, most of which benefited the household income, retail sales, lodging, and restaurant sectors. This spending also had substantial indirect effects, generating $5.25 million in added sales revenue for the Gunnison County economy (McKean et al. 1988). In Colorado as a whole, recreational fishing is estimated to add $600 million to the economy each year (Fielder and Pearson 1995). Similar types of multiplier effects as those found for fishing at Blue Mesa, whereby spending in one sector stimulates economic activity in other sectors of the county economy, likely accrue from other forms of nature-based tourism as well.

Services Provided by Gunnison County Ecosystems

Streams and rivers in Gunnison County provide valuable ecosystem services. "The Gunnison River . . . presently provides nearly all the anthropogenic benefits that can be extracted from a lotic system; and yet it retains surprising ecosystem integrity" (Stanford 1989, p. 5). Coal Creek, a tributary of the East River with its headwaters in the Elk Mountains west of Crested Butte, offers an excellent example of the services performed by a nondegraded watershed and riverine system. Coal Creek is a snowmelt-fed stream that drains a forested, undeveloped, 21-square-mile basin for much of its length before flowing through the town of Crested Butte and entering the Slate River.

Coal Creek offers a high-quality water supply for the town of Crested Butte. In addition, the creek and its adjacent riparian vegetation provide important habitat and food-chain support, both instream for trout and beavers and in riparian areas for elk, deer, bears, and numerous bird species (Cooper 1993). The Coal Creek watershed also regulates runoff and provides flood control. Water collects in the Coal Creek basin's snowpack throughout the winter and is gradually released in spring and summer, when the water is most needed downstream, with riparian vegetation, channel morphology, and beaver dams dissipating the stream's energy and slowing the movement of water downstream. In the spring of 1995, Coal Creek defied flood predictions by conveying runoff from a snowpack containing 415 percent of the basin's average snowpack water content (Hall 1995) through the town of Crested Butte without flooding.

The potential for degradation of Coal Creek's ecosystem services is substantial. Logging in the Coal Creek drainage, as has been proposed in the past by the U.S. Forest Service (USFS 1991), would increase peak flows by eliminating vegetation that provides runoff regulation functions, thereby exacerbating flooding hazards. Logging and other development activities, such as construction of new roads and buildings, would increase erosion and the risk of landslides. Increased sediment yield would impair fish habitat, accelerate downstream reservoir sedimentation, and reduce water quality, intensifying water treatment requirements for municipal water supply.

Large-scale diversions from Coal Creek have been proposed for molybdenum mining on Mt. Emmons. A plan advanced by the AMAX Corporation calls for processing of twenty thousand tons of ore per day for twenty-seven years. The water required for mining and milling of the molybdenum ore would amount to approximately ten times the current level of total consumptive use for municipal and domestic purposes in the East River basin (HDR Engineering 1989; USBR 1995). Although the value of the Mt. Emmons minerals is estimated at $3 billion, the corporation is attempting to patent the overlying land for $5 per acre under the 1872 Mining Law (Obmascik 1995). Such diversions, as well as mining-related heavy metal and acid pollution, would damage instream habitat and water quality, potentially forcing adoption of new water supplies for Crested Butte and impacting the recreation sector. A nondegraded Coal Creek system provides important economic benefits in the realms of fishing, hunting, tourism, water supply, and flood control. Degradation of the Coal Creek system of the sort described above would require technological substitions for the services provided by Coal Creek and impose substantial economic costs.

Owing to their transitional position between fluvial and terrestrial ecosystems, riparian areas in the Coal Creek watershed and elsewhere in the upper Gunnison River basin are rich in ecosystem services. Willow communities in the Slate River floodplain, for example, serve as critical bird habitat, increase streambank stability, and offer detrital input to aquatic food chains. Root networks of willows and other riparian plants increase the cohesiveness of bank materials, contributing to a bank's ability to resist erosion (Gregory et al 1991). Maintenance of bank integrity is an important service, since bank erosion increases turbidity levels, removes undercut bank areas relied on by fish for cover, reduces water quality, accelerates reservoir sedimentation, and causes property losses.

Gunnison County's upper East River valley also supports valuable wetlands. Throughout the upper East River valley, snowmelt nourishes wetlands by infiltrating and percolating as groundwater from mountain slopes into valley bottoms (Cooper 1993). Most of the county's wetlands are not pristine, having been subjected to grazing, vegetation removal, mine pollution, and other impacts. These wetlands, however, continue to filter pollutants, trap sediment, provide wildlife habitat, and deliver late-summer and autumn baseflows (Cooper 1993). Flood storage functions are moderately effective, but due to the above-mentioned wetlands impacts, do not match the performance levels of unimpaired floodplain systems, which are estimated to store two to five times as much water as modified floodplains. Beavers contribute to wetland integrity in this area by promoting storage of water on the floodplain and helping to maintain high water tables (Cooper 1993).

Other terrestrial ecosystems also perform important functions in Gunnison County. While the value of forests is often thought of in terms of "goods" such as wood supply, services provided by aspen and conifer forests are also significant in Gunnison County. Upland forests offer wildlife habitat for both game and nongame species, provide aesthetic value, contribute to hillslope stability and soil and runoff retention, and serve as carbon reservoirs. Soil and runoff retention purvey both local and regional benefits, since intact forests reduce the risk of landslides, downstream flooding (Burger 1922, as cited in Germann 1990), and reservoir and fish habitat siltation, while carbon storage in plants offers a global service by reducing atmospheric carbon dioxide concentrations. Montane meadows also furnish a global ecosystem service by acting as methane sinks, reducing the concentration of this potent greenhouse gas in the atmosphere through the action of methanotrophic bacteria (Torn and Harte 1995).

Past Ecosystem Service Degradation

The significance or contribution of ecosystem services can be discerned by considering the consequences of their degradation or loss. Although ecosystems in the county are generally healthy, other forms of degradation in addition to the above-mentioned wetlands impacts have been documented.

Willows in many parts of Gunnison County have been adversely affected by grazing and by lowering of the water table following channel incision and wetlands drainage. In the Slate River valley, approximately 40 percent of the original willow cover has been removed to expand grazing land (Cooper 1993). Loss of riparian vegetation in some areas has destabilized banks, allowing bank erosion and channel widening to occur in relatively low-magnitude streamflow events. Some ranchers in the county have become cognizant of these grazing-induced impacts to riparian systems and have undertaken innovative grazing management techniques such as hauling water to nonriparian areas. Efforts such as these have helped restore streambank stability, riparian vegetation health, and fish and wildlife habitat (Cairns and Pratt 1995).

In recent years, real estate development and population growth have also contributed to environmental degradation in the East River basin. A "boom in mountain living" affecting the entire Rocky Mountain region has accelerated the conversion of agricultural land to residential use in Gunnison County (Theobald and Riebsame 1995). The growth of subdivisions, condominiums, and ranchettes, along with associated road construction, has fragmented critical wildlife habitat and increased human populations in near-wildland areas. This process has had disproportionate impacts on ecologically important riparian and lower-forest ecotones, since private lands in the county are concentrated in valley bottomlands (Theobald and Riebsame1995).

The boom of real estate development in Gunnison County, which has been driven largely by the market for second homes, has not only created quality-of-life concerns related to aesthetic degradation, increased traffic, and affordable housing shortages, but has threatened the provision of ecosystem services by riparian, wetland, and forest ecosystems. Degraded water quality has already been observed; increased wastewater flows and insufficient treatment capacities have caused significant levels of nutrient loading in surface waters. This problem has been compounded by the fact that increased surface water diversions, particularly in winter, have diminished the pollution dilution capacity of river systems (USBR 1995). Despite these problems, overall water quality is still considered high.

Construction of Bureau of Reclamation dams on the Gunnison and Taylor rivers has impacted stream biota by reducing the seasonal variability of flow regimes and altering stream temperatures with cold hypolimnial releases in the summer (Stanford and Ward 1983). Elimination of many zoobenthic species from directly below the dams and a downstream shift in community structure has been observed (Stanford and Ward 1984). The Upper Gunnison River and its tributaries once supported outstanding cutthroat trout fisheries, and the lower Gunnison (which is outside of Gunnison County) supported abundant populations of squawfish and razorback suckers, which are both currently listed as endangered species. The demise of native fish species in the Gunnison River is likely due to dam-induced ecological discontinuities and water diversions (Stanford 1994).

Native fisheries have also been impacted by a large-scale program of exotic, hatchery-bred fish introduction. Rainbow and brown trout, which compete with native cutthroat for habitat and food supply, have been planted by the millions in the county's rivers, in an effort to "improve" fisheries for recreational purposes and to substitute for wild trout lost due to habitat degradation. Fears that fish raised in Gunnison County's Roaring Judy fish hatchery may be contributing to the spread of whirling disease, a nonnative disease that deforms young trout and has been found in Roaring Judy fish, raise questions about the sustainability of this effort. Whirling disease, which has been spread to many rivers of the Rocky Mountains by hatchery stocking and poses a substantial threat to fisheries throughout the region, exemplifies the pitfalls of attempting to replace ecosystem services with technological substitutes. Research efforts to develop a disease-resistant "designer" trout have found that native cutthroat trout may in fact carry the strongest resistance to whirling disease (Ring 1995), suggesting that restoration of biodiversity may protect fisheries against whirling disease.

Attempts to improve hatchery-stocked fisheries have backfired as well. In Taylor Reservoir, where non-native fish have been stocked for decades for recreational purposes, the fishery has been damaged by the intentional introduction in the 1970s of the possum shrimp (*Mysis relicta*) as a food supply for fish. Instead of improving Taylor Reservoir's fisheries, the possum shrimp has upset the reservoir's food web by outcompeting native invertebrate species, reducing the overall food supply for fish. Entry of the possum shrimp into the Gunnison River system, as would occur if water was spilled out of Taylor Reservoir, would potentially allow the shrimp to colonize and eventually overtake the downstream Blue Mesa Reservoir, whose valuable fishery could thereby be threatened. Possum shrimp disruption of food webs has also been documented in Montana's Flathead Lake (Susan Lohr, pers. comm.).

Mining has a history of stream system degradation in Gunnison County as well. Pollution from abandoned mines in the county continues to release heavy metals into the area's waterways, decades after the cessation of most mining. In some headwaters areas of the East River valley, heavy metal pollution limits species diversity and numbers of fish (USBR 1995).

Scenarios of Future Ecosystem Service Degradation

The benefits provided by ecosystem services in Gunnison County may be subtle and unappreciated. One way to highlight their importance is through analysis of currently proposed development projects in Gunnison County, which carry a significant potential for future degradation of ecosystem services and are typical of issues facing many areas in the Rocky Mountains. The county's clean water, scenic beauty, and healthy ecosystems make it attractive for many types of development but are in turn threatened by development forces. "There is good reason to fear that the region's natural wealth contains the virus of its ultimate impoverishment" (Younger 1971; this statement was originally made in reference to the Lake Tahoe basin, but it applies equally to Gunnison County and other areas of the West).

The county's abundant, high-quality surface water supplies have long been coveted both by agricultural and urban entities on the drier Eastern Slope of the Rockies, placing the Front Range's demand for fresh water at odds with instream and in-basin use of Gunnison River basin water. Numerous transmountain diversion schemes have been hatched over the years, including the current Union Park Reservoir Project, which would divert Gunnison County water to fuel growth in the Denver metropolitan area. The success of county residents in fending off past diversion plans has allowed the upper Gunnison River basin to retain substantial hydrologic integrity. The Union Park project would divert spring peak flows from headwater streams of the Taylor and East River basins to a 900,000-acre-foot reservoir at Union Park, which would become the second largest reservoir in Colorado, behind Gunnison County's Blue Mesa reservoir. Over 100,000 acre-feet per year, enough to supply approximately 500,000 people, would then be piped underneath the Continental Divide and eventually to the Front Range.

The economic and ecological impacts of this project would be substantial. Aquatic, riparian, and floodplain ecosystems would be harmed by reduced average flows, the loss of seasonal flood pulses, and submersion of land under the Union Park Reservoir. Reduced flows and aesthetic degradation would impact fishing, rafting, and other forms of instream flow-dependent recreation. Decreased water flows would adversely affect ranchers as well by reducing head and velocity of water in ditches and potentially forcing installation of new diversion or flow-control structures (Tyler Martineau, pers. comm., 1995; Pinnes 1995).

Expansion of the Crested Butte Mountain Resort ski area would also significantly alter the upper East River environment. Expansion plans include clearing of mature forest for new ski runs and ski lifts and real estate and commercial development, which would create substantial off-site impacts. This development would impact wetlands, wildlife, and the nearby Rocky Mountain Biological Lab and stress transportation, sewage, water, and parking infrastructure. Studies have also shown that ski area expansion would increase the area's avalanche hazard (Mears 1995), disrupt soil stability on the steep slopes of Snodgrass Mountain (RCE 1995), and change hydrologic flow patterns and the composition of existing vegetational communities.

Although the acute impacts associated with ski area expansions and transmountain water diversions may be more recognizable, the chronic effects of real estate development pose a substantial threat to ecosystem services. Many county residents fear that the upper East River valley will develop into another Aspen, where large vacation homes have replaced agricultural land, housing prices have forced locals to move elsewhere, and ecosytem values have been degraded. In the upper East River valley, build-out percentage, the ratio of occupied to vacant subdivided lots, was only 22 percent as of 1994, suggesting that more development is on the way (Theobald and Riebsame 1995).

Regional and global environmental impacts could also affect Gunnison County's economy in the future. Population growth in the Front Range is responsible for transmountain diversion proposals such as the Union Park project, while overcrowding of California and other areas stimulates real estate development in Gunnison County. Pollution from urban automobile use, coal-fired electricity generation, and copper smelting in areas geographically distant from Gunnison County pose threats to the county's air and water quality. Researchers at the Rocky Mountain Biological Laboratory have found evidence of acid precipitation in alpine lakes, which could reduce biodiversity and impact fish and amphibian populations (Harte and Hoffman 1989).

A climate change simulation experiment conducted in Gunnison County has yielded results with potential significance for the county's economy. This experiment, in which a montane meadow has been artificially heated for several years to simulate global warming, has found that warming stimulates a shift in the vegetation community in favor of sagebrush and to the detriment of forbs (Harte and Shaw 1995). Shifts in vegetation toward sagebrush dominance in a warmer climate could impact the grazing industry in Gunnison County and throughout the West, since sagebrush is lower-quality forage for cattle than forbs. Reductions in forb abundance could also impact the tourism sector, since the county's wildflowers attract many visitors. Climate change could therefore degrade the ecosystem service offered by a mixed-vegetation community structure and force substitution of irrigated pasture for natural meadow grazing areas.

Global warming induced by anthropogenic greenhouse gas emissions could also disrupt water supplies. A study of climate change impacts on the East River basin predicts that a two-degree temperature increase would reduce the East River's average annual flows by 10 percent, while a four-degree increase would cause a 17 percent flow reduction (Nash and Gleick 1991). Another study agrees that a temperature increase would reduce flows, but suggests that precipitation increases could counterbalance this effect (McCabe and Hay 1995). Seasonal shifts in streamflow distributions would also be expected with a warmer climate, as less precipitation fell as snow and snowmelt occurred earlier in the year (Carpenter et al. 1992). Runoff retention by snow would therefore be diminished, potentially increasing spring flood peaks and decreasing late-summer baseflows. This process could impact the irrigation regimes of Gunnison County ranchers as well as decreasing instream flows for fish and recreation.

Economic Analysis of Ecosystem Services

Assessing the economic impacts of ecological degradation requires an understanding of the value of ecosystem services. Supplementing qualitative descriptions of ecosystem-economy linkages with quantifications of the linkages between the human economy of Gunnison County and ecosystem services may be useful for some purposes but presents substantial methodological challenges. Instead of attempting to quantify the services provided by Gunnison County ecosystems that we have described above, we will discuss lessons from this case study about approaches to ecosystem service valuation.

Placing a dollar value on a good or service suggests that society would pay that amount of money for it or accept that amount to forgo it. While for market commodities, cost may coincide with value under certain circumstances, determining the exchange value of nonmarket goods and services is difficult. Many economists have attempted to determine willingness to pay for nonmarket commodities using the contingent valuation method. Contingent valuation may be able to assess some aspects of the value of ecosystem services, such as those associated with instream flows, and is capable of assessing existence value, which is the value derived from the mere knowledge that an ecosystem exists (see chapter 3). However, since many important ecosystem services, such as methane consumption by soils, are essentially invisible and are poorly understood, contingent valuation surveys would be unlikely to capture substantial portions of ecosystem service values. This points up the inability of markets, or pseudomarkets as the case may be in contingent valuation surveys, to be a good instrument for allocating resources in the absence of accurate information.

One method of translating the "cost" of ecosystem service degradation into monetary terms is through assessment of substitution costs, which are expenses associated with replacement of an ecosystem good or service by technological means. Substitution costs are incurred when, for example, tree farms are planted to replace natural forests, flood control dams are built to replace natural flood control by wetlands and other mechanisms, and natural pest control is replaced with pesticides (Ehrlich and Roughgarden 1987). Substitution costs do offer some notion of the economic importance of ecosystem services and make the worth of degraded services more evident but do not convey their full value. Substitution costs of ecosystem service degradation in Gunnison County could be assessed, for example, by examining costs associated with combating whirling disease, construction and maintenance of water and wastewater treatment plants, and other impact mitigation costs.

While costs may be incurred in substituting for lost ecosystem services, the costs or expenditures that are avoided due to the availability of production inputs like flood control or water filtration are also substantial if ecosystems are healthy. Calculating avoided expenditures is thus another method of measuring ecosystem service values (chapter 3).

Input-output analysis also can be applied to measurement of the economic importance of ecosystem services and the economic ramifications of ecological degradation. Input-output tables represent economic interactions with data on production and consumption flows between industries. Databases of input-output economic data containing figures for final demand and transaction multipliers for different sectors are available for every U.S. county. Once assumptions are made or empirical relationships derived about the level of a certain sector's output that is attributable to certain ecosystem services, one can use input-output analysis to model the multiplier effects of these services on the entire economy and to build scenarios assessing the economic effects of ecosystem service degradation. Traditional input-output analysis has examined economic benefits that growth in a certain sector might create, but this technique can also be used to analyze negative impacts, such as declines in the tourism and recreation sectors caused by the degradation of water quality and aesthetics a new mine might cause.

A number of workers have applied input-output techniques to analysis of interactions between ecosystem components (Hannon 1991). Such analyses can be used to model provision of ecosystem services and human impacts. For example, Hannon suggests a method of accounting for the service of bee pollination into which human impacts on bee populations could be incorporated. Ecosystem accounting frameworks can be combined with economic input-output analyses by adding matrices for the human inputs to ecosystems, such as pollution (Pederson 1992), and matrices for ecosystem services provided to the economy, such as water supply. This would allow development of estimates of the value of ecosystem services to the economy in question, of the interindustry effects of amelioration or degradation of these services, or of the human impacts on provision of ecosystem services (Hannon 1991).

Input-output analysis contains a number of weaknesses that may limit its effectiveness. Most input-output models are based on Leontief production functions, which use "fixed coefficient" technology, assume linearity, and do not allow for substitution or economies of scale (Silberberg 1990). Although it is difficult to quantify offsetting or substitution effects, it is important to recognize their existence. The coefficients in input-output tables representing interindustry relations are average coefficients, which are easier to determine but less useful than marginal coefficients, which are better able to show the significance of changes (Michael Hanneman, pers. comm., 1995). These difficulties, which are present in standard input-output analyses, are compounded by the problem of incorporating ecosystem services into input-output. Despite the shortcomings of input-output models, they offer reasonable predictions for many purposes.

Conclusions

In Gunnison County, healthy ecosystems provide the foundation for a viable local economy. The gradual release of the county's large winter snowpack sustains the county's recreation and ranching sectors in the summer months, and the aesthetic beauty of the county's mountains and rivers attracts tourists throughout the year. Riparian and wetland areas assist in flood control, water filtration, and bank stability, in addition to providing critical wildlife habitat. In spite of real estate development, grazing, mining, and dam construction impacts, the area's ecosystems continue to purvey benefits that contribute to the well-being of county residents. Exogenous pressures, such as calls for transmountain water diversion and demand for vacation homes, threaten the continued provision of ecosystem services at their present level.

Describing scenarios of future environmental degradation in Gunnison County is not difficult, but understanding and quantifying the feedbacks of the loss of ecosystem services to the local economy is a more vexing problem, which none of the valuation methods described above adequately addresses. Although a degree of economic activity is essential to human well-being, escalation of certain types of activity to the point where ecosystem functions on which humans depend are impaired may cause deterioration of well-being. In Gunnison County, economic activity related to molybdenum mining, for example, would provide quantifiable benefits in terms of jobs and income. But such activity may also generate negative economic feedbacks through its effect on ecosystem services. Degradation of the area's aesthetic beauty, wildlife habitat, and water quality could drive away tourists and reduce the quality of life of residents. The benefits from mining molybdenum are easier to measure than the ecosystem service benefits that derive from leaving land in its natural state (chapter 3), but striking a balance between economic activity and environmental quality in a manner that maintains well-being of current and future generations depends on an appreciation of these latter benefits.

In Gunnison County and elsewhere, efforts to establish sustainable economies may be hampered by lack of awareness of the value of ecosystem services and subsequent failure to incorporate this value into land-use decisions. In the Big Cypress Swamp example cited above, public understanding of ecosystem service benefits was crucial to mobilizing support for environmental protection. Likewise, in Gunnison County, public appreciation of the largely external benefits provided by healthy ecosystems will be needed to garner support for public policies necessary to protect important habitats (chapter 3). County residents displayed this type of awareness by voting in 1990 to raise their own property taxes in order to fund litigation opposing the Union Park transmountain water diversion project.

Further efforts to protect ecosystem functions will be necessary in the face of the development pressures outlined above. Land purchases, conservation easements, and stricter planning laws will help protect against the conversion of ranches and wildlands into ranchettes or subdivisions, thereby mitigating against impacts associated with real estate development; and expanded instream-flow protection efforts will be necessary to protect the recreation sector against the increasing demand for freshwater diversions. Such efforts will need to be accompanied not only by policy decisions outside of Gunnison County to curb emissions of greenhouse gases and other pollutants and to limit population growth, but also by replacement of antiquated nineteenth-century laws governing water use and mining with laws representing modern values and understanding of environmental benefits.

Decision making that is blind to ecosystem values and to the negative externalities of development can lead to environmental degradation, negative economic feedbacks, and a decline in well-being. In contrast, policy choices incorporating an appreciation of the benefits offered by ecosystems will allow establishment of sustainable levels and types of economic activity and assure continued provision of ecosystem services. In Gunnison County, the weight given to environmental values may spell the difference between a future in which ecosystem services continue to sustain the recreation and ranching sectors and the well-being of county residents versus a future in which little water is left flowing in streams; valley bottomlands are filled with vacation homes and mountains are covered with ski slopes or mining operations; and technological substitutes are required for the services once provided by the county's forests, wetlands, and streams.

Acknowledgments

We would like to thank Lisa Micheli and Yasemin Biro for research assistance, Susan Lohr and the Rocky Mountain Biological Lab for logistical support, Tyler Martineau and Ralph Clark III for valuable discussions about Gunnison County, and our reviewers for helpful comments on an earlier draft.

References

Burger, H. 1922. "Physikalische Eigenschaften der Wald—und Freilandboeden." *Mitteilungen der Schweizerischen Centralanstalt fur das Forstliche Versuchswesen* 23(1): 1–221.

Cairns, J., Jr., and J. R. Pratt. 1995. "Ecological restoration through behavioral change." *Restoration Ecology* 3(1):51–53.

Carpenter S. R., S. G. Fisher, N. B. Grimm, and J. F. Kitchell. 1992. "Global change and freshwater ecosystems." *Annual Review of Ecology and Systematics.* 23:119–139.

Cooper, D. J. 1993. "Wetlands of the Crested Butte region: Mapping, functional evaluation, and hydrologic regime." Prepared for the Town of Crested Butte. March 1993.

County Information Service. 1992. "Gunnison County." Colorado State University Cooperative Extension Service, Fort Collins.

Ehrlich, P., and J. Roughgarden. 1987. *The Science of Ecology*. New York: Macmillan.

Fielder, J., and M. Pearson. 1995. *Colorado: Rivers of the Rockies*. Englewood, CO: Westcliffe Publishers.

Germann, P. F. 1990. "Macropores and hydrologic hillslope processes." In M. G. Anderson and T. P. Burt, eds., *Process Studies in Hillslope Hydrology*. New York: John Wiley & Sons, pp. 327–363.

Gregory, S. V., F. J. Swanson, W. A. McKee, and K. W. Cummins. 1991. "An ecosystem perspective of riparian zones." *BioScience* 41(8): 540–551.

Hall, D. 1995. "Area snowpack at 415% of normal." Crested Butte Chronicle & Pilot. May 26, 1995, p. 1.

Hannon, B. 1991. "Accounting in ecological systems." In R. Costanza, ed., *Ecological Economics: The Science and Management of Sustainability*. New York, Columbia University Press, pp. 234–252.

Harte, J., and E. Hoffman. 1989. "Possible effects of acidic deposition on a Rocky

Mountain population of the tiger salamander *Ambystoma tigrinum*." *Conservation Biology* 3(2): 149–158.

Harte, J., and R. Shaw. 1995. "Shifting dominance within a montane vegetation community: Results of a climate-warming experiment." *Science* 267:876–880.

Harte, J., and R. Socolow. 1971. *Patient Earth*, pp. 181–202. New York: Holt, Rinehart, Winston.

HDR Engineering, Inc. 1989. *Final Report: Upper Gunnison-Uncompaghre Basin Phase I Feasibility Study*, 2 vols. Prepared for Colorado Water Resources and Power Development Authority. Denver, CO.

McCabe, G. J., and L. E. Hay. 1995. "Hydrological effects of hypothetical climate change in the East River basin, Colorado, USA." *Hydrological Sciences* 40(3): 303–318.

McKean, J. R., D. M. Johnson, and R. G. Walsh. 1988. "Gunnison County Interindustry Spending and Employment Attributed to Fishing at Blue Mesa Reservoir." Colorado Water Resources Research Institute. Technical Report No. 53. December 1988. Fort Collins.

Mears, A. I. 1995. "Snow Avalanche Technical Report: Environmental Impact Statement Considerations, Crested Butte Mountain Resort." Prepared for Pioneer Environmental Services. Gunnison, CO.

Nash, L. L., and P. H. Gleick. 1991. "Sensitivity of streamflow in the Colorado basin to climatic changes." *Journal of Hydrology* 125: 221–241.

Obmascik, M. 1995. "Land in Crested Butte for $5 an Acre." *Crested Butte Chronicle & Pilot* 35(32):6.

Peckarsky, B. L, and K. Z. Cook. 1981. "Effect of Keystone Mine effluent on colonization of stream benthos." Environmental Entomology 10(6): 864-871.

Pederson, O. G. 1992. "An input-output satellite model for Danish CO_2, SO_2, and NO_2 emissions." Paper for the Israeli-Danish Binational Symposium: Environmental Resources in National Income Accounting. Ein-Gedi, Israel. Nov. 29–Dec. 4, 1992.

Pinnes, E. 1995. "Dry year options for water supply." Presented at 1995 Water Workshop, Gunnison, CO. Aug. 2–4, 1995.

RCE (Resource Consultants and Engineers). 1995. "Geologic hazard assessment and mitigation planning for Crested Butte Mountain Resort, Gunnison County, Colorado." Prepared for Pioneer Environmental Services. Fort Collins, CO.

Ring, R. 1995. "The West's fisheries spin out of control." *High Country News* 27(17): 1, 10–13.

Silberberg, E. 1990. *The Structure of Economics: A Mathematical Analysis*, 2nd edition. New York: McGraw Hill.

Sparling, E. W., D. A. Harpman and J. Booker. 1994. "Final Report: Upper Gunnison Basin In-stream Flow Project." Department of Agriculture and Resource Economics, Colorado State University. Fort Collins, CO. September 20.

Stanford, J., and J. V. Ward. 1983. "The effects of mainstream dams on physicochemistry of the Gunnison River, Colorado." In V.D. Adams and V.A. Lamarra, eds., *Aquatic Resources Management of the Colorado River Ecosystem,* pp. 43–56. Ann Arbor Science, Ann Arbor, MI.

Stanford, J., and J.V. Ward. 1984. "The effects of regulation on the limnology of the Gunnison River: A North American case study." In A. Lillehammer and S.J. Saltveit, eds. *Regulated Rivers* pp.467–480, Oslo University Press, Norway.

Stanford, J. 1994. "Instream Flows to Assist the Recovery of Endangered Fishes of the Upper Colorado River Basin." Washington, D.C.: U.S. Department of Interior, National Biological Survey. Biological Report 24 (July).

Stanford, J. 1989. "Flow alteration and ecosystem stability in the Gunnison River, Colorado: A perspective." March 6, 1989. Unpublished.

Theobald, D., and W. Riebsame. 1995. "Land use change on the Rocky Mountain forest fringe." Unpublished.

Theobald, D., H. Gosnell, and W. E. Riebsame. 1995. "Land use and cover change in the Rocky Mountains II: Case study of the East River Valley, Colorado." *Mountain Research and Development,* in review.

Torn, M., and J. Harte. 1996. "Methane consumption by montane soils: Implications for positive and negative feedback with climatic change." *Biogeochemistry* 32(1):53–67.

U.S. Bureau of Reclamation (USBR). 1995. "Preliminary Draft, East River Water Supply and Water Quality Study." Task memorandum no. 2 (July) 1995.

U.S. Forest Service (USFS). 1991. "Amended Land and Resource Management Plan and Final Supplemental EIS. Grand Mesa, Uncompaghre, and Gunnison National Forests." U.S. Department of Agriculture, U.S. Forest Service, Delta, CO.

U.S. Geological Survey. 1994. "Water Resources Data: Colorado. Water Year 1994. Vol. 2. Colorado River Basin." *U.S. Geological Survey Water-Data Report* CO-94-2.

Vandenbusche, D. 1980. *The Gunnison Country,* 1st edition. Vandenbusche, Gunnison, CO.

WRC Engineering, Inc. 1991. "Water Availability Report for the Union Park Reservoir Project Water Rights. Case No. 86 CW 226, Case No. 88 CW 178." Submitted to Arapahoe County.

Younger (People ex rel.) v. County of El Dorado. 1971. California Supreme Court. 5 C.3d 480; 96 Cal. Rptr. 553, 487 P.2d 1193.

Chapter 18

WATER QUALITY IMPROVEMENT BY WETLANDS

Katherine C. Ewel

During the last twenty-five years, recognition of the services that wetlands provide to humanity has increased significantly (table 18.1), and we now know that wetlands are useful in both direct and indirect ways. We have also come to understand, however, that not every wetland can provide all the services listed in table 18.1, and that exploiting a wetland for one service may compromise its ability to perform another. This chapter describes the diversity of wetlands that exist and the services they provide, and then focuses on the conflicts engendered by exploiting one of these services: water quality improvement.

Four major groups of wetlands are recognized (Brinson 1993). *Fringe wetlands* include salt marshes and lakeside marshes in which water typically flows in two opposite directions, influenced by lunar and/or storm tides. In *riverine wetlands*, which occupy floodplains, water generally flows in one direction. *Depressional wetlands*, such as prairie potholes in the north-central United States, usually receive much of their water from runoff and/or groundwater seepage rather than from surface water bodies, so that water residence times are much longer. *Extensive peatlands* also have long water residence times, but the accumulated peat creates a unique hydrologic regime that differs from the previous three types of wetlands. Geomorphic setting, water source, and hydrodynamics generate considerable variation within each of these major categories.

Although wetlands account for only a small portion of the earth's surface, they are often concentrated in a particular area, where they dominate the

Table 18.1. Importance of wetlands in providing ecosystem services

Biodiversity: Sustenance of Plant and Animal Life
- Evolution of unique species
- Production of harvested wildlife:
  - Water birds, especially waterfowl
  - Fur-bearing mammals (e.g., muskrats)
  - Reptiles (e.g., alligators)
  - Fish and shellfish
- Production of wildlife for nonexploitative recreation
- Production of wood and other fibers

Water Resources: Provision of Production Inputs
- Water quality improvement
- Flood mitigation and abatement
- Water conservation

Global Biogeochemical Cycles: Provision of Existence Values
- Carbon accumulation
- Methane production
- Denitrification
- Sulfur reduction

landscape. Most of the wetlands in North America north of Mexico are in Canada and Alaska (table 18.2), where most are freshwater bogs (extensive peatlands). In the forty-eight contiguous states in the United States, nearly two-thirds of the wetlands are in the South, where both depressional and riverine wetlands are common.

None of the values listed in table 18.1 is specific to any single wetland type or region, and each of the major types of wetlands provides more than one main class of services. For instance, besides providing production inputs through flood control, river swamps (riverine wetlands) also contain very diverse assemblages of plants and animals that can be harvested (such as bottomland hardwood trees and fish) or observed (such as birds viewed from a canoe or a boardwalk at a nature center). Because of the multiplicity of ways in which an individual wetland can be valued, exploiting it for one value may infringe on other values, compromising our ability to realize its full worth. It is therefore useful to explore the consequences of such exploitation, comparing an ecosystem's service with the service that technology can provide, and considering also the effect of that exploitation on other services and even other ecosystems.

Water quality improvement is a service that is widely attributed to wetlands, which have absorbed and recycled nutrients from human settlements

Table 18.2. Distribution of current wetland area in United States and Canada

	Wetland Area (1,000 ha)	Percentage of Area in Wetlands	Percentage of Nation's Wetland Area
Canada	127.2	14	
Manitoba	22.5	35	18
Ontario	29.2	31	23
United States	111.1	12	
Alaska	70.7	43	63
Florida	4.5	29	4
Louisiana	3.6	28	3
Minnesota	3.5	16	3
Texas	3.1	4	3

Sources: Much of this information is summarized in Mitsch and Gosselink (1993); some data were obtained from Dahl et al. (1991) and Hall et al. (1994).

since the dawn of civilization. However, recognition of this service in the United States did not come until the 1970s, after increasing environmental awareness mandated reversing eutrophication of the nation's waterways and led to the Clean Water Act in 1972. At this time, demands for increased funding for wastewater treatment collided with escalating fossil fuel costs caused by the energy crisis. The expense of providing additional wastewater treatment therefore soared beyond the reach of many communities, in spite of the best of intentions. The search for alternate, low-energy ways of treating wastewater was already underway; golf courses, lawns, and even forests were being irrigated experimentally with wastewater (e.g., Sopper and Kardos 1973). Public health concerns and unwanted changes in vegetation prevented widespread adoption of any of these alternatives. By the mid-1970s, the concept of discharging wastewater to wetlands, in which soils, vegetation, and animals were already adapted to flooding, seemed attractive.

Wastewater is treated to remove suspended solids, dissolved nutrients, pathogens, and other toxic compounds such as heavy metals. Three stages of treatment were developed over a period of several decades. Primary treatment was a physical process, usually including screening and settling, that removed most of the suspended solids and, along with them, most of the associated nutrients. Adding the secondary level of treatment reduced most remaining suspended solids, biological oxygen demand, and nitrogen and phosphorus (sometimes by as much as 90 percent), mainly through biological processes. Both primary and secondary treatment included some mechanism of disinfection such as chlorination. In the United States and many other developed countries, secondary wastewater treatment was standard by the 1970s. Removing the last pollutants and effecting more thorough removal of bacteria, viruses, and other toxins through tertiary sewage treatment (or more likely an advanced wastewater treatment process that combines primary, secondary, and tertiary treatment) can more than double the cost. Twenty years ago, the concept of allowing wetlands to provide this tertiary treatment, enabling communities to install or maintain secondary treatment plants but upgrade the effluent at an affordable cost, was very appealing.

Using Natural Wetlands for Wastewater Recycling

Deciding whether and how to use a wetland for wastewater discharge requires that several issues be addressed:

1. *Availability of an appropriate wetland.* What wetlands are available, and can they provide this service?
2. *Input to the wetland.* What are the quality, quantity, and rate of delivery of wastewater available (or desired) for discharge into the wetland?
3. *Effect on the wetland.* How will wastewater discharge change species composition, growth rates, and demographic characteristics of plants, animals, and microbes in a wetland? How will these changes affect the wetland's long-term ability to sustain treatment?
4. *Costs.* How do the costs of pretreatment, transportation, and other necessary features compare with those of more conventional methods?
5. *Effects of wastewater discharge on wetland landscapes.* How will discharging wastewater into a wetland affect communities and ecosystems in the surrounding landscape? What are the risks for human society?

Availability of an Appropriate Wetland

Wastewater treatment systems that use depressional wetlands, riverine wetlands, fringing wetlands, and extensive peatlands have been put into service during the last two decades; several hundred wetlands in the United States and Canada are now used for wastewater treatment, many of them in the Southeast and in the states and provinces surrounding the Great Lakes (Knight et al. 1993). Because the descriptions of many of these projects are anecdotal, this chapter draws extensively on results from large-scale, long-term research projects based on two kinds of wetlands that seemed most promising in the 1970s for wastewater disposal. One was based on an extensive peatland near Houghton Lake, Michigan (Tilton and Kadlec 1979), and the other on depressional forested wetlands dominated by pondcypress (*Taxodium distichum* var. *nutans*) near Gainesville, Florida (Ewel and Odum 1984).

Vast peatlands occupying ancient lakebeds and scoured-out basins in northern glaciated landscapes are remote from cities but are often found near small communities and vacation resorts. The extensive peat deposits (often several meters deep), sealed basins, and remoteness offered a substantial nutrient storage reservoir isolated from human contact. The small cypress swamps are isolated, stillwater wetlands in which standing water fluctuates dramatically during most years, seeping slowly through 1–3 m of peat and then sand before reaching clay (Spangler 1984). Although these cypress ponds are usually 1–10 ha, much smaller than a northern peatland, they are found throughout the southeastern United States coastal plain, often in high densities, and they may be common near towns and cities.

Input to the Wetland

Discharge to wetlands in the United States is regulated by the Environmental Protection Agency through the National Pollutant Discharge Elimination System permit program of the Clean Water Act. The major criteria are effluent quality, limits to acceptable changes within the treatment wetland, and adherence to water quality standards for discharge from wetland to downstream water bodies. These standards were intended to allow natural wetlands to be used without sacrificing their unique values and functions or the water quality of receiving waters (Bastian et al. 1989). The state of Florida is the only state to have chemical and biological standards specifically for wetlands used for wastewater treatment (Kadlec and Knight 1996). At both state and federal levels, permits are issued on a case-by-case basis.

Full-scale wastewater discharge to the extensive peatland near Houghton Lake, Michigan, has averaged 485,000 m^3/season (serving a seasonal community of approximately five thousand people) during a study that began in the mid-1970s and continues to be monitored after two decades of full-scale operation (EPA 1993). The wastewater is distributed from May through September along a gated irrigation pipe. Water ponds above the discharge and then follows a very shallow gradient to the stream that drains the peatland 3 km away.

In cypress ponds and other depressional wetlands, water seeps downward, eventually forming a groundwater mound so that most water infiltrates around the perimeter of the pond (Heimburg 1984). Heavy rainfall in some seasons may limit the quantity of water that can be discharged. In the Florida study, wastewater from a small treatment plant in a mobile home community of approximately 250 people was added through a single pipe to the center of a cypress pond at the rate of approximately 39 m^3/day (2.8 cm/wk) for seven years. Loading rates used in both the peatland and the cypress swamp study are now considered conservative; 1–2 cm/day is currently recommended for natural wetlands (Knight 1990).

Natural wetlands are not likely to be used for wastewater from industrial communities, because treatment of heavy metals is neither easily controlled nor predictable. Some heavy metals are taken up readily by peat or mineral soils, but they may not be retained, particularly if pH and other water quality measures change with wastewater application (Giblin 1985). A wastewater treatment plant can tailor processes to specific pollutants, but a natural wetland cannot be manipulated so finely.

Effect on the Wetland

Vegetation in a natural wetland changes dramatically after wastewater discharge begins. Changes in vegetation in the Michigan peatland marked the advance of high concentrations of dissolved phosphorus and nitrogen as exchange sites in the soil became saturated (Kadlec 1987). In the vicinity of the discharge, initial increases in biomass of sedges (*Carex* sp.) and grasses (particularly *Calamagrostis canadensis*) were followed by development of thick carpets of duckweed (*Lemna* sp.) and stands of cattails (*Typha* sp.). This visually detectable change in species composition developed over several years and now characterizes 70 ha of the 600-ha peatland (Robert Kadlec, pers. comm., 1995). In addition, many of the trees outside the area of visual impact have died because of changes in the hydrologic regime. Because of the large size of this peatland, effects of nutrient enrichment are contained within it.

Over a fourteen-year period, nutrient retention has averaged 96 percent for dissolved inorganic nitrogen and 97 percent for total phosphorus in this wetland (EPA 1993). Insect, bird, and mammal populations have changed, but the decreases in numbers and diversity of vertebrates recorded from other sites with long histories of discharge have not occurred (Kadlec 1987). Both muskrats and waterfowl are now more common because of sustained flooding (EPA 1993).

Tree growth rate nearly doubled in the Florida cypress ponds (Brown and van Peer 1989); this response has been sustained for more than two decades in other swamps receiving wastewater (e.g., Nessel et al. 1982). A layer of duckweed (primarily *Lemna* spp., *Spirodella oligorhiza*, and *Azolla caroliniensis*) also developed, and both invertebrate and vertebrate fauna changed as well (Brightman 1984, Harris and Vickers 1984).

Throughout a five-year monitoring period, concentrations of organics, nutrients, and minerals remained at background levels in shallow wells around the perimeter of the experimental cypress pond (Dierberg and Brezonik 1984a). In spite of the growth response by cypress trees, soils are clearly the main nutrient storage unit in this ecosystem (Dierberg and Brezonik 1984b). Laboratory column leaching studies suggested that the level of removal observed in the experimental ponds could continue for at least twenty years (Dierberg and Brezonik 1984a).

Uptake by organic matter followed by accumulation in sediments and deposition with iron and aluminum seem to be the major ways in which phosphorus is retained in a wetland (Richardson 1985, Cooke 1992, Cooke et al. 1992). Although peat is generally perceived as having substantial storage capacity for nutrients, it is in fact better at transforming nutrients than retaining them, and annual retention capacity is not particularly high (Richardson and Davis 1987). Instead, wetlands flooded with wastewater accumulate new sediments; some of these sediments become a fine microdetritus and remain as suspended solids, some of the organic phosphorus may be mineralized, and some phosphorus-rich sediments are buried (Kadlec 1995). Accumulation of these new sediments alters the hydrology of a basin, particularly in wetlands where water flows laterally above or below the surface rather than percolating downward, but the continual inflow of water and the tendency for microdetritus to be pushed away from the discharge site prevent the basin from "filling in."

Nitrogen removal is most expediently accomplished by denitrification, which can average 75 percent of inputs (Richardson and Davis 1987) and can continue with no time limits. Denitrification in the Florida cypress ponds accounted for only 14 percent of nitrogen inputs because of low concentrations of nitrate and nitrite in the wastewater; vegetation took up a similar amount, and the rest remained in the sediments (Dierberg and Brezonik 1984b). Denitrification rates can be increased by controlling the concentration of nitrate in the discharge from the treatment plant. One difficulty in reducing the small remaining concentrations of phosphorus and nitrogen in this tertiary stage of treatment is the need to retain a ratio between the two nutrients in order to prevent nitrogen fixation from occurring. An interesting characteristic of the mat of floating vegetation in the cypress ponds receiving wastewater was an occasional bloom of *Azolla carolinensis*, which is a source of nitrogen fixation.

Costs of Using Wetlands for Wastewater Treatment

Natural wetlands have been adopted for tertiary wastewater treatment primarily by small communities (e.g., table 18.3). Capital costs include the secondary treatment facility, land, and irrigation pipe; operating costs include pumping and maintenance costs in addition to monitoring expenses. A cost analysis of wetland treatment using cypress ponds included operating and capital costs except for the secondary sewage treatment facility (Fritz et al. 1984). Using cypress ponds cost $0.14/m³ compared with $0.28/m³ for advanced wastewater treatment; this differential disappeared with higher land costs. Costs also rose with the length of pipe needed and with wetland area, particularly as the number of individual wetlands increased to meet a community's needs. These relationships suggest that the use of nearby wetlands is feasible for a community that is too small to be able to afford advanced wastewater treatment or too remote to be able to lay and maintain a pipeline to a regional facility.

Table 18.3. Major features of wastewater treatment systems dependent on natural and constructed wetlands

Location	Population	Inflow to Wetland (m^3/day)	Capital Costs	Annual Operating Costs	
				Total	Per m^3
Natural					
Houghton Lake, Michigan	5,000 (May–September)	9,800 (1978)	$397,900	$15,300	10¢
Cannon Beach, Oregon	1,200–4,000 (seasonal)	2,600	$1.5 million (1983)	$72,000	8¢
Constructed					
Martinez, California	16,000	5,000	$300,000	$40,000	2¢
Vermontville, Michigan	825	300	$395,000 (1972)	$4,200	4¢

Source: EPA (1993).

The wastewater treatment facilities described in table 18.3 have found additional use in environmental education for elementary schools through graduate and professional training programs. Willingness to pay for the services provided by those wetlands is surely increased by the educational benefits derived, although these benefits may be unique to a particular wetland and the community it serves, and they may not be appreciated until after the facility has been put into operation. This additional service may counterbalance any marginal loss of value perceived by the community when a wetland is first adopted for wastewater treatment.

Using natural wetlands for wastewater treatment is becoming more difficult, however, and the utilitarian approach of a cost-benefit analysis is becoming less satisfactory. Wetlands are no longer considered wastelands, and their contributions to regional biodiversity, such as through the habitat they provide for endangered species, are less and less likely to be treated as externalities. Their non-use value has also increased, as Goulder and Kennedy (chapter 3) point out; many people prefer having them left untouched. The success of wetlands in achieving recognition at last for the many services they provide is making the service of water quality improvement more difficult to exploit.

Effects of Wastewater Discharge on Wetland Landscapes

The effects of wastewater discharge beyond the boundaries of a wetland have not yet been adequately evaluated. Diseases associated with human wastes, such as cholera, were common near wetlands in early years (see discussion in Purseglove 1988 of the correlation between wetlands and diseases in early England), suggesting that treatment capacities were frequently exceeded. Removal of pathogens now depends on both chlorination and long detention times, particularly for bacteria (Gerba et al. 1975). The experimental cypress ponds proved effective in containing coliform bacteria (Fox et al. 1984), but viruses were detected in groundwater outside the ponds (Wellings et al. 1975).

As human populations grow, contamination of surface water bodies and groundwater supplies may increase, exceeding public health standards. Although the threat is probably no greater than exists with many standard wastewater treatment plants, the unique nature of each wetland that treats wastewater focuses attention on it.

A wetland used for wastewater treatment can also affect other ecosystems through animal vectors. Arboviruses circulate naturally in Florida swamps, and there was concern that migrating birds, which began using the experimental cypress swamp more intensively, would spread them. This concern was unwarranted, because peak virus activity and peak migration times did not coincide (Davis 1984). However, this fortuitous asynchrony may not occur at other latitudes, and birds or mammals that use wastewater-treatment wetlands may be exposed to pathogens or trace elements that are not normally found in undisturbed wetlands. For instance, wading birds that catch fish from polluted canals and ditches in Florida may infect nestlings with a nematode that can cause high mortality (Spalding et al. 1993); wastewater-treatment wetlands seem to attract great egrets and snowy egrets especially (Frederick and McGehee 1994). Outbreaks of avian botulism and high concentrations of trace elements have been recorded in birds that use high-nutrient wetlands formed after phosphate mining (summarized by Marion 1989), suggesting that similar outbreaks associated with wastewater-treatment wetlands are possible. No such events have been documented, but there have been no studies conducted to determine explicitly whether use of wastewater-treatment wetlands affects reproduction and mortality of birds and other terrestrial vertebrates.

Another risk incurred when a natural wetland is used for water-quality improvement is the diminution or loss of other services (Ewel 1990). The wastewater-treatment wetlands described above have experienced substantial changes not only in species composition but in functional relationships as well. Wetland plants that take up excess nutrients, particularly in a previously low-nutrient wetland, will generate more labile organic matter; this in turn alters the substrate for microbes and other detritivores and disrupts relationships among interdependent organisms in the wetland (Wetzel 1993).

Impacts of changes in a wetland can therefore be propagated well beyond its boundaries. When a wastewater-enriched wetland discharges into a downstream ecosystem, such as another wetland, changes in hydroperiod as well as water quality may eventually have a serious impact. Two important lessons have been learned from long-term observations of greentree reservoirs, which are bottomland hardwoods in the southern United States that were impounded during the dormant season to increase wintering waterfowl populations (summarized by Guntenspergen et al. 1993). First, in spite of careful plans and the best of intentions, manipulations of natural ecosystems may be too subtle to monitor effectively, allowing substantial degradation to occur before change is finally detected; and second, because vegetation changes can be caused by changes in timing and degree of water level fluctuation, small changes in water quality may also have significant cumulative effects.

Ecological risk factors are often interdependent and hence far-reaching (chapter 3), and some wetlands are at greater risk than others for affecting other ecosystems. Because of the extensive contiguity between fringe wetlands and open bodies of water, together with the usually high frequency of flooding, few coastal marshes have been considered for wastewater discharge, even though short-term studies suggest that these wetlands may serve as sinks for both nitrogen and phosphorus (e.g., Simpson et al. 1978, Valiela et al. 1985). Similarly, riverine wetlands may have short-term nutrient removal capabilities (e.g., Brinson et al. 1984) but not offer sufficient protection of adjacent water bodies, particularly during flooding.

Riverine wetlands are now being used to provide a water-quality improvement service that probably contains fewer risks than tertiary wastewater treatment: buffering rivers from non-point-source pollution by fertilizers and other agricultural chemicals (e.g., Peterjohn and Correll 1984, Cooper and Gilliam 1987). Although the threats to public health appear to be fewer, measures of success in these studies so far have been restricted to removal of phosphorus and nitrogen from runoff and groundwater flow from the fields. Changes in understory species composition and foliage nutrient concentrations, and therefore attractiveness to animal populations, have not been reported. Studies of the potential for using Everglades peatlands to take up the nutrients released by mineralization of organic cultivated soils show that changes in nutrient concentrations and species composition of macrophytes will occur and could have significant impacts on aquatic and terrestrial vertebrates (Craft et al. 1995).

Solution: Constructed Wetlands

Long before final research results evaluating the use of natural wetlands for wastewater discharge were available, design and construction of wetlands to meet this need had begun. The concept of constructing wetlands specifically for wastewater treatment is now advocated for expediency (Reed et al. 1995) and cost (table 18.3). Wetlands are constructed such that water flows primarily over the sediment and through vegetation, or as vegetated submerged bed systems in which water flow is engineered for contact with plant roots (see Knight 1990 for more detailed descriptions). These wetlands are excavated with a shallow gradient in soils of low permeability (or lined with an impermeable barrier and then filled with an appropriate soil). They are then either planted or allowed to be vegetated naturally. They usually comprise several cells that can operate in series or parallel, allowing flows to be redistributed for greater control and easier maintenance.

Nutrient retention and processing features that are characteristic of natural wetlands can be exploited in constructing wetlands (Wetzel 1993). For instance, macrophytes can be kept in a rapid growth phase by intentional, programmed disturbances. Maintaining at least a moderate species diversity makes the system more responsive to variations in loading rates of different nutrients. Anaerobic conditions and large areas of vegetation-free sediment surface can maximize retention of both organic matter and nutrients. Manipulation of water turnover time and addition of other electron acceptors besides oxygen can also be better accomplished in constructed wetlands. Linking constructed wetlands (where flows and vegetation can be controlled) with natural wetlands (for supplying soluble iron and aluminum needed for phosphorus removal) may be a mechanism for taking advantage of both strategies when they are available (e.g., Cooke et al. 1992).

Constructed wetlands can serve the same small communities as natural wetlands and can probably be incorporated into treatment systems for larger communities as well. They may be especially well suited to degraded wetlands, surface-mined areas, and borrow pits, taking advantage of both low land costs and freedom from the need to consider conflicting wetland values. Costs of the constructed wetland in Martinez, California, over an eighteen-year period were estimated to be less than one-third the cost of having wastewater treated in a neighboring deep-water diffuser (EPA 1993).

Wetlands can also be constructed to treat agricultural runoff or other nonpoint sources of pollution. A 4-ha wetland can cost as little as $300,000 or more than $1 million, depending on the quality of the liner used and the need for gravel (Reed et al. 1995). This is an added cost for an agricultural operation and wasn't needed when fewer fertilizers and other chemicals were used and when fields were more distant from water bodies. It also provides an estimate of the avoided costs of the use of natural wetlands for filtering agricultural runoff (see chapter 3).

Gren (1995) demonstrated that restoring wetlands to reduce nitrogen loading to the Stockholm, Sweden, archipelago is considerably less expensive than construction of wastewater treatment plants. Ancillary benefits, such as nitrogen and saltwater filtering, supply of water and nutrients, production of food, and support of endangered species, may increase threefold the economic advantage over construction of wastewater treatment plants, although it is not clear that all can be realized simultaneously.

The faster turnover rate of constructed wetlands, as expressed in both detention time and hydraulic loading rate, together with their grassy or herbaceous vegetation, make them similar to fens, which are peatlands that are less acidic and more productive than bogs. Forcing water to flow laterally rather than percolating into the groundwater provides greater control over the processes and allows cells to be used or bypassed as needed, and also provides easily accessed monitoring points. Constructed wetlands are therefore based more on the extensive peatland model than on depressional wetlands, for both ecological and engineering reasons.

Conclusions

Many natural wetlands may indeed be able to provide a service to humanity by filtering nutrients and improving the quality of wastewater discharged into them, thereby protecting downstream water bodies from pollution at less cost than an advanced wastewater facility would entail. However, impacts on the wetlands themselves may not be inconsequential, and increases in the perceived value of other wetland services in recent years make exploiting the wastewater treatment service to the detriment of others less defensible. Constructing wetlands *de novo* for wastewater treatment appears to get around this conundrum. The most enduring service that wetlands provide may therefore be the lesson of how to combine engineering and ecological knowledge to accommodate human population growth. As the demands of population growth that are being placed on our ecosystems conflict with each other more frequently, conscious decisions must be made on how to assess the risk for obtaining an ecosystem service.

Acknowledgments

I thank Kathleen Stearns Friday and Robert L. Knight for thoughtful comments on the manuscript. Robert L. Knight and Robert H. Kadlec generously provided information from unpublished sources.

References

Bastian, R. K., P. E. Shanaghan, and B. P. Thompson. 1989. "Use of wetlands for municipal wastewater treatment and disposal—Regulatory issues and EPA policies." In D. A. Hammer, ed., *Constructed Wetlands for Wastewater Treatment*, pp. 265–278. Lewis Publishers, Chelsea, Mich.

Brightman, R. S. 1984. "Benthic macroinvertebrate response to secondarily treated wastewater in north-central Florida cypress domes." In K. C. Ewel and H. T. Odum, eds., *Cypress Swamps*, pp. 186–196. University Press of Florida, Gainesville.

Brinson, M. M. 1993. *A Hydrogeomorphic Classification for Wetlands*. Wetlands Research Program Technical Report WRP-DE-4. U.S. Army Corps of Engineers Waterways Experiment Station, Vicksburg, Miss.

Brinson, M. M., H. D. Bradshaw, and E. S. Kane. 1984. "Nutrient assimilative capacity of an alluvial floodplain swamp." *Journal of Applied Ecology* 21: 1041–1058.

Brown, S., and R. van Peer. 1989. "Response of pondcypress growth rates to sewage effluent application." *Wetlands Ecology and Management* 1: 13–20.

Cooke, J. G. 1992. "Phosphorus removal processes in a wetland after a decade of receiving a sewage effluent." *Journal of Environmental Quality* 21: 733–739.

Cooke, J. G., L. Stub, and N. Mora. 1992. "Fractionation of phosphorus in the sediment of a wetland after a decade of receiving sewage effluent." *Journal of Environmental Quality* 21: 726–732.

Cooper, J. R., and J. W. Gilliam. 1987. "Phosphorus redistribution from cultivated fields into riparian areas." *Journal of the Soil Science Society of America* 51: 1600–1604.

Craft, C. B., J. Vymazal, and C. J. Richardson. 1995. "Response of Everglades plant communities to nitrogen and phosphorus additions." *Wetlands* 15: 258–271.

Dahl, T. E., C. E. Johnson, and W. E. Frayer, eds. 1991. *Status and Trends of Wetlands in the Conterminous United States Mid-1970's to Mid-1980's*. U.S. Department of the Interior, Fish and Wildlife Service, Washington, D.C.

Davis, H. 1984. "Mosquito populations and arbovirus activity in cypress domes." In K. C. Ewel and H. T. Odum, eds., *Cypress Swamps*, pp. 210–215. University Press of Florida, Gainesville.

Dierberg, F. E., and P. L. Brezonik. 1984a. "Effect of wastewater on water quality of cypress domes." In K. C. Ewel and H. T. Odum, eds., *Cypress Swamps*, pp. 83–101. University Presses of Florida, Gainesville.

Dierberg, F. E. and P. L. Brezonik. 1984b. "Nitrogen and phosphorus mass balances in a cypress dome receiving wastewater." In K. C. Ewel and H. T. Odum, eds., *Cypress Swamps*, pp. 112–118. University Press of Florida, Gainesville.

EPA (Environmental Protection Agency). 1993. *Constructed Wetlands for Wastewater Treatment and Wildlife Habitat*. U.S. Environmental Protection Agency, EPA832-R-93-005, Washington, D.C.

Ewel, K. C. 1990. "Multiple demands on wetlands." *BioScience* 40: 660–666.

Ewel, K. C., and H. T. Odum, eds. 1984. *Cypress Swamps*. University Press of Florida, Gainesville.

Fox, J. L., D. E. Price, and J. Allinson. 1984. "Distribution of fecal coliform bacteria in and around experimental cypress domes." In K. C. Ewel and H. T. Odum, eds., *Cypress Swamps*. University Press of Florida, Gainesville.

Frederick, P. C., and S. M. McGehee. 1994. "Wading bird use of wastewater treatment wetlands in central Florida, United States." *Colonial Waterbirds* 17: 50–59.

Fritz, W. R., S. C. Helle, and J. W. Ordway. 1984. "The cost of cypress wetland treatment." In K. C. Ewel and H. T. Odum, eds., *Cypress Swamps*, pp.239–248. University Press of Florida, Gainesville.

Gerba, C. P., C. Wallis, and J. L. Melnick. 1975. "Fate of wastewater bacteria and viruses in soil." *Journal of the Irrigation and Drainage Division* (American Society of Civil Engineers) 157: 11572.

Giblin, A. E. 1985. "Comparison of the processing of elements by ecosystems. II. Metals." In P. J. Godfrey, E. R. Kaynor, S. Pelczarski, and J. Benforado, eds., *Ecological Considerations in Wetlands Treatment of Municipal Wastewaters*, pp. 158–177. Van Nostrand Reinhold, New York.

Gren, I.-M. 1995. "Costs and benefits of restoring wetlands: Two Swedish case studies." *Ecological Engineering* 4: 153–162.

Guntenspergen, G. R., J. R. Keough, and J. Allen. 1993. "Wetland systems and their response to management." In G. A. Moshiri, ed., *Constructed Wetlands for Water Quality Improvement*, pp. 383–390. CRC Press, Boca Raton, Fla.

Hall, J. V., W. E. Frayer, and B. O. Wilen. 1994. *Status of Alaska Wetlands*. U.S. Fish and Wildlife Service, Alaska Region, Anchorage, Alaska.

Harris, L. D. and C. R. Vickers. 1984. "Some faunal community characteristics of cypress ponds and the changes induced by perturbations." In K. C. Ewel and H. T. Odum, eds., *Cypress Swamps*. University Press of Florida, Gainesville.

Heimburg, K. 1984. "Hydrology of north-central Florida cypress domes." In K. C. Ewel and H. T. Odum eds., *Cypress Swamps*. University Press of Florida, Gainesville.

Kadlec, R. H. 1987. "Northern natural wetland water treatment systems." In K. R. Reddy and W. H. Smith, eds., *Aquatic Plants for Water Treatment and Resource Recovery*. Magnolia Publishing, Orlando, Fla.

Kadlec, R. H. 1995. "Overview: Surface flow constructed wetlands." *Water Sciences and Technology* In press.

Kadlec, R. H., and R. L. Knight. 1996. *Treatment Wetlands*. Lewis Publishers, Boca Raton, Fla.

Knight, R. L. 1990. "Wetland systems." In *Natural Systems for Wastewater Treatment*, pp. 211–260. Manual of Practice FD-16. Water Pollution Control Federation, Alexandria, Va.

Knight, R. L., R. W. Ruble, R. H. Kadlec, and S. Reed. 1993. "Wetlands for wastewater treatment: Performance database." In G. A. Moshiri, ed., *Constructed Wetlands for Water Quality Improvement*, pp. 35–58. CRC Press, Boca Raton, Fla.

Marion, W. R. 1989. "Values of phosphate-mined wetlands for birds: A balanced approach." In R. R. Sharitz and J. W. Gibbons, eds. *Forested Wetlands and Wildlife*. CONF-8603101, Department of Energy Symposium Series No. 61, U.S. Department of Energy Office of Scientific and Technical Information, Oak Ridge, Tenn.

Mitsch, W. J., and J. G. Gosselink. 1993. *Wetlands*. 2nd edition. Van Nostrand Reinhold, New York.

Nessel, J. K., K. C. Ewel, and M.S. Burnett. 1982. "Wastewater enrichment increases mature pondcypress growth rates." *Forest Science* 28: 414–417.

Peterjohn, W. T., and D. L. Correll. 1984. "Nutrient dynamics in an agricultural watershed: Observations on the role of a riparian forest." *Ecology* 65: 1466–1475.

Purseglove, J. 1988. *Taming the Flood*. Oxford University Press, New York.

Reed, S. C., R. W. Crites, and E. J. Middlebrooks. 1995. *Natural Systems for Waste Management and Treatment*, 2nd edition. McGraw-Hill, Inc., New York.

Richardson, C. J. 1985. "Mechanisms controlling phosphorus retention capacity in freshwater wetlands." *Science* 228: 1424-1427.

Richardson, C. J., and J. A. Davis. 1987. "Natural and artificial wetland ecosystems: Ecological opportunities and limitations." In K. R. Reddy and W. H. Smith. eds., *Aquatic Plants for Water Treatment and Resource Recovery*, pp. 819–854. Magnolia Publishing, Orlando, Fla.

Simpson, R. L., D. F. Whigham, and R. Walker. 1978. "Seasonal patterns of nutrient movement in a freshwater tidal marsh." In R. E. Good, D. F. Whigham, and R. L. Simpson, eds., *Freshwater Wetlands: Ecological Processes and Management Potential*, pp. 243–257. Academic Press, New York.

Sopper, W. E., and L. T. Kardos, eds. 1973. *Recycling Treated Municipal Wastewater and Sludge through Forest and Cropland.* Pennsylvania State University Press, University Park.

Spalding, M. G., G. T. Bancroft, and D. J. Forrester. 1993. "The epizootiology of eustrongylidosis in wading birds (Ciconiiformes) in Florida." *Journal of Wildlife Diseases* 29: 237–249.

Spangler, D. P. 1984. "Geologic variability among six cypress domes in north-central Florida." In K. C. Ewel and H. T. Odum, eds., *Cypress Swamps*, pp. 60–66. University Press of Florida, Gainesville.

Tilton, D. L., and R. H. Kadlec. 1979. "The utilization of a freshwater wetland for nutrient removal from secondarily treated wastewater effluent." *Journal of Environmental Quality* 8: 328–334.

Valiela, I., J. M. Teal, C. Cogswell, J. Hartman, S. Allen, R. Van Etten, and D. Goehringer. 1985. "Some long-term consequences of sewage contamination." In P. J. Godfrey, E. R. Kaynor, S. Pelczarski, and J. Benforado, eds., *Ecological Considerations in Wetlands Treatment of Municipal Wastewaters*, pp. 301–315. Van Nostrand Reinhold, New York.

Wellings, F. M., A. L. Lewis, C. W. Mountain, and L. V. Pierce. 1975. "Demonstration of virus in groundwater after effluent discharge onto soil." *Applied Microbiology* 29: 751–757.

Wetzel, R. G. 1993. "Constructed wetlands: Scientific foundations are critical." In G. A. Moshiri, ed., *Constructed Wetlands for Water Quality Improvement.* CRC Press, Boca Raton, Fla.

Chapter 19

SERVICES SUPPLIED BY SOUTH AFRICAN FYNBOS ECOSYSTEMS

Richard M. Cowling, Robert Costanza, and Steven I. Higgins

Perched at the southwestern tip of Africa is the world's smallest and, for its size, richest floral kingdom, the Cape Floristic Region. This tiny area, occupying a mere 90,000 km^2, supports 8,500 plant species (of which 68 percent are endemic), 193 endemic genera, and 6 endemic families (Bond and Goldblatt 1984). Because of the many threats to this region's spectacular flora, it has earned the distinction of being the world's "hottest" hot spot of biodiversity (Myers 1990).

The predominant vegetation in the Cape Floristic Region is fynbos (figure 19.1), a hard-leafed and fire-prone shrubland that grows on the highly infertile soils associated with the ancient, quartzitic mountains (mountain fynbos) and the windblown sands of the coastal margin (lowland fynbos) (Cowling 1992). Owing to the prevalent climate of cool, wet winters and warm, dry summers, fynbos is superficially similar to California chaparral and other Mediterranean-climate shrublands of the world (Hobbs et al. 1995). Fynbos landscapes are extremely rich in plant species (the Cape Peninsula has 2,554 species in 470 km^2), and narrow endemism ranks among the highest in the world (Cowling et al. 1992).

What services do these species-rich fynbos ecosystems provide and what is their value to society? The valuation of ecosystem services is fraught with problems (see chapters 3 and 4), and very few studies provide a comprehensive economic valuation. We know of no ecological-economic studies from unusually species-rich ecosystems. Here we review recent research in the fynbos that demonstrates unequivocally the substantial economic value

Figure 19.1. Fynbos (foreground) with invasive alien plants (*Pinus* spp. from the Mediterranean Basin and *Acacia* spp. from southwestern Australia) in the background. Note the greater height, and hence biomass, of the aliens. *Photo:* R.M. Cowling.

of ecosystem services such as sustained supply of clean water, wildflowers, recreational opportunities, and biodiversity storage. These studies provide convincing economic incentives for management interventions aimed at conserving and restoring biodiversity. We hope that this review will be useful to researchers and policy makers working in other species-rich ecosystems that are threatened by extensive transformations.

Our chapter is divided into four parts. First, we outline the nature and value of ecosystem services provided by fynbos. Then we briefly discuss the major threat to fynbos ecosystems and the biodiversity they harbor. The next section reviews static and dynamic models that quantify the economic value of fynbos ecosystems under different management scenarios. Finally, we discuss the policy implications of these studies.

Fynbos Ecosystem Services: Their Nature and Value

Fynbos ecosystems provide a diverse array of services (Cowling and Richardson 1995). However, only recently have there been explicit attempts to provide an economic valuation of these (Burgers et al. 1995, van Wilgen et al. 1996, Higgins et al. 1996a; table 19.1). The major services are derived from consumptive use (wildflower harvesting), nonconsumptive use (hiker and ecotourist visitation), indirect use (water runoff), future use or option value (plant biodiversity), and the existence value of fynbos landscapes (see also chapters 3 and 4). We describe the nature and value of these services below.

Wildflowers

The fynbos flora is widely harvested for cut and dried flowers (van Wilgen et al. 1992). The combined value for 1993 of these enterprises, much of which was made up of export earnings, was US$18–19.5 million and provided a livelihood for twenty to thirty thousand people in an otherwise agriculturally marginal zone (Cowling and Richardson 1995). Most wildflowers are harvested on the lowlands and lower mountain slopes of the southern fynbos region (Greyling and Davis 1989). The unit value of wildflowers varies considerably but may exceed US$10,000/km^2 in certain areas (Higgins et al. 1996a; table 19.1).

Hiker and Ecotourist Visitation

The fynbos region includes a comprehensive network of nature reserves and wilderness areas (Rebelo 1992) with an excellent infrastructure of hiking trails and overnight facilities. At present, the use of most of these facilities is considerably lower than potential visitation limits of 2.8 hikers/km^2/month and one ecotourist/km^2/month (Higgins et al. 1996a). However, tourism is

Table 19.1. Nature and unit value of services provided by South African mountain fynbos ecosystems

Service	Value[a]
Native plant species maintenance ($/sp)	55–5,500
Endemic plant species maintenance ($/sp)	27,400–274,000
Hiking opportunities ($/visitor/day)	3.5–7.0
Ecotourism opportunities ($/visitor/day)	22–274
Unit value wildflowers ($/km^2)	543–11,415
Unit value water ($/m^3)	0.04–0.12

[a] Figures represent the lower and upper limits of attempts at valuation.
Source: Higgins et al. 1996a.

a major growth industry in the fynbos region, and improved marketing of its beautiful landscapes and exceptional biota is likely to result in an upsurge in ecotourism.

Water

Fynbos-clad mountain watersheds yield about two-thirds of the region's water requirements (Burgers et al. 1995). Runoff from these watersheds is very high, owing to the generally high rainfall in the mountains, the porous, sandy soils, and the low water use by fynbos plants (Le Maitre et al. 1996). Furthermore, water quality is excellent and minimal treatment for domestic use is required (Burgers et al. 1995). Mountain-derived water plays a crucial role in the fynbos region's economy, which is centered in the semi-arid lowlands; in 1992 it generated a gross domestic product of US$15.3 billion (Bridgeman et al. 1992). For example, the deciduous fruit industry, which is entirely dependent on water derived from adjoining mountain watersheds, generated a gross export earning of US$560 million in 1993 and provided employment for about 250,000 people (van Wilgen et al. 1996). The minimum unit value of water to society is normally taken as the tariff for bulk untreated water from state supply schemes (Burgers et al. 1995; table 19.1).

The fynbos region is also home to large and rapidly growing numbers of economically marginalized people who live in informal settlements on the periphery of urban centers. Most of these communities do not have access to reliable sources of clean water. The Reconstruction and Development Programme (RDP) of South Africa's government of national unity endorses the principle that all South Africans have a right to "convenient access to clean water" (African National Congress 1994). This right will be realized only if watersheds are optimally managed to ensure the delivery of this ecosystem service in a cost-effective manner.

Biodiversity

Fynbos landscapes are exceptionally rich in plant species and typically include many narrow endemics (Cowling et al. 1992). Many fynbos plants have been developed as food and drug plants (e.g., rooibos tea—*Aspalathus linearis;* honeybush tea—*Cyclopia* spp.; buchu oil—*Agathosma crenulata*) (Donaldson and Scott 1994), and numerous others, including proteas, pelargoniums (geraniums), heaths, gladioli, freesias, and restios, have been developed as horticultural crops (Cowling and Richardson 1995). There are undoubtedly many as yet undiscovered plants that have economic potential or option value.

Fynbos ecosystems provide a storage service for this plant biodiversity. Higgins et al. (1996a) estimated the value of this biodiversity storage service as the cost of maintaining indigenous plant gene banks (the two values in table 19.1 reflect the cost of two South African schemes). The value of a narrow endemic species was estimated as the cost of producing a new floricultural variety (i.e., the cost of creating a novel combination of genes).

Existence Value

The existence value to society of fynbos ecosystems is very difficult to quantify in economic terms (see chapter 3). Nonetheless, the exquisite beauty of fynbos plants and the grandeur of fynbos landscapes are of considerable aesthetic and cultural value to the people of the southwestern Cape and, increasingly, elsewhere in South Africa and the world (Cowling and Richardson 1995).

Human-Induced Disruptions of Fynbos Ecosystem Services

Alien invasive plants, all shrubs and trees from other fire-prone Mediterranean-climate ecosystems (Richardson et al. 1992), are the major human-induced threat to fynbos biodiversity and ecosystem services. These weeds invade rapidly after fires and soon displace the fynbos flora (Richardson et al. 1992; figure 19.1). Alien plants increase the biomass of fynbos ecosystems by between 50 and 1,000 percent (Versfeld and van Wilgen 1986), resulting in a decrease in runoff from watersheds of between 30 and 80 percent (van Wilgen et al. 1992, Le Maitre et al. 1996; figure 19.2). Despite the obvious economic costs of a lack of effective management to counter the threats posed by alien plants (Burgers et al. 1995, Le Maitre et al. 1996), funding for fynbos watershed and reserve management, which is largely absorbed by clearing alien plant infestations, has been inadequate for several years. Recent estimates are that 31 percent of the area of proclaimed mountain fynbos watersheds is invaded by alien plants (Burgers et al. 1995); the situation is much worse in the lowlands (Richardson et al. 1992). Rates of invasion are exceptionally rapid, and models have predicted that without management, pristine watersheds will have alien plant cover of between 80 and 100 percent after one hundred years (Le Maitre et al. 1996; Richardson et al. 1996; figure 19.3).

In conclusion, invasive alien plants eliminate native plant biodiversity and reduce substantially water production from fynbos ecosystems. Therefore,

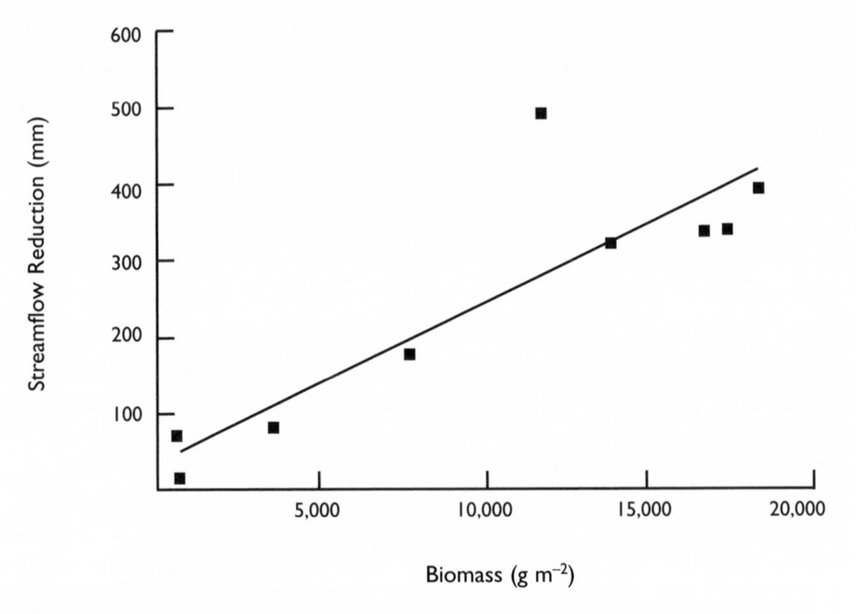

Figure 19.2. Relationship between biomass and reduction in streamflow from nine gauged watersheds in the southwestern Cape with varying degrees of invasion by alien plants.
Source: From Le Maitre et al. (1996), reproduced with permission from Blackwell Science Ltd.

they pose a direct and serious threat to all of the services provided by these ecosystems.

A Lack of Management

Most studies that have attempted ecosystem valuation adopt a static approach (e.g., Costanza et al. 1989; but see Krysanova and Kaganovich 1994). However, there is a growing recognition that effective articulation of ecosystem services will require a dynamic approach that combines, in integrative models, both ecological and economic processes (Costanza et al. 1993, Bockstael et al. 1995), and assesses the marginal costs and benefits of policy interventions (see chapter 3). In this section we discuss static and dynamic models that evaluate the economic consequences of a lack of management (chiefly alien plant control) in mountain fynbos ecosystems.

Van Wilgen et al.'s (1996) static model compared the costs of developing water supply schemes, water yield, and unit cost of water in two identical fynbos watersheds, with and without the management of alien plants (table 19.2). Both watersheds cover an area of 10,000 ha and have a mean annual

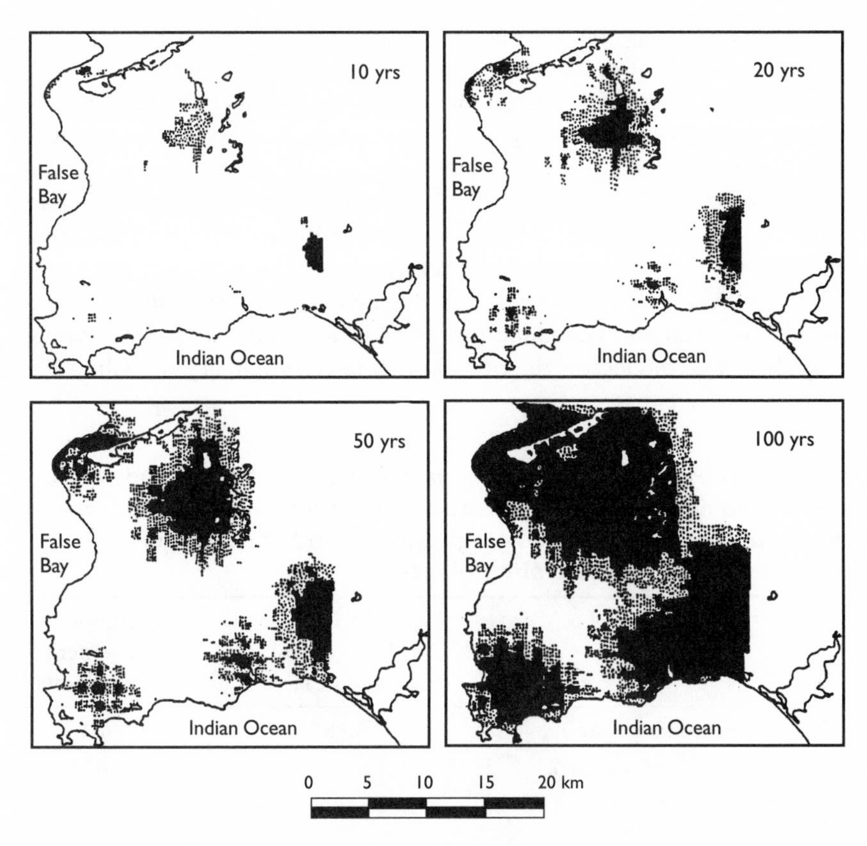

Figure 19.3. Maps showing the extent of infestation by alien plants in the Kogelberg mountain watershed (which supplies the city of Cape Town) in the southwestern Cape at various stages after the start of simulations of spread. Solid areas represent dense infestations; shaded areas represent lower-density classes, while unshaded areas are free of alien plants. If alien plant invasions are left unchecked, 86 million cubic meters of water could be lost annually. This represents 34 percent of the present annual water use by the city. *Source:* From Le Maitre et al. (1996), reproduced with permission from Blackwell Science Ltd.

rainfall of 1,500 mm. At a post-fire age of fifteen years, the managed watershed supports about 3,800 gm^{-2} of fynbos vegetation; the same watershed, if fully invaded by alien trees, supports a biomass of about 11,000 gm^{-2} at the same stage (Le Maitre et al. 1996). Runoff from the invaded catchment would be reduced by approximately 30 percent.

The unit cost of water for the two hypothetical watersheds was calculated by assuming an annual interest cost on capital outlays (the building of a water supply scheme in both cases, and the initial clearing of alien plants in one) and combining this with the annual operating costs (table 19.2). Although total annual costs are 11 percent higher for the watershed where alien trees are cleared and managed, the unit cost of water production is 14 percent lower, owing to the larger volumes of water that would be produced from a watershed where alien trees are controlled. Furthermore, such a watershed would yield an additional 14.1 million m^3yr^{-1}. This last point is particularly important in view of the limited opportunities for establishing new water supply schemes in the fynbos region (Little 1995).

Higgins et al. (1996a) developed a dynamic simulation model that integrated ecological and economic processes in a hypothetical 4 km^2 fynbos watershed. The model was developed in an interdisciplinary workshop set-

Table 19.2. Assumptions for the parameters and costs and water yields associated with two identical, hypothetical watersheds in the mountains of the fynbos region, with and without the management of invasive alien plants

	With Management of Alien Plants	Without Management of Alien Plants
Aboveground biomass (gm^{-2})	3,867	10,964
Reduction in streamflow due to plant biomass at 15-yr post-fire (mm rainfall equivalent)	114	256
Capital cost of clearing initial infestation (US$ ha^{-1})	830	0
Annual cost of alien plant management (US$ ha^{-1})	8	0
Capital cost of developing water supply scheme and initial clearing of aliens (US$ $\times 10^6$)	76	67.7
Annual interest on capital cost at 8% (US$ $\times 10^6$)	6.1	5.4
Operating costs (US$ $\times 10^6$ yr^{-1})	1.36	1.27
Total annual costs (interest plus operating) (US$ $\times 10^6$ yr^{-1})	7.46	6.67
Water yield ($m^3 \times 10^6$ yr^{-1})	62.7	48.6
Unit cost of water (cents m^{-3})	11.8	13.7

Sources: Biomass and streamflow reductions were calculated from relationships given in Le Maitre et al. (1996). Table adapted from van Wilgen et al. (1996).

ting using STELLA (1993, High Performance Systems, Inc.), an icon-based simulation language that facilitates collaborative model construction. A monthly time step was selected in order to simulate the seasonal and fire-related dynamics of the fynbos watershed ecosystem over a fifty-year period. The model comprises five interactive submodels, namely hydrological, fire, plant, management, and economic valuation (figure 19.4). Parameter estimates for each submodel were either derived from the published literature or established by workshop participants and consultants (they are described in detail in Higgins et al. 1996a). The plant submodel included both native and alien plants. Simulation provided a realistic description of alien plant invasions and their impacts on river flow and runoff (figure 19.5). Our discussion below will focus mainly on the output of the economic submodel in relation to different alien plant management scenarios (table 19.3). The model enabled us to quantify the marginal costs and benefits to society as a result of different management policies (scenarios). It is important to note that the model does not provide a realistic estimate of the total value of the watershed, since we lack information on what society would be willing to pay for ecosystem management as a function of alien plant invasion (see chapter 3). The nature and unit value of ecosystem services quantified in the model are given in table 19.1.

Under management scenario M1 (present management, table 19.3), inadequate clearing of aliens had a major negative impact on the ecosystem services provided by the hypothetical watershed (Higgins et al. 1996a). The

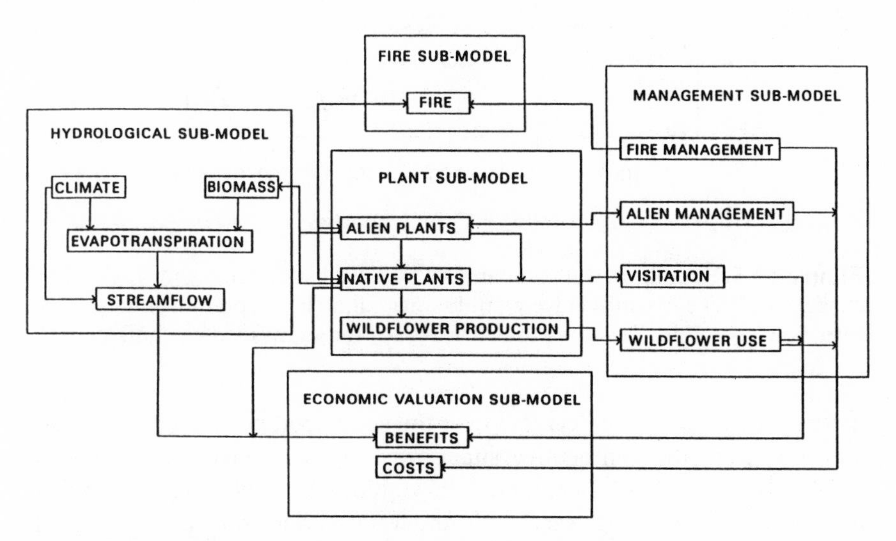

Figure 19.4. Conceptual diagram of the dynamic ecological-economic model of the fynbos mountain watershed model.

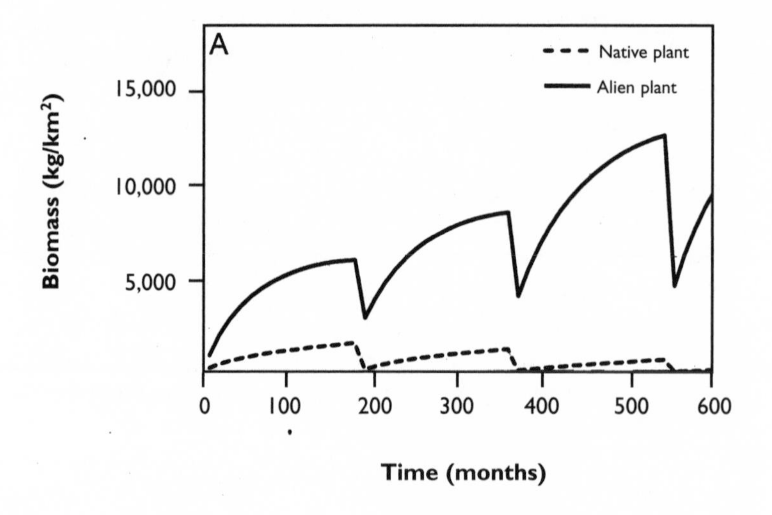

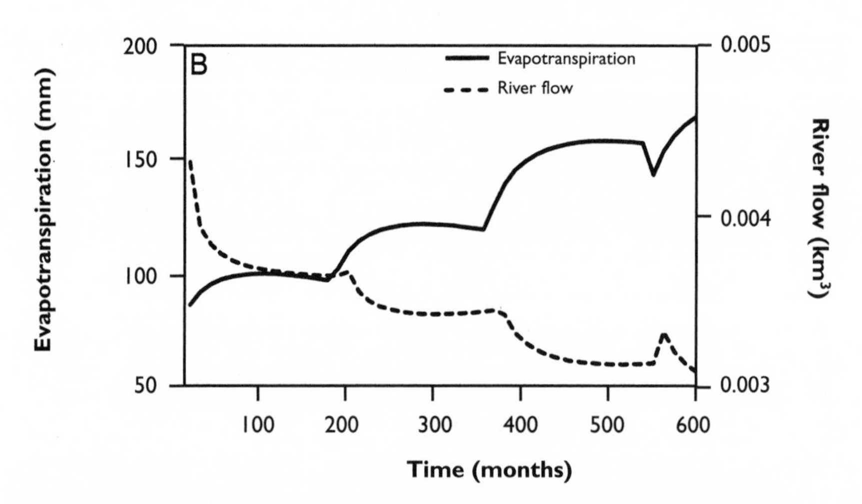

Figure 19.5. Output of the plant and hydrological submodels showing changes in: (A) alien and native plant biomass and (B) evapotranspiration and river flow over time (including four fire cycles) for an invaded catchment.

model predicts a steady decrease over the simulation period in water yield and, owing to the competitive superiority of alien versus native plants (Richardson et al. 1992), a reduction in wildflower production and a decline in the biodiversity storage service. As the watershed is invaded, it is avoided first by ecotourists and later by hikers, thereby diminishing revenues from recreational activities. When this scenario was combined with the low end of the range of unit value estimates for the four main categories of ecosystem

Table 19.3. Management scenarios used for modeling the economic outputs from the fynbos model[a]

Code	Description
M1	Present management: 50% invaded by aliens; inadequate clearing (0.003 km^2 mo^{-1}).
M2	Proactive management: 50% invaded by aliens; adequate clearing (0.01 km^2 mo^{-1}).
M3	Pristine management: uninvaded; no clearing required.

[a]Codes appear in the text as well as in figures 19.6 and 19.7, where they are combined with two levels of economic valuation of ecosystem services: E1 (lower limit of unit values in table 19.1) and E2 (upper limit of unit values).

Source: Higgins et al. 1996a.

services considered (E1; see table 19.3), the net value (i.e., with management costs deducted and discounted at 3 percent over fifty years) of the watershed was US$5.2 million (figure 19.6). At the high end of the range of unit values (E2) the value estimates totaled US$29.3 million.

When the management strategy involves the eradication of aliens from an invaded watershed (scenario M2 in table 19.3), the net value over the simulation period increased to US$7.1 million under low-end valuation, to US$55.3 million under high-end valuation (figure 19.6). This occurs in spite of increased management costs associated with a greater effort at clearing aliens (see also van Wilgen et al. 1996).

Pristine watersheds are free of alien plants and require little management other than controlled burns and hiking trail maintenance (scenario M3 in table 19.3). Under this scenario, the net value of the watershed under low-end valuation (E1) was US$7.7 million (figure 19.6). This was only marginally higher than under the proactive management scenario (M2,E1) and demonstrates that alien clearing can restore the value of key ecosystem services, especially water production (see also Burgers et al. 1995). Predictably, the highest net value (US$82.5 million) was recorded for the pristine management (M3) and high-end valuation (E2) combination.

Water dominates the value of the watershed under low-end valuation (E1), whereas the biodiversity storage service has the greatest value under high-end valuation (E2) (figure 19.7). Endemic plants contribute little to this service, owing to the small size of the watershed; the empirical relationship between watershed area (4 km^2) and species richness predicts a value 465 plant species and only 1.46 endemics (Higgins et al. 1996a). In relative terms, recreational activities (hiking and ecotourism) contribute little to the gross value of the watershed, but absolute values are substantial in well-managed situations: the derived revenues could make an important contri-

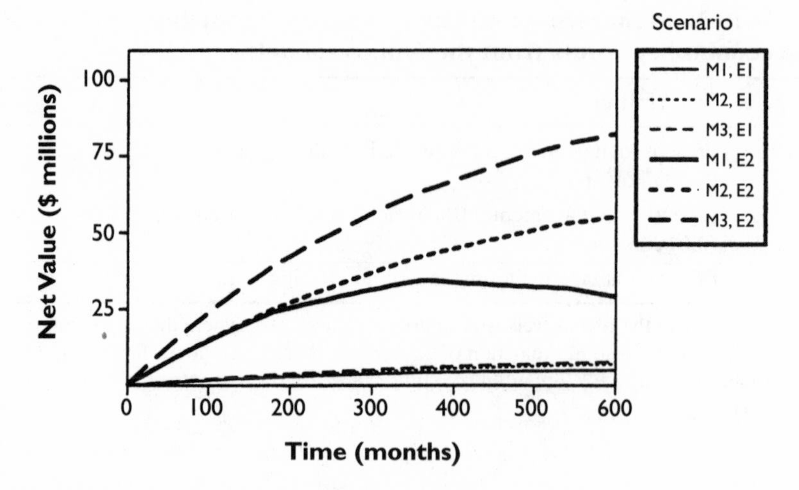

Figure 19.6. Changes in net value with time for three management scenarios (M1–M3) and two economic valuations (E1 and E2) for a 4 km^2 fynbos mountain watershed. Scenarios are described in table 19.3.

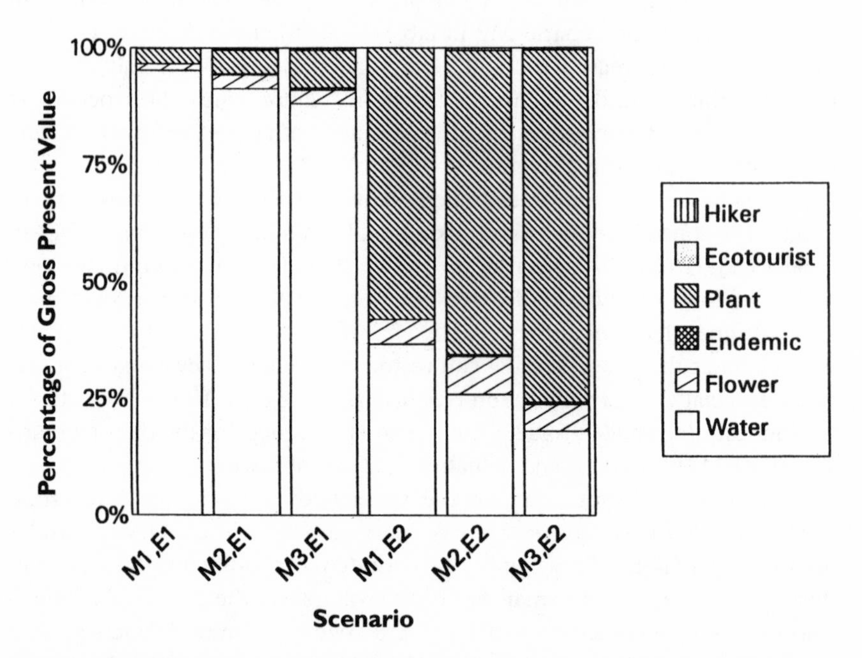

Figure 19.7. The proportional contribution of ecosystem services to gross present value of a 4 km^2 fynbos mountain watershed under three different management scenarios (M1–M3) and two economic valuations (E1 and E2). Scenarios are described in table 19.3.

bution to management costs. In this respect it is important to note that the costs of proactive management (principally alien clearance) amount to only 0.6 percent (high-end valuation, E2) and 4.8 percent (low-end valuation, E1) of the value to society of the ecosystem services supplied by the watershed.

Policy Implications

This chapter has shown that species-rich fynbos ecosystems provide a wide array of services that are of considerable economic value to society. We have also shown that the costs of optimal ecosystem management, aimed largely at eradicating and preventing invasions by alien shrubs and trees, are minuscule when compared to the benefits provided by pristine ecosystems. Our chapter raises several important implications for policy formulation. We discuss some of these in this section.

The deleterious impact of alien trees and shrubs on native plant biodiversity has been known to fynbos ecologists for several decades (e.g., see Richardson et al. 1992 and van Wilgen et al. 1992). Indeed, an appreciation of the importance of fynbos-clad watersheds for the economic development of the southwestern Cape resulted in the proclamation in the early 1970s of most of the mountains of the region as protected watersheds. Management plans were drawn up for each watershed, and alien plants were vigorously combated. However, the past five years have seen a substantial decline in funding for watershed management, and this has resulted in an alarming increase in the extent and density of alien plant infestations (Burgers et al. 1995). Given the considerable economic value to all sectors of society of the services supplied by pristine watersheds, these budgetary shortfalls make little sense. The benefits of optimal watershed management in terms of increased water yield—water being a critical limiting resource in the fynbos region—is a sufficiently persuasive argument for a policy that would provide sufficient funds for the eradication and prevention of alien plant invasions.

With this in mind, a group of fynbos ecologists, under the auspices of the Foundation for Research Development's Fynbos Forum, developed an audio-visual "road show" that described the economics of water and watershed management. The road show was presented to the minister of water affairs and forestry in May 1995. So impressed was the minister that he declared alien plant removal from watersheds as a project within the Reconstruction and Development Programme (RDP). The RDP is aimed at kick-starting socioeconomic development in post-apartheid South Africa. The watershed project is envisaged to run for twenty years (the time estimated for complete restoration) and, in the process, will create thousands of jobs, involve numerous training programs, stimulate scientific research, and safeguard many thousands of plant and animal species. The project was launched in late 1995 at several sites in the fynbos region.

With that battle apparently won, we need to turn our attention to lowland fynbos, where water is not a significant ecosystem service. There are three issues here that are important for policy development. First, lowland fynbos ecosystems have better wildflower resources than the mountains and are also more accessible. It is not surprising, therefore, that they support the bulk of the wildflower industry (Greyling and Davis 1989). Alien plants pose a major threat to wildflower resources throughout the lowlands (Cowling 1990). Although alien plants are an economically valuable fuelwood resource for the subsistence sector in lowland areas adjacent to urban centers, Higgins et al. (1996b) have shown that wildflower harvesting is more important economically and more sustainable in the longer term. Landowners and policy makers need to be apprised of the long-term economic benefits of controlling alien plant invasions for the benefit of the rapidly growing wildflower trade in the fynbos lowlands.

Second, tourism is the fastest-growing industry in the fynbos region, and there is enormous potential to market the fynbos as an ecotourist resource of international significance (Cowling and Richardson 1995). As a result of their spectacular flora, accessibility, scenic qualities, and proximity to the ocean, lowland fynbos ecosystems are prime ecotourist destinations. Alien plant invasions, which degrade scenery and eliminate native plant and animal biodiversity, represent a major threat to the ecotourism industry. There are no realistic estimates of the contribution of ecotourism to the economy of fynbos lowland regions. Entry fees charged by nature reserves (table 19.1) are not good indicators of this contribution, since users actually pay much more (e.g., transportation, time, and accommodation costs) to visit these facilities (see chapter 3). Techniques such as the travel cost method (chapter 3) urgently need to be employed to provide a realistic assessment of the contribution of ecotourism to the economies of lowland fynbos regions.

Finally, lowland fynbos ecosystems provide a storage service for many thousands of fynbos plants and animals, including numerous threatened taxa and local endemics (Cowling et al. 1992, Rebelo 1992). Many lowland species have been developed as horticultural crops, medicinal plants, and foodstuffs (Donaldson and Scott 1994). The option value of the remaining species represents a considerable economic resource. Policy makers must be made aware of the valuable service provided by these ecosystems and allocate funds to counter threats that may disrupt it.

Ecological economics is a rapidly developing discipline. However, it already provides the concepts, theories, and tools to explore the economic implications of habitat destruction and biodiversity loss (e.g., Jansson et al. 1994). The explicit economic valuation of the services provided by mountain fynbos ecosystems has resolved conflicts regarding the allocation of limited funds and provided the incentives for policies aimed at ecosystem restoration and optimal management. The fynbos model developed by Higgins et al. (1996a), which has a user-friendly interface and can be played as an interactive game by a wide range of interested parties with no prior modeling experience, could be used to communicate the value of lowland fynbos ecosystems in terms of wildflowers, ecotourism, and biodiversity storage, and thereby contribute to the resolution of management and land-use conflicts. We are confident that such initiatives will lead to improved conservation and utilization of lowland fynbos ecosystems. Such interventions will be to the benefit of all South Africans and indeed to citizens of the entire planet.

Summary

The species-rich fynbos ecosystems of South Africa represent the world's foremost hot spot of plant diversity and endemism. These ecosystems provide a diverse array of services to society, such as wildflowers, ecotourism opportunities, water supplies, biodiversity storage (including many actually and potentially valuable horticultural, food, and drug plant species), and, owing to their beauty, immense existence value. Fynbos ecosystems are severely threatened by alien plant invasions, which substantially reduce the quantity and quality of the services they provide. This chapter reviews recent research that evaluates the impact on the economic value of fynbos ecosystem services of a lack of alien plant control. Results show that the marginal costs of management to remove alien plants range between 0.6 percent (high-end valuation) and 4.8 percent (low-end valuation) of the marginal value to society of the services provided by a fynbos mountain watershed. Although water dominates the value under low-end valuation, the biodiversity storage function has the greatest value under high-end valuation. This emphasizes the importance of providing a realistic value of the latter service. The economic impact of reduced water supplies from fynbos mountain watersheds invaded with alien plants has provided the incentives for a development project aimed at restoring these ecosystems. We discuss potential policy interventions for restoring and maintaining native plant biodiversity in lowland fynbos regions, where water supply is not a major ecosystem service.

Acknowledgments

Much of the research reported here was undertaken at a workshop on the valuation of fynbos ecosystem services, held at the University of Cape Town in July 1995. We thank all workshop participants, especially Dave le Maitre, Christo Marais, Guy Midgley, and Jane Turpie, who helped develop the fynbos watershed model. Funds were provided by the Pew Charitable Trusts, the University of Cape Town, and the Foundation for Research Development.

References

African National Congress. 1994. *The Reconstruction and Development Programme*. Johannesburg: Umanyano Publications.

Bockstael, N., R. Costanza, I. Strand, W. Boynton, K. Bell, and L. Wainger. 1995. "Ecological economic modeling and valuation of ecosystems." *Ecological Economics* 14:143–159.

Bond, P., and P. Goldblatt. 1984. "Plants of the Cape Flora." *Journal of South African Botany* (Supplement) 13:1–455.

Bridgeman, D.H.M., I. Palmer, and W.H. Thomas, eds. 1992. *South Africa's Leading Edge: A Guide to the Western Cape Economy.* Cape Town: Wesgro.

Burgers, C.J., C. Marais, and S.J. Bekker. 1995. "The importance of mountain catchments for maintaining the water resources of the Western Cape Province and the need for optimal management." In *Managing Fynbos Catchments for Water,* C. Boucher and C. Marais, eds., pp. 99–123. Pretoria: Foundation for Research Development Programme Report Series 24.

Costanza, R., S.C. Farber, and J. Maxwell. 1989. "Valuation and management of wetland ecosystems." *Ecological Economics* 1: 335–361.

Costanza, R., L. Wainger, C. Folke, and K.G. Maler. 1993. "Modeling complex ecological economic systems." *BioScience* 43: 545–555.

Cowling, R.M. 1990. "Farming fynbos—reconciling conservation with exploitation." UCT News 17, no. 1: 8–10.

Cowling, R.M., ed. 1992. *The Ecology of Fynbos. Nutrients, Fire and Diversity.* Cape Town: Oxford University Press.

Cowling, R.M., P.M. Holmes, and A.G. Rebelo. 1992. "Plant diversity and endemism." In *The Ecology of Fynbos. Nutrients, Fire and Diversity,* R.M. Cowling, ed., pp. 62–112. Cape Town: Oxford University Press.

Cowling, R.M., and D.M. Richardson. 1995. *Fynbos. South Africa's Unique Floral Kingdom.* Cape Town: Fernwood Press.

Donaldson, J.S., and G. Scott. 1994. "Aspects of human dependence on plant diversity in the Cape Mediterranean–type ecosystem." *South African Journal of Science* 90: 338–342.

Greyling, T., and G.W. Davis, eds. 1989. *The Wildflower Resource: Commerce, Conservation and Research.* Occasional Report 40. Pretoria: Council for Scientific and Industrial Research.

Higgins, S.I., E.J. Azorin, R.M. Cowling, and M.J. Morris. 1996a. "An ecological economic simulation model of mountain fynbos ecosystems: Dynamics, valuation and management." *Ecological Economics* (submitted).

Higgins, S.I., E.J. Azorin, R.M. Cowling, and M.J. Morris. 1996b. "A dynamic ecological-economic model as a tool for conflict resolution in an invasive-alien plant, biological-control and native-plant scenario." *Ecological Economics* (submitted).

Hobbs, R.J., D.M. Richardson, and G.W. Davis. 1995. "Mediterranean-type ecosystems: Opportunities and constraints for studying the function of biodiversity." In *Mediterranean-Type Ecosystems. The Function of Biodiversity,* G.W. Davis and D.M. Richardson, eds., pp. 1–42. Berlin: Springer-Verlag.

Jansson, A., M. Hammer, C. Folke, and R. Costanza, eds. 1994. *Investing in Natural Capital: The Ecological Economics Approach to Sustainability.* Washington, D.C.: Island Press.

Krysanova, V., and I. Kaganovich. 1994. "Modeling of ecological and economic systems at the watershed scale for sustainable development." In *Investing in Natural Capital: The Ecological Economics Approach to Sustainability,* A. Jansson, M. Hammer, C. Folke, and R. Costanza, eds., pp. 215–232. Washington, D.C.: Island Press.

Le Maitre, D.C., B.W. van Wilgen, R.A. Chapman, and D.H. McKelly. 1996. "Invasive plants and water resources in the Western Cape Province, South Africa: Modelling the consequences of a lack of management." *Journal of Applied Ecology* 33, no. 1:161–172.

Little, P.R. 1995. "Water resources of the Western Cape." In *Managing Fynbos Catchments for Water,* C. Boucher and C. Marais, eds., pp. 3–8. Report Series 24. Pretoria: Foundation for Research Development Programme.

Myers, N. 1990. "The biodiversity challenge: Expanded hot-spots analysis." *The Environmentalist* 10: 243–255.

Rebelo, A.G. 1992. "Preservation of biotic diversity." In *The Ecology of Fynbos. Nutrients, Fire and Diversity,* R.M. Cowling, ed., pp. 309–344. Cape Town: Oxford University Press.

Richardson, D.M., I.A.W. Macdonald, P.M. Holmes, and R.M. Cowling. 1992. "Plant and animal invasions." In *The Ecology of Fynbos. Nutrients, Fire and Diversity.* R.M. Cowling, ed., pp 271–308. Cape Town: Oxford University Press.

Richardson, D.M., B.W. van Wilgen, S.I. Higgins, T.H. Trinder-Smith, R.M. Cowling, and D.H. McKell. 1996. "Current and future threats to plant biodiversity on the Cape Peninsula, South Africa." *Biodiversity and Conservation* (in press).

van Wilgen, B.W., W.J. Bond, and D.M. Richardson. 1992. "Ecosystem management." In *The Ecology of Fynbos. Nutrients, Fire and Diversity,* R.M. Cowling, ed., pp. 345–371. Cape Town: Oxford University Press.

van Wilgen, B.W., R.M Cowling, and C.J. Burgers. 1996. "Valuation of ecosystem services: A case study from South African fynbos." *BioScience* (in press).

Versfeld, D.B., and B.W. van Wilgen. 1986. "Impacts of woody aliens on ecosystem properties." In *The Ecology and Control of Biological Invasions in South Africa,* I.A.W. Macdonald, F.J. Kruger, and A.A. Ferrar, eds., pp. 239–246. Cape Town: Oxford University Press.

Part V

Conclusion

Chapter 20

Valuing and Safeguarding Earth's Life-Support Systems

Gretchen C. Daily

Unless humanity is suicidal, it should want to preserve, at the minimum, the natural life-support systems and processes required to sustain its own existence. Many would argue that society should do more than the minimum, preserving for anthropocentric reasons the material basis of a richer, fuller life and preserving for deeper spiritual and ethical reasons "the Creation"—our only known living companions (and the only known habitable planet) in the universe. However the increasingly vexing tradeoffs between natural ecosystem preservation and conversion to other uses are eventually resolved, their analysis clearly requires, above all, the explicit establishment of a basis for value. This is not an academic issue but a matter of social choice today in the context of humanity's cultural heritage.

The core analyses presented in this book attempt to value ecosystems and their component species only insofar as they confer benefits, in the form of life-support goods and services, to human beings. This focus does not in any way preclude making decisions on the basis of other values as well, such as existence values of nonhuman organisms and their habitats; aesthetic, historical, religious, or other cultural significance; recreational values; etc.—all independent of any contribution to the fundamental material ingredients of human well-being. As a group, the contributors to this volume advocate a decision-making framework that considers a multitude of values, and most chapters discuss many types of value. Our primary focus here is on ecosystem service values because they are both very large and greatly underappreciated if, indeed, they are recognized at all.

Challenges in Valuation

The contributors tackled a series of difficulties in assigning value to ecosystem services, difficulties that are worth some reflection before reviewing the general findings and implications of the book (see also chapter 16). To begin with, the problem of merely identifying relevant components of value can hardly be overstated. Consider how important is the role played by organisms, by definition, in determining the character of ecosystems, yet how little is known about the role of the diversity of life—at any level, from genes to ecosystems—in ecosystem functioning (explored in chapter 6). The economic analyses presented in this book often required that organisms be lumped into broad groups according to their presumed roles (e.g., plants as photosynthesizers and soil protectors; microorganims as decomposers and nutrient cyclers; insects as pollinators, crop pests, and pest enemies; marine fish and shellfish as marketable seafood; etc.). Then the relationship between human well-being and the ecosystem functions performed by these general groups was characterized and valued.

The lack of information on the role, and value, of biodiversity in the supply of ecosystem services necessarily renders the characterizations overly simple and the valuations lower-bound, conservative estimates. A high-diversity system is assigned the same total value as a low-diversity system with the same number of functional groups because the probable added value of diversity remains poorly known and thus accounts for nothing at present. This lack of information also has the effect of diminishing the marginal value imputed to natural ecosystems. Larger areas of natural habitat tend to have higher levels of biodiversity, but this value of increased area, above and beyond that conferred by simply having a larger area generating services, cannot presently be quantified.

This gets to a second critical challenge, namely to determine the marginal value of ecosystems and the services they supply. It is, of course, eminently clear that the total value of ecosystem services is infinite—we could not possibly live without them. But establishing sound ecosystem conservation policies requires determining the costs of destroying the next unit of relatively intact natural habitat. By how much would the destruction of a particular 1 km^2 of a 100 km^2 forest disrupt the hydrological cycle, the retention of soil, or the natural pest control provided by the forest ecosystem to adjacent cropland? As the chapters make clear, such marginal analyses are difficult at best.

A third complication encountered in assigning economic value to ecosystem services is their context-dependency, both geographical and temporal. Consider savanna ecosystems suited to grazing livestock. The service of supplying forage would be valued only in those geographic areas (now a substantial portion of the land surface) where human societies graze livestock. Moreover, livestock have different economic (not to mention cultural) values in different parts of the world—one must thus specify a particular value of livestock being used to make the calculation. Similarly, the erosion-control service will appear less valuable in regions where low-priced subsistence crops are grown than in areas where high-priced cash crops for export are produced, all else equal. Likewise, sea level rise matters (in this context) only if it rises to a level that damages property, freshwater aquifers, or other utilized sources of material well-being. If no people lived near enough to a given coastline to be directly affected, and there were no deleterious indirect effects of sea level rise, then the ecosystem service value of carbon sequestration as protection from sea level rise would be zero. Finally, the striking diversity in value systems among human cultures poses an additional challenge, especially where services operating on regional or global scales are concerned.

The monetized value of a service also depends on current prices and current social preferences, and may thus vary through time quite independently of any changes in the quality or rate of flow of the service. As livestock and grain prices fluctuate, for instance, so would the calculated value of ecosystem services that underpin livestock and grain production. Similarly, if anthropogenic greenhouse gas emissions were to cause (for the sake of argument, beyond a shred of doubt) deleterious climatic change in the future, the value imputed to natural ecosystems for carbon sequestration would increase. Just as geographic context dependency precludes assigning any one value to ecosystem services everywhere and summing up globally, temporal context dependency precludes assigning a single value to an ecosystem service for all time. Future generations may impute different values to ecosystem services than does the current generation; such changes are difficult to forecast. Human societies are now so interdependent, however, that undervaluation by one group may imperil all.

A fourth challenge is that market prices, to the extent that they are distorted by externalities, subsidies, barriers to trade, etc., will be poor indicators of the value of ecosystem goods and services. For example, to the extent that the price of food does not incorporate the social costs of unsustainable agricultural practices, it underestimates the value of ecosystem services sustaining food production.

A fifth challenge is in quantifying services for which there is no easy translation into market value, such as stability (low variance) in ecosystem productivity. How much is stability worth, all else equal? Even for a single individual, stability typically takes on different values through different stages of life. Suppose that there is an intrinsic tradeoff between the mean and variance in the productivity of a pasture, such that a pasture planted in a monoculture has a higher average annual yield, but is more susceptible to drought, than a pasture with relatively high plant diversity. The more diverse pasture does quite well in drought, having a suite of drought-tolerant species that thrive under dry conditions, free from the competition of drought-intolerant species. The relative values of productivity and stability in such pasture will vary as a function of the user's access to other sources of forage during drought conditions. If access is low, stability will be essential to maintaining a herd of animals and a livelihood. Many real-world cases involve a multitude of factors and are less clear-cut. Economists have various tools for assaying the value of an entity like stability, such as, in this case, willingness to pay for certain kinds of insurance. Fortunately, one does not always need to calculate an absolute value. Here, for instance, a relative comparison of the value of higher mean yield and lower variance might allow a preference for pasture type to be determined.

Finally, the interdependence and the arbitrary categorization of services preclude obtaining a grand total value of all ecosystem services in any given area by simple summation. There are two related problems. First, as our characterizations indicate, it is impossible to classify the services into entirely distinct, independent conditions and processes; however they are classified, many could not operate without others and would be worthless in isolation. Second, it thus follows that the number of services contributing to a given source of human benefits is necessarily arbitrarily specified. For example, climatological, hydrological, soil fertility, pest control, pollination, and many other interdependent services sustain agricultural production. Without soil and the vegetation it supports, for instance, the hydrological cycle could not function as needed; conversely, without water, many of the processes that maintain soil fertility would terminate. One could calculate the value of each of these services on a given farm as the value of the food grown there, say $X. (Since food can't be grown without water, the value of hydrological services is $X; similarly, the value of soil services would also be $X; assuming (see chapter 9) that a total lack of natural pest control would decimate crops, the value of pest control is $X as well; and so on.) The total value of ecosystem services contributing to crop production, however, is not the number of arbitrarily categorized services multiplied by the value of the crop. It is simply the value of the crop.

Technical difficulties such as these, exacerbated by imperfect information, pervade attempts to evaluate rationally the tradeoffs that involve the fate of natural ecosystems. Academicians and policy makers must accept this and devise ways to deal with it that ensure against making bad, irreversible choices.

Conclusions Regarding Ecosystem Service Value

In spite of these difficulties, several conclusions stand out clearly from our characterizations and valuation analyses. First, the services operate in intricate and little-explored ways that would be very difficult to substitute for using technology. Their marginal value will therefore almost certainly only go up as their supply dwindles. Second, the total value of ecosystem services is very large. The services examined in this volume amount to many trillions of dollars annually. Third, marginal values of ecosystem services appear high in areas where we have an indication of them. Loss of soil, pest control, pollinators, and ecosystem goods are all presently impoverishing local or regional groups of people. Finally, safeguarding ecosystem services represents one of the wisest economic investments society could make.

Research Needs and Policy Implications

Ideally, a constructive interplay between basic research and policy needs exists, whereby each influences the development of the other in devising solutions to important problems facing society. The primary needs of society with respect to ecosystem services are their identification, characterization, valuation, monitoring, and safeguarding. Below I discuss very briefly the motivation for, and some of the key issues in, each of these major areas, recognizing that each could be the subject of an entire book or more.

Identification

Obviously, the safeguarding of critical ecosystem services requires that they first be identified. At a variety of scales, from local communities to nations and the entire globe, an explicit cataloging of important ecosystem services is needed. For given geographic locations, one would also like to know which of these services are supplied locally, which are imported from elsewhere, and which are supplied globally. Such identification should represent an important part of national and local "green plans." It might also eventually form a basis for international agreements on the management of earth's life-support systems, building on the slight framework of those already formulated on the oceans, atmosphere, forests, and biodiversity.

One way of making publicly available scientific information on the sources, supply, and importance of ecosystem services would be the professional development of an Ecosystem Services Site on the World Wide Web.

Such an effort not only would enhance general knowledge of ecosystem services but, perhaps more important, could help build a constituency for their preservation. In local communities and nations around the world, appreciation of the value of and need to safeguard ecosystem services could be fostered by bringing representatives of different sectors of society together to jointly build models of the natural underpinnings of the economy (with or without sophisticated technology such as Geographic Information Systems). Many, if not most, of the problems societies face in formulating sound environmental policies stem quite simply from the lack of recognition of the crucial roles that natural ecosystems play in maintaining their health and happiness.

Characterization

The identification of ecosystem services is a necessary but clearly insufficient step in supplying information needed to make sound policy choices regarding the fate of natural ecosystems. Rational evaluation of the tradeoffs faced by society requires the development of a sound understanding of how ecosystem services work. While more than enough is known to guide policy at the present, further information would permit more efficient allocation of effort and material resources. A brief list of some of the broad research questions that require investigation includes (see also Holdren 1991):

- Which ecosystems supply what services?
- What is the impact of various human activities upon the supply of services?
- What are the relationships between the quantity or quality of services and the condition of the ecosystem (e.g., relatively pristine vs. heavily modified) supplying them?
- To what extent do the services depend upon biodiversity (from the genetic to the landscape level)?
- To what extent have various services already been impaired? How are impairment and risk of future impairment distributed geographically?
- How interdependent are the services? How does exploiting or damaging one influence the functioning of others?
- To what extent, and over what time scale, are the services amenable to repair?
- How effectively, and at how large of a scale, can existing or foreseeable human technology substitute for ecosystem services?

- Given the current state of technology and scale of the human enterprise, what proportion and spatial pattern of land must remain relatively undisturbed, locally, regionally, and globally, to sustain the delivery of ecosystem services?

Valuation

Valuation is critical to incorporating the importance of ecosystem services into decision-making frameworks, which are largely structured in economic terms. Having reviewed above some of the major challenges in valuation, I turn briefly to other important issues.

As Osvaldo Sala has pointed out, the undervaluation of ecosystem services has serious consequences not only for policy but also for the prioritization and undertaking of scientific research. To a first approximation, research is driven by two strong forces, intellectual curiosity and practical, economic needs. The study of ecosystem services has not fit well into either of these ends of the spectrum of scientific research. How ecosystems confer benefits on humanity represents too applied a topic to qualify as an area of "pure" research; at the same time, ecosystem services have neither been sufficiently recognized nor valued to attract the funds that support "applied" research. This lack of interest and lack of funding is sustained in a detrimental positive feedback.

Sala has suggested, as a thought exercise, comparison of the numbers of publications on the efficient production of cattle or timber versus those that report on the efficiency of carbon sequestration or water purification services of natural ecosystems. This comparison reveals a level of research on ecosystem goods and services with no direct market value that is orders of magnitude less than that on the production of major commodities, such as beef. In many important areas concerning the supply of ecosystem services, there is virtually no research at all. What is known about ecosystem services so far has been learned largely incidentally through their unintended disruption (although the recently established Beijer International Institute of Ecological Economics and International Society for Ecological Economics have helped pioneer work in this area; see, e.g., Folke et al. 1991; Dasgupta et al. 1994; Gren et al. 1994; Prugh et al. 1995).

Another hindrance to the characterization and valuation of ecosystem services is that much of the work is inherently interdisciplinary. Neither ecologists nor economists can do it all alone—even working in interdisciplinary teams. The expertise of individuals in numerous other disciplines, such as medicine and engineering, is required to answer key questions. For instance, epidemiologists are needed to elucidate the influence of changes in natural and managed ecosystems on human susceptiblity to disease (e.g., Daily and Ehrlich 1996)—an area of ecosystem services not addressed in this volume. Engineers are needed to develop and deploy technologies that minimize damage to and, where possible and desirable, substitute for aspects of ecosystem services.

The investigation of ecosystem services actually offers a powerful combination of intellectual stimulation and challenge with such societal importance that it is increasingly attracting the interest of scientists from all parts of the spectrum. Economists are making rapid progress in valuation methods, including the paramount issue of discounting, central to valuing an entity, such as an ecosystem service, the vast majority of whose value lies in the future and will always lie in the future (Chichilnisky et al. 1995, 1996).

Monitoring

In addition to knowing what services are delivered, how they are delivered, and how important they are (in economic or other terms), it is critical that society be able to track trends in the quality and rate of supply of ecosystem services, just as it tracks similar trends in its financial capital. Sustainable management of earth's life-support systems requires widespread, systematic monitoring of services all over the world, measured at appropriate scales. Since not everything can be monitored, indicators of various sorts need to be developed, tested, and refined. Recent changes in national accounting show that many societies are beginning to do this. There already is some monitoring of such things as fish stocks, soil conditions, water flows, atmospheric conditions, and the like—but no coordinated, comprehensive effort in terms of the delivery of services. This is not the place for a treatise on monitoring, but as a preliminary suggestion, the Intergovernmental Panel on Climate Change might serve as a partial model for a similar Intergovernmental Panel on the Status of Ecosystem Services.

Safeguarding

The safeguarding of ecosystem services will require that their value be explicitly incorporated into decision-making frameworks. In many cases, however, ecosystem service value is, and will remain, highly uncertain. The pace of ecosystem destruction, and the typical irreversibility thereof on a time scale of interest to humanity, warrants substantial caution. Just as it is impossible to quantify the full value of a human being, it is likewise impossible to determine the full value of natural ecosystems. Yet just as societies have established fundamental human rights, the establishment of fundamental ecosystem protections may be the most prudent approach in the face of uncertainty. New institutions and agreements—international and subnational—that encourage fair participation will be needed to do this (see, e.g., Heal 1994). Governments need to take a flexible, adaptive management approach, in which policy decisions to establish research priorities are made in consultation with the appropriate expert communities and the acquisition of new information is considered in honing the future policy course.

It is, of course, obvious that the underlying forces driving the impairment and destruction of ecosystem services include unsustainable growth in the scale of the human enterprise (overconsumption, overpopulation, and the use of environmentally destructive technologies and cultural practices). An important driver of these is the pattern of gross inequity within and between nations (e.g., Ehrlich et al. 1995). Sustainability and environmental security (sensu Myers 1993) cannot be achieved unless these fundamental problems are addressed much more effectively than they are today. It is a daunting challenge.

Nonetheless, I'd like to end this volume on two optimistic notes. Even though many of our results are preliminary, they show that the overall value of ecosystem services is so gigantic that, once recognized, their maintenance is bound to move toward the top of the international political agenda. And even though ecologists, economists, and others now concerned with maintaining the services still lack the knowledge of them they would like to have, there was virtual unanimity in our group that more than enough is known already to accelerate today's nascent movements toward safeguarding the very basis of our existence on earth.

Acknowledgments

This chapter benefitted from discussions with David Layton, Paul Ehrlich, Harold Mooney, and Osvaldo Sala.

References

Chichilnisky, G., G. Heal, and A. Beltratti. 1995. "The green golden rule." *Economic Letters* 49: 175–179.

Chichilnisky, G., G. Heal, and T. Lovejoy. 1996. "Sustainable cost-benefit analysis: A practical proposal" (in review).

Daily, G. C., and P. R. Ehrlich. 1996. "Global change and human susceptibility to disease." *Annual Review of Energy and the Environment* 21: 1–20.

Dasgupta, P., C. Folke, and K.-G. Maler. 1994. "The environmental resource base and human welfare." In *Population, Economic Development, and the Environment*, K. Lindahl-Kiessling and H. Landberg, eds., pp.25–50, Oxford: Oxford University Press.

Ehrlich, P. R., A. H. Ehrlich, and G. C. Daily. 1995. *The Stork and the Plow: The Equity Solution to the Human Dilemma*. New York: Putnam Press.

Folke, C., M. Hammer, and A.-M. Jansson. 1991. "Life-support value of ecosystems: A case study of the Baltic Sea Region." *Ecological Economics* 3: 123–137.

Gren, I.-M., C. Folke, K. Turner, and I. Bateman. 1994. "Primary and secondary values of wetland ecosystems." *Environmental and Resource Economics* 4: 55–74.

Heal, G. 1994. "Formation of international environmental agreements." In C. Carraro, ed., *Trade, Innovation, Environment*, pp. 301–332. Dordrecht: Kluwer Academic Publishers.

Holdren, J. P. 1991. "Report of the Planning Meeting on Ecological Effects of Human Activities." National Research Council, 11–12 October, Irvine, California, mimeo.

Myers, N. 1993. *Ultimate Security: The Environmental Basis of Political Stability*. New York: W.W. Norton.

Prugh, T., R. Costanza, J. Cumberland, H. Daly, R. Goodland and R. Norgaard. 1995. *Natural Capital and Human Economic Survival*, ISEE Press, Solomons, Md.

INDEX